Irrigation Institutions under Large Multi-Village Tanks

Irrigation Institutions under Large Multi-Village Tanks

Dr. K. Sivasubramaniyan
M.A., M.Phil., Ph.D.
Professor, Department of Economics

KALPAZ

Madras Institute of Development Studies

First Published, 2021

Published by

Kalpaz Publications
C-30, Satyawati Nagar,
Delhi – 110052
Ph.: 011-47034999, 9811692060
www.kalpazpublications.com
E-mail: kalpaz@hotmail.com

Madras Institute of Development Studies
79 II Main Road, Gandhi Nagar
Adyar, Chennai–600 020
www.mids.ac.in

Printed at: G. Print Process, Delhi

Cataloging in Publication Data—DK
Courtesy: D.K. Agencies (P) Ltd. <docinfo@dkagencies.com>

Sivasubramaniyan, K., author.
Irrigation institutions under large multi-village tanks / Dr. K. Sivasubramaniyan.
pages cm
Study conducted in Vellore and Tiruvannamalai districts of Tamil Nadu, India.
ISBN 9789353240462

1. Tank irrigation—India—Vellore (District) 2. Tank irrigation—India—Tiruvannamalai (District) I. Madras Institute of Development Studies, publisher. II. Title.

LCC S619.T36S58 2019 | DDC 631.587095482 23

The book has been published on the remembrance of the Golden Jubilee Celebrations (2021) of MIDS and Devoted to my Teacher Prof. A. Vaidyanathan Who taught me the nuances of Irrigation

Contents

Foreword

Historically the tank irrigation systems had played a significant role in rural life and particularly in promoting the irrigated agriculture. Irrigation tanks in India are century old water harvesting structures. There are about 208,000 tanks in the country irrigating about 2.3 m.ha and benefitting mostly marginal and small farmers besides the multiple uses from these tanks benefitting the landless households in the tankfed villages. After India's independence, Government's investment priorities shifted towards the major canal systems and groundwater development resulting in the neglect of these precious irrigation structures. As a result, tank foreshore encroachment, poor tank management, and government appropriation of the tanks have been the main outcomes of the failure of local authority system to enforce the institutional arrangements under common property resources management regime. In fact, the inefficient management of the tanks had also resulted in lower land productivity compared to canals and wells. However, still tanks in the rural regions help mobilize the livelihood for the poor. The social capital of tank irrigation has now been morphing in response to demographic pressures, out-migration, and development activities in terms of non-agrl investment. The resilience of the small tank system is thus explained by the existing synergy between the tank ecology, social capital and institutional structures.

In the recent years, groundwater has emerged as a supplemental source in the tank irrigation and the stabilization value of groundwater supplementation is observed to be high thus necessitating the importance of conjunctive use of tank and well water. However, due to constraints in the future expansion of the canal and well irrigated areas, tank irrigation still offers more scope for revival. Several interventions such as revenue mobilization, augmenting groundwater resources, strengthening the multiple uses, tank sluice management, reviving tank farmers organization, integrating social forestry, desilting etc., are incorporated in the tank modernization programs being implemented in different states with national and international funding.

In this context, irrigation tanks are considered still as neglected opportunities. Hence, looking at the future of tank irrigation potential, the issue is considered as one of increasing 'access' to tank water supplies and technology adoption and 'management'. Improving the availability of tank water supplies will minimize the burden on groundwater pumping due to poor electricity supplies and increasing well drilling costs. It is expected that by conjunctive use of the surface and groundwater supplies along with the enhanced multiple uses will improve the livelihood of the households and enhance the sustainability of the tanks. Also, the impact of climate change on future water storages can also be easily handled by strengthening the tanks as storage structures because of their wider geographical distribution.

As such, limited publications are available on tanks focusing on these critical issues. In this context, the research works undertaken by Dr Sivasubramaniyan are timely and much warranted. The book has interesting chapters derived from in-depth field research. I am sure this book will be very useful to researchers, NGOs, planners and policy makers dealing with small scale irrigation tanks in India and elsewhere.

I congratulate Dr Sivasubramaniyan for bringing out this very valuable and timely book at a time when the tank irrigation is getting the attention of the Government departments, NGOs, corporates and national and international funding agencies.

09, December, 2019.

K. Palanisami
Ph.D., P-Doc (USA), FNAAS.,
Agricultural and Water Expert, ISWRD,
Govt of Telangana, Hyderabad, &
Emeritus Scientist,
International Water Management Institute (IWMI),
New Delhi.

Preface

In India, agriculture still occupies the prime place in providing both employment to huge population and solving hunger to all the people, apart from exporting considerable quantum of its output. This has been possible mainly due to expansion of irrigation facilities, especially after Independence. Although four sources of irrigation (canals, tanks, wells and other sources) are responsible for this achievement, one of the surface irrigation sources in India such as tanks have been gradually, steadily and continuously declining since the 1970s. Consequently, tank irrigated agriculture in south India, especially in Tamil Nadu is not prosperous even now. Why this is happening and what are the contributing factors for this decline? This is the core aspect discussed in this book with the evidence from two major irrigation tanks in Tamil Nadu.

In the process of publishing this book many have helped to achieve the target. At the outset, I wish to thank Prof. Shashanka Bhide, then Director of MIDS, who helped to strengthen the project proposal initially for funding from the Indian Council of Social Science Research (ICSSR) in June 2015 and sustained it to get the project sanctioned. I thank the ICSSR for liberally granting me two extensions, three months each, to effectively complete the project report. It is my duty to remember and thank Prof. A. Vaidyanathan, my PhD Supervisor in the late 1980s, who introduced me the subject of 'irrigation institutions' and helped me to take up the study under tank irrigation for my PhD thesis and encouraged me in several ways to complete it in 1995. Actually, he taught me the important concepts of irrigation and the linkages between irrigated agriculture and the society at large. I am extremely thankful to him for his initiation of this subject of study that helped me now tremendously to study further and further on 'irrigation institutions' related issues more deeply. This book is an evidence of that sustenance.

My sincere thanks to Tamil Nadu Public Works Department officials such as the Chief Engineer, Executive Engineers, Assistant Executive Engineers and Assistant Engineers for their various helps especially for collection of secondary data from their offices during the course of the study. I wish to record my thanks to the District Revenue Officers at Vellore and Tiruvannamalai who permitted to collect the time consuming secondary data on cropping pattern, pattadars landholdings and caste details, and well details etc., from the Village Administrative Offices located in all the 32 villages and a detailed information for the 9 sample villages.

At village level many have helped to get proper details for the study. The village leaders, knowledgeable persons on tank related issues, big land owners, village assistants working with the VAOs, the former irrigation functionaries such as Nattanmaikars, Kavaimaniyam and Kammukutties who were then responsible for tank irrigation management. Also, the presently functioning Territorial Constituency Members (TCMs) of the Water Users Associations are contacted to know the present system of maintenance of channels and management of irrigation supplies in their village jurisdiction of both the tanks. For all of them my sincere thanks for sharing their time and giving enriched details. I express my deep gratitude to the ayacutdars and various levels of village leaders who were kind enough in allowing us to interview them and who readily responded to numerous questions.

I Thank Mr. P.G. Pattabi, and Mr. P. Anbazhagan who are responsible for collection of farmer level survey data in four sample villages served by Dusi-Mamandur Tank, both have helped me in conducting the long lasted field work initially in the 32 villages as a first stage coverage and again in 9 sample villages in the two study tanks during 2016-17. I express my thanks to them. I am thankful to the library staff of Madras Institute of Development Studies by placing at my disposal the latest literature related to the area of my research. I also acknowledge the help of the administrative staff of MIDS in various ways during the course of this work.

I owe special thanks to Mr. R. Dharumaperumal and Mr. P. Thyagarajan for their computer assistance and Mr. A. Arivazhagan, in the publication section, for his help in typing the Tamil Mamulnamas and preparing detailed figures for the study. Finally, I am grateful to Dr. V. Rengarajan, Senior Consultant, MIDS and Mr. M. Teeka Raman another consultant of my Tribal Study project who gave moral support in completing the work.

13 March 2020

K. Sivasubramaniyan
Chennai

List of Tables

List of Figures

Abbreviations

AE	=	Assistant Engineer
AEE	=	Assistant Executive Engineer
CE	=	Chief Engineer
CI	=	Cropping Intensity
DI	=	Direct Irrigation
DMT	=	Dusi-Mamandur Tank
EE	=	Executive Engineer
EL	=	Earth Level
FTL	=	Full Tank Level
FSL	=	Full Supply Level
GCA	=	Gross Cropped Area
GCM	=	Gross Cropped Months
GIA	=	Gross Irrigated Area
GOI	=	Government of India
GOTN	=	Government of Tamil Nadu
GVO	=	Gross Value Output
HA	=	Hectare
HHs	=	Households
HR	=	Head Reach
HYV	=	High Yielding Variety
IB	=	Irrigation Board
IIs	=	Irrigation Institutions
IWs	=	Irrigation Workers
JE	=	Junior Engineer

Kg	=	Kilogram
KPT	=	Kaveripakkam Tank
KM	=	Kilometer
LA	=	Less Active
MCFT	=	Million Cubic Feet
MCM	=	Million Cubic Metre
MI	=	Millimeter
MR	=	Middle Reach
MS	=	Murai System
NIA	=	Net Irrigated Area
NSA	=	Net Sown Area
OH	=	Operational Holdings
OMP	=	Oru Madai Paichal
PAS	=	Palar Anicut System
PWD	=	Public Works Department
Rs	=	Rupees
SDO	=	Sub-Divisional Officer
SE	=	Superintending Engineer
SO	=	Section Officer
TR	=	Tail Reach
VAO	=	Village Administrative Officer
WI	=	Work Inspector
WS	=	Water Sales

Glossary

Aadi	=	Fourth Month of the Tamil Year
Adangal	=	Cultivation accounts register maintained by the Village Administrative Officer
Adi	=	Bottom portion
Anicut	=	A permanent masonry structure to store and/or divert river flows for irrigation
Aathu kalvai	=	An excavated river channel for irrigation
Ayacut	=	Area commanded by an irrigation work
Collector	=	The Head of the Revenue Administration Department at the district level
Diversion vents	=	Vents with specified dimension constructed to divert water from the main/branch channels
Dry land	=	Revenue classification of land which does not have access to surface irrigation sources
Ettram	=	Man operated irrigation lift
Eravai Paichal	=	Irrigation by lifting water manually from the main channels
Guttee mulee	=	Term used for the payment in kind
Inam	=	Land held free or on concessional rates of revenue
Jally water	=	Uncontrolled leakage gushing from the tank sluices
Jamam	=	Period of three hours duration
Kammukutti	=	Hereditary village level irrigation worker
Kani	=	Equivalent to 1.33 acres or 0.54 hectare
Karai	=	Murai among kinship group
Karnam	=	Village accountant
Kasam	=	An excavated river channel to provide water to the ayacut during the monsoon season

Kavaimaniyam	=	Village level specialist in canal management
Keel	=	Lower portion
Kolmurai Kanakku	=	A method to measure the length of channels for maintenance by using five feet wooden rod
Kondam	=	A temporary earthen embankment to raise the water level to irrigate elevated lands
Korambu	=	A temporary earthen embankment to divert water by using wood and sand-bags
Kudimaramath	=	Traditional community labour participation for the maintenance of irrigation channels
Lascar	=	Lower level functionary of PWD
Madagu	=	Sluice
Mamulnamas	=	Documents written in 1815 setting down the traditional rules and regulation for augmentation and distribution of tank water
Mamul right	=	The customary right
Maniagar/Maniyam	=	Village Headman
Marakkal	=	Local measure for grains, equivalent to 4.52 kg of paddy
Maramat nesuff	=	Term used for payment in kind, to the persons engaged in maintenance of irrigation channels and other related works like construction of new tanks
Mattu	=	A masonry structure from which a specified dimension of two or more channels take off to supply water to the ayacut
Mel	=	Upper portion
Mettukkal Paichal	=	Lands irrigated by higher level supply channel
Mirasidari	=	A land tenure system in which the ryots paid the land Tax to the mirasidar
Munsiff	=	A village administrator
Murai	=	A system of fixed time-bound rationing of tank water supply adopted during scarcity period
Nadu	=	Middle portion
Naligai	=	A period of 24 minutes
Nanjai Kalani	=	Wet land under tank irrigation

Nattanmaikar	=	One of the village Headmen
Neerkatti	=	A government appointed irrigation functionary to supervise the tank water supply
Oru madai Paichal	=	Channel wise turn distribution of water supply, practised during the deficit tank supply period
Oru suli malai	=	Rainfall equivalent to one inch depth
Pallakkal Paichal	=	Lands irrigated by lower level supply channel
Panchayat	=	A village council traditionally consisting of five members
Pangu	=	Share
Patta	=	A written document settling title and land revenue of a land holder
Picottah	=	A shallow water lift by using human power
Ryotwari	=	A land tenure system in which the farmer directly paid the land tax to the government
Samba	=	Main paddy crop grown during July-December
Seer	=	A measure equivalent to 0.933 Kg
Spring Channel	=	An excavated river channel formed along the contour line to irrigate the low-lying lands adjacent to the river
Talaiyari	=	A village revenue functionary to assist the VAO
Tahsildar	=	Taluk level Head of the Revenue Department
Thotti or Toti	=	Hereditary village level peon
Thulir	=	Top portion
Toom	=	A measure of grains equivalent to 48 seers
Wet land	=	A revenue classification of land cultivated by surface irrigation sources
Yadas	=	A written slip, for release of a specified quantity of tank water to the ayacut, given by the village representative to the Secretary of the irrigation Board or to the JE of the PWD
Zamindar	=	The owner of an estate under the Permanent Settlement.
Zamindari	=	A land tenure system in which the ryots paid the land tax to the Zamindar.

Chapter - 1

Introduction

Indian agriculture has come a long way since Independence. The foodgrains output increased from 50.8 million tonnes (mt) in 1950-51 to 265 mt in 2013-14 (indiaagristat.com) to feed an Indian population of 1.28 billion in 2014 and 1.368 billion in 2019. Yet, ensuring food security and sustainability remains a continuing challenge. Among the three sectors of the economy – agriculture, industry and services – the primary sector is the key to several states as well as India's economy, because it provides employment for a large chunk of the population even now. In India, in 1970 agriculture provided employment to about 75% of the workforce, which declined to 55% in 2006 and further to 49% in 2013. The contribution of agriculture to gross domestic product (GDP) stood at 48% in 1970 reducing to 18% in 2006 and further down to 14% in 2013 (Sivasubramaniyan 2014: 117). Likewise in Tamil Nadu, three-fifths of employment (91 lakh out of 151 lakh workers) was generated by agriculture in 1970, which declined to 50.3% (146 lakh out of 290 lakh workers) and further to 42% in 2011 (139 lakh out of 328 lakh workers). The contribution of agriculture to gross state domestic product (GSDP) has steadily declined from 35% in 1970 to 26% in 1980, to 17% in 2000 and further down to 6% in 2013 (Tamil Nadu an Economic Appraisal, Various issues).

1.1 Role of Irrigation In Raising Agriculture

Given the agriculture scenario, water is one of the crucial inputs to enhance agricultural production. Crop yields depend critically on the amount of water available to crops at different stages of their growth. Rainfall alone is an inadequate and undependable source of moisture to sustain high productivity of agriculture. In India, a greater part of the year is practically dry; and the available rainfall even during the monsoon is often inadequate and highly variable in its distribution. While some regions have excess rainfall often resulting in floods, others suffer drought. These are among the principal reasons for the low level of yield in Indian agriculture. Under these conditions irrigation plays a crucial role in supplementing rainfall. It augments the supply of water to crops, makes it more assured, extends supply into the dry season, helps raise crop yields, adopt more productive crop patterns and increase the intensity of cultivation. Good quality of irrigation in terms of adequate, assured and controlled supply of water according to crop needs is especially important to realise the full potential of new technology based on fertiliser intense high yielding varieties.

The expansion and improvement of irrigation facilities has been an important element of India's agricultural development programme. During the six decades 1950-2012 the public sector has spent more than Rs. 4940 billion on irrigation development (including 368 billion of institutional finance almost exclusively to support groundwater exploitation by the farmers) mostly on major and medium reservoir-based projects (Sivasubramaniyan, 2016: 6-7). However, public investments in minor surface irrigation works (tanks and small diversion works serving relatively small areas over a few villages utmost) have been limited. The exploitation of groundwater has been growing rapidly: The number of wells and tube wells has increased manifold. At the time of Independence most wells were operated manually or with animal power. But now most of them are operating with electric (about 95%) and diesel (5%) pumps. The proportion of net irrigated area (NIA) served by wells has risen from 29.8% in the 1950s to 62.1% during 2010-11 to 2014-15 (see Table 1.1). Except for public tube wells in north India this expansion has been mostly in the private sector with liberal financial assistance (mainly loans at relatively low interest) from the public sector. Likewise, in Tamil Nadu, the proportion of NIA served by wells has risen from 23.5% in the 1950s to 60.5% during 2010-11 to 2014-15.

More importantly, among the two surface irrigation systems, viz. canals and tanks, both the sources have recorded a declining trend in percentage terms from the 1950s to 2000s. The canal irrigation in absolute term at all-India level performed much better (from 91.9 lakh ha. to 159.6 lakh ha) than in Tamil Nadu (8.0 lakh ha to 6.8 lakh ha). The more deplorable situation of irrigation source, both absolute and percentage terms, is noticed

Table 1.1: Trends in Net Area Irrigated by Sources in India and Tamil Nadu from 1950-51 to 2014-15.

(Area in Lakh Hectares)

Source	*1950-51 to 1959-60*		*1960-61 to 1969-70*		*1970-71 to 1979-80*		*1980-81 to 1989-90*		*1990-91 to 1999-2000*		*2000-01 to 2009-10*		*2010-11 to 2014-15*	
	Area	*% to NIA*	*Area*	*% to NIA*	*Area*	*% to NIA*	*Area*	*% to NIA*	*Area*	*% to NIA*	*Area*	*% to NIA*	*Area*	*% to NIA*
INDIA														
Canals	91.9	41.2	111.9	41.9	137.7	40.1	163.1	38.3	173.4	32.7	156.9	26.4	159.6	24.0
Tanks	41.5	18.6	44.5	16.6	38.1	11.1	29.9	7.0	31.1	5.9	19.8	3.3	18.4	2.8
Wells+ tubewells	66.3	29.8	87.1	32.6	144.1	41.9	207.8	48.7	292.5	55.2	364.1	61.2	412.8	62.1
Other sources *	23.2	10.4	23.9	8.9	23.8	6.9	25.4	6.0	33.1	6.2	53.9	9.1	73.5	11.1
Total NIA	222.9	100.0	267.3	100.0	343.6	100.0	426.3	100.0	530.1	100.0	594.7	100.0	664.3	100.0
Total NSA	1270.0		1369.0		1402.0		1404.0		1423.0		1401.0		1407.0	
TAMIL NADU														
Canals	8.0	37.6	8.8	35.6	8.9	33.2	8.2	33.0	8.3	29.3	7.3	26.8	6.8	24.4
	(8.7)		(7.9)		(6.5)		(5.0)		(4.8)		(4.7)		(4.3)	
Tanks	7.8	36.8	9.1	36.8	8.5	31.5	6.2	24.7	6.3	22.4	5.1	18.6	4.2	14.9
	(18.7)		(20.5)		(22.3)		(20.6)		(20.4)		(25.6)		(22.8)	
Wells+ tubewells	5.0	23.5	6.5	26.0	9.2	34.1	10.4	41.6	13.5	47.7	14.8	54.2	16.9	60.5
	(7.5)		(7.4)		(6.4)		(5.0)		(4.6)		(4.1)		(4.1)	
Other sources *	0.5	2.2	0.4	1.6	0.4	1.3	0.2	0.8	0.2	0.6	0.1	0.4	0.07	0.2
	(2.0)		(1.6)		(1.5)		(0.7)		(0.5)		(0.2)		(0.1)	
Total NIA	21.2	100.0	24.8	100.0	27.0	100.0	25.0	100.0	28.4	100.0	27.3	100.0	28.0	100.0
% to All-India	(9.5)		(9.3)		(7.8)		(5.9)		(5.4)		(4.6)		(4.2)	
Total NSA	55.45		60.26		61.35		56.22		56.33		50.22		48.04	

Note: Figures in brackets indicate source wise percentage compared to India. NIA = Net Irrigated Area. NSA = Net Sown Area.

Other sources * Indicates Anicuts, Bhandaras, Springs, Kuttai, Thangal, Small diversion networks and so on.

Source: Indian Agricultural Statistics, 1985-86—1989-90, Vol.I, Ministry of Agriculture, GOI. New Delhi.

CMIE, Agriculture, Various Issues. Indian Agriculture in Brief 27th Edn. January 2000.

GOTN, Tamil Nadu-An Economic Appraisal, Various Issues, Evaluation and Applied Research Department, Chennai.

GOTN, Season and Crop Reports, Various Issues. Chennai. www.IndiaAgristat.com

only under Tank Irrigation both at all-India and Tamil Nadu. The maximum absolute area under tank irrigation is recorded only in the 1960s and this was 44.5 lakh ha (16.6% of NIA) at all-India and 9.1 lakh ha (36.8% of NIA) in Tamil Nadu. This share of tank irrigation declined tremendously both in India and Tamil Nadu during 2010-11 to 2014-15. The former has registered a decline to 18.4 lakh ha (2.8% of NIA) and the latter to 4.2 lakh ha (14.9% of NIA).

1.1.1 Nuance of Irrigation under Different Sources

The quality of irrigation that is, the ability to regulate timing and quantity of water supply to particular plots according to crop needs varies between different type of projects. Large reservoirs, which harness rainfall over extensive areas, provide a larger volume, and more assured supply, than minor irrigation works like tanks which depend mostly on local rainfall. By contrast, in the case of wells which usually serve small areas, often belonging to a single farmer, it is easier toregulate water supply according to specific circumstances and needs of individual farmers. So,the quality of irrigation is better in the case of wells. However, wells are not always reliable in all areas, especially, where there is no surface source of supply available. In those areas, well water mostly depends upon the availability of rainfall and this situation prevails in most parts of the country. In order to overcome this difficulty, the rainwater harvesting method was encouraged by the government. This method is very useful to store water that is available in a smaller catchment area through construction of tanks. But these tanks are not altogether feasible to store water which is available from the larger catchment through big river basins and perennial streams. To utilise the river water the governments took initiative to build dams and reservoirs to store available water which was mainly used for irrigation other than Power generation. Hence the growing role of reservoir-based surface irrigation and wells have helped improve the quality of irrigation.

1.2 Irrigation Institutions

The quality of irrigation is not merely a function of the nature of the water source but of how well the source is managed. Partly this is a matter of engineering: For example, the ability to regulate the amount of water supplied to different segments of the command depends on how well the canals and distributaries are designed and constructed, and whether they are equipped with regulatory devices. But to a large extent these matters also depend on having a clearly defined and consistent set of rules to guide the allocation between various uses and users, and the ability to enforce these rules. All surface systems typically serve numerous users ranging from a few dozen farmers in the case of small tanks to hundreds of thousands in projects like Bhakra Nangal. In these cases, the scheduling of supplies between time and space is especially complex. This is where institutions come in. Apparently, the reservoir-based irrigation attracts a large number of communities to be involved to utilise the available water. In order to regulate and manage the surface irrigation systems (whether it is small or big) separate irrigation institutions (IIs) have to build up.

The early work on IIs was done by Wittfogel (1957), who concerned with the relation of irrigation with the emergence of strong centralised state systems. He argued that hydraulic agriculture, which requires great cooperative effort for organising capital and labour to build dams and canals, leads to greater political integration and centralisation. Following Wittfogel there have been several studies which have questioned the universal applicability of his irrigation hypothesis (large scale irrigation system requires great cooperative efforts to build dams and canals which in turn leads to greater political integration and centralisation). Thus, Leach (1961), Millon (1962), and Maass and Anderson(1978) found no clearcut relationship between the degree of centralization and the size of irrigation system; nor any evidence that irrigation works invariably result in a hydraulic bureaucracy or strengthening of the centralised power. However, there is some confusion in the concept of centralisation. Hunt and Hunt (1976: 129) have argued the need to distinguish between two distinct types of centralisation:

"One refers to authority in terms of irrigation systems, who exercised different decisions of the social and technical needs of irrigation. The other refers to generalised political authority, who exercised water as one aspect of a complex political role, which may involve other functions beyond water control." Whether and to what extent irrigation development contributes to centralisation of the state is evidently a debatable issue. But our concern here is with the way irrigation operations are structured and function and how they affect agriculture.

As it is indicated at the outset water is the basic input to raise crop productivity. Here it is important to point out the peculiarities of water and its use in agriculture. Water is in the nature of common resource serving numerous uses and users. Harnessing and using the resource for common benefit of users is generally done (or regulated) by some form of public organisation. Private ownership and operation is rare except in the case of wells which serve small areas and are within the means of individual farmers.

In surface irrigation the arrangements range from exclusive user management of construction maintenance and water allocation to exclusively government managed systems, with a wide variety of mixtures in between.

In principle, since irrigation makes a significant increase in productivity, the potential beneficiaries have an interest in collective efforts to develop and manage it. The world over, there are innumerable examples of such efforts in developing small works (like tanks, diversion of streams and rivers) which involve relatively simple engineering and serve relatively small, compact community of users. But in larger systems involving complex engineering, large outlays and vast number of users, the state tends to play a greater role in financing, constructing and managing them. The organisations in such cases are necessarily larger, more complex, more dependent on formal rules and bureaucracies. Indeed, in several countries, including India the state has taken over the full responsibility for all these activities in the case of modern reservoir-based irrigation. Even in the case of groundwater, which is exploited privately, regulation to prevent exploitation requires public (State) intervention.

The relative roles of the government, the user community, individual farmer, and theinterrelation among them obviously vary a great deal. This is reflected in differences in the form of the organisation, its internal structure (including the distribution of responsibilities and powers within the organisation; the locus and modalities of decision making; the manner in which functionaries are appointed and controlled) and the relation between irrigation organisations and the wider sociopolitical framework. However, it is inappropriate, and misleading, to view irrigation organisation as if it were a single entity: rather it has to be viewed in terms of arrangements for performing various tasks such as facilities construction, operationmaintenance, water allocation and conflict resolution in different phases of irrigation management (Kelly 1980: 1420).

These variations are a function of several factors: Rainfall, geology, topography and other features of the physical environment have a significant bearing on the kind of systems (both in terms of water source and scale) which are desirable and feasible in a particular context. The scale and complexity of the system affects whether or not it can be handled on an individual basis or by cooperative effort on the part of users or needs State involvement. The extent of State, users in terms of their sharing in the costs of setting up and operating systems have influence their interest and involvement. Often the interests of politicians and the bureaucracy dictate a high degree of direct role for the state in funding and managing irrigation[1].

Comprehensive descriptions of the structure and functioning of particular systems in a wider, historical perspective are rare; systematic comparative studies are even more so[2]. The literature on IIs has focused more on issues as centralisation vs decentralisation, bureaucracy vs user management, privatisation, and cost recovery. This preoccupation has been strengthened by the growing concern about the poor quality of service, depletion of groundwater; salinity, water logging and environmental problems; low cost recovery and similar features of State managed systems, which have become increasingly important in surface irrigation all over the world.

However, it is not, as if users have been indifferent to the way systems are managed. Studies in India have shown the existence of collective efforts among users of even Staterun canal systems. Wade (1987a) found that scarcity of water, cooperation among farmers and specialised irrigation were characteristic of the Tungabhadra canal system of Andhra Pradesh. Further he stated that there are strong and active village committees dealing with irrigation problems. These committees collectfunds, appoint common irrigators and sluice guards to patrol distributaries at the top end of canals, arrange representations to officials and politicians and have a sizeable common fund for different expenses. Many other cases of local mobilisation to persuade and/or pressure the state to construct/improve irrigation works, to change operational rules, ensure adequate supplies from the main system especially in times of scarcity have been reported (Vaidyanathan and Janakarajan 1989; Rajagopal 1991 and Sivasubramaniyan 1995). But, instances of the users actively participating in the design, framing of operational rules or continuing management of the system as a whole in India are few and far between.

1.3 User Participation in Management

There is good reason to believe that centralised bureaucratic management with little or no user participation leads to inefficient irrigation which contributes to the poor quality of irrigation provided by staterun systems. This has long been emphasised in irrigation management literature (see Coward 1980; Downing and Gibson 1974). Recent studies underscore this point even more forcefully (see Vaidyanathan 1991b; Chambers 1988; Sengupta 1991; Ostrom 1992; Tang 1992; GOI 1992; IIMI 1994, Sivasubramaniyan 2006, 2011).

A work by Tang (1992), based on 47 case studies of irrigation systems around the world, highlights wide diversity of experiences in the relationship between institutions and collective action in resource management. The results suggest that systems that include elements of selfgovernance (by users) can work well. As Tang points out, how well each system works depends on the physical, social and institutional factors operating in each case. These factors interact to create the unique set of needs and problems that must be handled by each governing institution.

On the basis of a comparative study of India and Philippines Sengupta (1991) has argued that farmers' participation is feasible in practical terms, and beneficial whenever applied. In a subsequent study (Sengupta 1993) based on data for over fifty such systems in India, and their aggregate effects, environmental impact, multipurpose uses and salient management features, he has suggested that a suitable alternative strategy can be developed based on the existing traditional techniques of irrigation and water management. He has also suggested that such an approach would promote efficient water management, better flood control and soil conservation, farmers' participation and balanced agricultural development.

In India, the need for wider involvement and participation of users for effective management has been long recognised. Encouraging users' involvement is now part of official policy. The necessary enabling legislation has also been enacted in several States. A number of experiments in this direction have been made under the auspices of nongovernmental organisations in different parts of the country. However, a recent assessment of this experience has concluded thus:

> "Efforts to actually organise farmers' groups and make them participate in management have been patchy and can claim only a limited success so far. Most State Governments have not really pursued the idea seriously" (GOI 1992: 125). Even after more than two and half decades this observation still persists in all the states of India.

The best example for the government not really pursued to organise farmers' groups could be seen through the enactment of Tamil Nadu Government's Act (Act no.7/2001) called, "Tamil Nadu Farmers Management of Irrigation Systems Act – 2000" (TNFMIS ACT 2000), to enable all surface irrigation systems, especially the tank irrigation that should form Water Users Associations (WUAs) to manage the tank system efficiently through active participation of both users and officials of Public Works Department (PWD). But this notion has not yet yielded appreciable results so far in Tamil Nadu (Brief summary of TNFMIS ACT 2000 is given in Appendix 1.1).

However, Maharashtra and Gujarat have done considerable pioneering work, both through nongovernmental and by the government organisations. Even then, the Planning Commission evaluated on this aspect thus:

> "... the area covered by these initiatives is very small, less than one per cent of area irrigated at present... For the most part, the outlet and canal committees are there only in name; their functions are vague; they seldom meet; they are not consulted on substantial issues; nor are department officers required to follow the advice" (GOI 1992: 126).

Elsewhere in the world, similar experiments[3] have been made or are under way. They vary in scope some have set up local water users associations; some have tried to involve them in the management of the system as a whole; and in a few cases more radical reforms seeking to get users to participate in system design, contribute to the costs and take responsibility for maintenance and operation, are being attempted.

The conference strengthened the growing consensus to move towards greater involvement of users and to substantially reduce the role of government in constructing and managing irrigation systems. But such reform not only requires the creation of Water Users Associations but also to strengthen the Associations at the tertiary level. Major role for users in decision making and greater responsibility for carrying out various management tasks such as construction of irrigation structures, operation and maintenance, water allocation and resolution of conflicts in all four phases of irrigation system such as water source control, water delivery, water use and water drainage is necessary. As Ostrom (1992) has pointed out that involving users and suppliers directly in all stages (including design) helps ensure that institutions are well matched to the particular physical, economic and cultural environment of each system. Such "selfgoverning" systems are more flexible, distribute water more equitably, and neutralize the "perverse incentives" that lead managers and consumers to abuse and neglect irrigation facilities.

The question of moving towards selfgoverning form of irrigation organisation with user involvement is important in the case of relatively large systems now under State control and even smaller, multicommunity systems where the State plays an important role. But the problem is rather different in the case of local surface irrigation works which have a tradition of community management. These works are very numerous and account for a sizeable proportion of the irrigated area in India. While their problems are different, their experience in user management the manner in which they are managed, the way they have changed over time, and their problems in ensuring efficient and equitable water management are of considerable practical interest to institutional reform in larger irrigation systems.

1.4 Management of Local Systems

India has a long tradition of using small storage and diversion canals drawn from local streams and rivers. Before the introduction of large reservoirbased canal irrigation and energised pumping of groundwater, tanks were the dominant form of irrigation in several parts of the country. In 195051 tanks and other minor sources irrigated 6.6 million hectares, which constituted about 3132 per cent of the country's total irrigated land. Since then the bulk of the extension of irrigation has been through large reservoirs and groundwater. The relative importance of minor irrigation, according to official statistics, even the absolute area under this category has heavily declined. This is attributed partly to the concentration of investment on large storages and groundwater; and partly to the deterioration of traditional local systems due to neglect (Vaidyanathan 1992: 25).

Among minor irrigation works tanks are the most important category. As per the 5th Minor Irrigation Census (2013-14), India has 21.70 million Minor Irrigation (MI) structures in 6,46,784 villages. In which 94.5% (20.51 million) accounts for groundwater depended well sources. In the remaining (1.19 million) surface flow schemes, MI tanks / ponds have the largest share of 41% which is estimated 2,37,000 tanks are in use in the country. The reported area under tank irrigation heavily declined to 1.97 million hectares in the 2000s that was reported about 3.0 million hectares in the eighties as against 4.5 million hectares recorded in the 1960s as the maximum tank irrigated area in the country. Tank irrigation is particularly widespread in South India. Nearly 60 per cent of the area under tanks is in Tamil Nadu, Andhra Pradesh and Karnataka. In these three states, during the 1980s tanks irrigate a much higher proportion of the total cultivated land (seven per cent against the national average of two per cent) and of the total irrigated land (24 per cent as against the national average of eight per cent) (GOI 1989). However, this trend has changed in the 2000s where the three states tank irrigated area to the total cultivated land declined to 6% against the national average of 1.5 percent and of total irrigated land it was 16% as against the national average of 3.4% (GOI Nov. 2017. Report of 5th Census of MI Schemes, Ministry of Water Resources, New Delhi).

1.4.1 Tanks in Tamil Nadu

Tamil Nadu has 41,127 tanks: of which 80.6% (33142) are small (serving less than 40 hectares each) and depend mostly on local rainfall. The remaining 7985 tanks are large category serving more than 40 ha each. In the latter category, some 3600 tanks are linked to larger systems fed by canals drawn from large rivers and storages (GoTN 1979; GoTN 2014 – TN an Economic Appraisal: S-43). Though the tanks are generally small, each serving one village or a part of it; some are large spread over several villages serving several thousand hectares. (The tanks which are the subject of this present study belong to the latter category). However, the volume and duration of water supply relative to the area irrigated, as well as its assurance of water supply vary widely among them.

There are some sketchy accounts available of their historical origin and evolution (see Raju 1941; Ludden 1978; Reddy 1990; Sivasubramaniyan 1995, 2006). For the most part tanks were constructed long ago (in some cases centuries ago) by efforts of local chieftains, temples and village communities. These tanks were maintained and managed by user communities. The *Kudimaramath* (a system of voluntary labour participation to upkeep and maintenance of irrigation channels) system provided for collective maintenance by community efforts: Under the *Mirasi* system of land tenure, a part of the village produce was earmarked for tank maintenance.

The wars led by local chieftains preceding the advent of British rule caused considerable physical damage to the irrigation works. Changes in land tenure and in particular the introduction of the Ryotwari system in place of *Mirasi* tenure and the recognition of exrulers/chieftains as *Zamindars* have further weakened the traditional socioeconomic structures which took care of the tanks. The deterioration in irrigation adversely affected government revenues land revenue was the dominant source and this led the colonial government to assume direct responsibility for maintenance of system facilities (other than distribution networks) and subsequently for repair, renovation and improvement, construction of new tanks (Raju 1941; Krishnaswami 1947; Dikshit et al. 1994).

The government takeover of maintenance did not, by all accounts, make much of a difference, partly because of inadequate allocation of resources. Attempts to revive *Kudimaramath* were also unsuccessful. However, the government did undertake substantial investments from the mid-19th to the early 20th centuries for major repairs, new construction/remodeling of anicuts to augment supplies to tanks, increasing the number of system tanks and construction of new tanks. Such works were taken up in Palar, Vaigai and Tambrabarani basins.

In the early 1900s, the government had initiated a survey of existing tanks to ascertain whether they were up to required standards and if not taken up works needed to bring them to standard. This survey made relatively slow progress. There was some quickening of pace in the postIndependence period but outlays have been rather small and results moderate. During the 1980s and 2000s a large tank Modernisation project was started. This programme focused mainly on engineering works to improve tank structures and distribution networks more or less entirely under the Public Works Department (PWD). Lately the programme has emphasised participation of users in the tank modernisation projects. Surprisingly, however, **neither government reports on minor irrigation nor the modernisation schemes, recognised that tank communities generally have informal local institutions for maintenance and operation.** Nor do they explore the role which they might play in improving management. In the 2007-15 Word Bank funded Tamil Nadu Irrigated Agriculture Modernization and Water Bodies Restoration and Management (TN-IAMWARM) project the World Bank Team has specifically insisted to strengthen the institutional aspects of the communities' participation in tank management. However, before getting the loan the PWD attempted some extent to strengthen this aspect by preparing basin wise beneficiary water users list to form the Water Users Associations as stipulated by the World Bank, but later, once the WUAs were formed no involvement of users in the modernization program was entertained.

1.4.2 Community Management of Tanks

The local irrigation works have traditional community-based institutions for their management is well known and documented by studies in different parts of the world[4]. There are also several studies relating to India[5]. These studies, many of which relate to Tamil Nadu, are rather limited in scope and coverage, but nevertheless useful.

There are few properly documented accounts of the way these user communities worked before the advent of the British rule. It is known that they were governed by customs and tradition; in some cases (notably North Arcot district of Tamil Nadu) these have been codified in writing and published (Water *Mamulnamas*) in the early 19th century. Many accounts in the latter part of the 19th century attributed the collapse of the

Kudimaramath to the decline of these traditions. Currently too there is a widespread belief that traditional arrangements for tank management have disintegrated and become ineffective. But recent studies cited above show that **while local tank management institutions are not uniformly effective, and there have been considerable changes, there is no warrant for the belief that they have declined everywhere**.

Most communities studied are found to have well defined rules for the contributions by *ayacutdars* for maintenance and managing water allocation, and mechanisms for enforcing the same. In many studies, it is found that community was to be actively involved in collective action for appropriation of water (by participating in channel maintenance to get more supply), in maintenance of distribution networks in the *ayacut* where water is given, and in regulating water use especially in times of shortage through functionaries specifically appointed for the purpose.

The general pattern observed in Tamil Nadu (for details see Vaidyanathan and Janakarajan 1989; Rajagopal 1991; Sengupta 1991, 1993; Ramanathan unpublished; Sivasubramaniyan 1995, 2006) is that the village leaders (*Nattanmaikars*) drawn from the large landholders of the dominant, usually upper, castes take active part in tank management. In some villages, the task is entrusted to a specialist functionary (*Kavaimaniyam*) experienced and knowledgeable about tank matters. Decisions made by them are implemented by irrigation workers (*Kammukatties*) mostly from the Scheduled Castes by rotation. They are expected to monitor water flows to the tank, watch the bunds and other tank structures, open and close sluices and regulate water flow in the *ayacut* when required by the *Nattanmai/ Kavaimaniyam*. The *Kammukatties* are paid by the *ayacutdars*, usually in kind, on per acre basis. In the Tambrabarani basin, the tank level organisations are more formal and elaborate (Sengupta 1991, 1993; Ramanathan unpublished).

There are well recognised procedures for maintenance of the distribution networks, including the contributions expected from the *ayacutdars*, penalties for nonparticipation and the organisation of work. The *ayacutdars* are expected to contribute labour in proportion to their holding; sometimes cash contributions are allowed. In some cases, the *ayacutdars* participate in cleaning inlet channels, when the PWD fails to do so. In some villages detailed accounts of contributions, default and fines are maintained.

There are also conventions and rules for regulating the flow of water in the *ayacut*. Based on a survey of about a dozen tanks in North Arcot district, Chambers (1977) noted the existence of systematic though varying rules for allocating water in times of scarcity. One village distributed water to fixed acreages during scarcity period; in five villages the usual distribution system of the "topenders first" was practised by the organisation, while in three others a special provision for "tail enders first" in times of scarcity was practised. Chambers reported however that water distribution under tanks was usually both inequitable and inefficient in terms of productivity. Because of conveyance losses water is less productive to the tail end of a channel. Scarcity of water also mostly affects the tail end farmers. The traditional rules of allocation according to the conditions prevailing then may not be applicable to the present. New rules may therefore emerge or changes may take place in the structure and functioning of earlier tank irrigation organisations.

Similar variations are recorded by Vaidyanathan and Janakarajan (1989) in their study on Palar Anicut System (PAS). According to them, there are considerable variations in the working of tank organisations: They found that while almost all villages had *Nattanmaikars* and *Neerkatties*, only five had *Kavaimaniyam*. Eight of the 15 villages covered by the study reported regular maintenance of inlet channels, distributaries and field channels. Six reported that regular maintenance was not done in the last five years. The system of rationing by turn in terms of scarcity was in vogue only in seven villages. The main features of villages where the institution functioned effectively include concentration of land control in one or two upper castes; absence of factions; tank being the only source of irrigation; low level of technology, and low absentee ownership (Janakarajan 1993b). The development of well irrigation and the fact that the 1980s were marked by a series of droughts and low tank supply must also have affected their functioning at the time of the survey.

Another study on functioning of tank institutions in two large system tanks in North Arcot district, Sivasubramaniyan (1995) reveals considerable variations in the functioning of irrigation institutions and also maintenance of irrigation channels and management of tank water supply. He observed: Although Kaveripakkam Tank (KPT-39) and Dusi-Mamandur Tank (DMT-31) had 69 *Nattanmaikars* in (KPT 14 and DMT 18) 32 villages, the functioning of number of *Kavaimaniyam* was only 3 each in both the tanks in 1990. Number of irrigation board delegates working in DMT (57) was more than that of KPT (14). Also, in both the tanks, the lower level irrigation workers such as *Kammukutties* and *Thotties* functioned more or less same (KPT – 43 and DMT -33) as per the total extent of ayacut in each tank. Regarding maintenance of irrigation channels, almost all the ayacut villagers contributed mostly their own labour and to some extent cash contribution for maintenance in normal tank supply year. However, the extent of contribution was only from head and middle reach villagers when the tank water supply is decreased to three-fourth to half the full tank level. During less than full tank supply years, supplemental well irrigation helps to achieve assured yield in the ayacut but non-well farmers are affected. Further, well irrigation is more favourable for KPT ayacut than DMT ayacut due to geological conditions. On the whole, functioning of local institutions is satisfactory to manage the tank water for better crop production.

A comparative study of canal and tank irrigation Systems in Cauvery Basin in Tamil Nadu has shown stronger institutional set up for collective action to secure supply in the systems with relatively poor, uncertain water availability compared to the system with abundant supply; greater uncertainty of supplies combined with greater difference between irrigated and unirrigated yield has stimulated greater interest in collective action in the lesser water availability system. The study also reported that the role of institutions in water management to be significant in the lesser water availability system where the degree of scarcity and consequently the potential productivity of water is high, but it is relatively insignificant as the system is not affected much by water scarcity. Even within an irrigation system, irrigation organisation is much stronger in the tail reach as it is more affected by water scarcity than others. This indicates that the extent of involvement of irrigation organisation in water management is linked to the extent of benefits likely to accrue to them by averting crop losses (Rajagopal 1991: 324-8).

The aforesaid studies also show that the characteristics of village level organisation have undergone changes because of vast increase in the number of *ayacutdars*, shift in land control from a relatively small number of upper caste landowners to numerous middle and lower castes, the attitude changes of farmers, and the spread of well irrigation in the tank *ayacut* of several villages. As a result, **the arrangements have got modified but have not broken down. There is still a strong tendency to invoke customs and tradition in managing irrigation supplies.**

1.4.3 Importance of Studying Large Tanks

Most of the available studies on tank irrigation deal with relatively small tanks whose *ayacuts* fall mostly within one village. Larger tanks such as Kaveripakkam (KPT) and DusiMamandur (DMT), the concern of the study, represent a considerably more complex structure. Both the KPT and

DMT are multisluice and multivillage tanks integrated into the larger Palar Anicut System (PAS). (For details of multisluice and multivillage system tank see figure 1.1). Their functions depend on the way the system as a whole is managed including the tank and individual villages served by the tank.

The concern of the selected tanks such as KPT and DMT are of special interest because these tanks involve a considerably more complex usermanaged organisational arrangements and greater interaction with the PWD in matters of water management, than the small, single village tanks. The PWD is not only responsible for deciding water releases in the channels drawn from river (through the anicut) but the control over the operation of the sluices also vests with its functionaries. They, however, work with and operate on the advice of a body representing the *ayacutdars*. Below the tank level organisation, which has a formal status, there are informal village level organisations to take care of maintenance and water allocation within their territories. There is reason to believe that **all these arrangements, evolved over centuries, have undergone significant changes due to a variety of factors and yet are working more or less smoothly**.

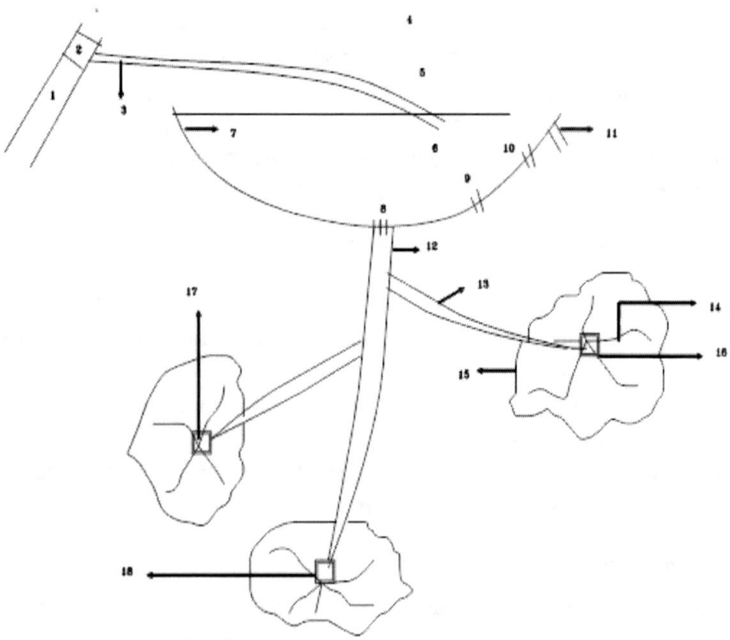

River 2.Anicut 3.Inlet Channel 4.Catchment Area 5. Tank Foreshore Area 6. Tank Water Spread Area 7. Tank Bund 8. Deep Sluice 9. Middle Sluice 10. Elevated Sluice 11. Surplus Weir 12. Main Channel 13. Branch Channel 14. Field Channel 15. Village Boundary 16.Head Reach Village 17. Middle Reach Village 18. Tail End Village

Figure 1.1: Structure of a System Tank and its Distribution Network

The objectives of this study are:

- To get a detailed picture of the structure and functioning of the management of Kaveripakkam Tank and Dusi-Mamandur Tank.
- To understand the roles of the PWD, the tank level organisation, and the villages and the interrelations between them.
- To examine the characteristics of water supply and distribution networks.
- To study the criteria by which water releases are regulated: the mechanisms of regulation adopted by the functionaries.
- To evaluate the combined impact of all these on the availability of water and productivity in different segments of the command.

An attempt is also made to bring out the changes in respect of each of the above dimensions which have occurred in the past, and how they have affected the organisation and the way it functions.

1.5 Methodology and Data Sources

The study was done in three stages: The tank as a whole; the villages in the *ayacut*; and a detailed survey of farms in selected villages.

1.5.1 System Level

System level inquiry in the first stage was to ascertain the characteristics of the tank systems and their management as a whole; how they were designed to work, how they had been actually working, and how the deviations had been accommodated. For this purpose, provision was made to understand the history of the systems, changes which have occurred and the causes thereof.

Data and information for this phase of study were derived from several sources: A limited amount of Archival material relating to Revenue and Public Works Departments; PWD records of rainfall data, storage levels of the tanks for about four decades, and water releases by channel and the tank restoration *Memoir* which gives tank specific technical data. The *Mamulnamas*, unique to North Arcot district, set down in written form, the rules, conventions and mechanisms which had evolved over a long time were in vogue in the early 19th century, are the basic historical documents to understand more about the tank specific details. (The original Tamil version and a translated version of the text of the KPT and DMT *Mamulnamas* is given in Appendix 3.1). Village level data on land use, irrigation and cropping at different parts of the *ayacut* villages are taken from registers relating to the Settlement (1882), Resettlement (1913), Updated Settlement (1983) and Adangal registers (2016). Chitta and Adangal registers of the villages were used to identify the current owners of *ayacut* land and wells.

1.5.2 Selection of Villages

The location of the villages in a tank is widely believed to be a major factor influencing the access to water which has in turn a bearing on the way institutions work and how users interact with them. However, there are also other factors, like topography, extent of groundwater use, and lack of supplemental source of water other than the tank affect the functioning of IIs. The variations and their implications can be captured only by detailed interviews in the villages served by the tanks. Accordingly, at the second stage, a village level survey was conducted in 2016 in all the 32 villages (KPT 14 and DMT 18).

Data for the second stage were derived from interviews held with (a) government officials in charge of managing the PAS as a whole; (b) government functionaries, delegates representing ayacut villages, and villagers knowledgeable in the functioning of the tank level organisation; and (c) interviews with village leaders who previously functioned as *Nattanmaikars, Kavaimaniyam, Thalaiyaries*, and *Neerkatties*, in each of the ayacut villages. Interviews were conducted with the help of a check list of questions on aspects relevant to the study. More importantly a similar survey was conducted in 1991 in the same sample villages. Hence wherever required that year details are also discussed for proving a comparative picture of changes in village / tank level organisations.

The government functionaries, village elders and knowledgeable persons were interviewed individually. The number of persons interviewed was conditioned by constraints of the project's resources and the difficulties of locating the persons. In every case attempt was made to collect information by interviewing at least two or three persons. In each village, the lower level irrigation functionaries who functioned a decade ago such as *Kammukatties* and *Thotties* were usually interviewed in a group (ranging from two to six persons). As said above, the earlier survey (1991) has done a good work on this and relevant details on this aspect are referred in this report.

The checklist sought to elicit information on the following aspects: (1) the number and type of personnel involved in managing irrigation, the basis of their remuneration and the approximate quantum received during the latest two supply years; (2) the organisation of maintenance, form and magnitude of labour contribution, frequency of maintenance and participation by farmers; (3) the basis of water allocation during normal and deficit periods, the procedures for regulating the water use in scarcity periods; and (4) the effect of wells on water management.

A census of all the villages was not feasible. So, the following methodology was adopted to conduct the farm survey. A priori, the conditions of water supply are expected to be important in determining the working of institutions for community management and people's attitude to it. The initial village survey suggested that when tank water supply was assured collective institutions were less active. They were more active when water supply was accessible but uncertain. The extent of development of wells also seemed to affect the functioning of the tank management institutions. Therefore, the villages were classified by sluice and reach. Deepest sluice/ head reach villages are relatively good in water supply conditions and elevated sluice/tail reach villages the worst.

In Kaveripakkam tank, on the basis of elevation of the sill level, the ten sluices have been grouped into three categories: deep, middle and elevated. The villages served by these sluices are categorised into head, middle and tail reaches depending on the location in the channels fed by the sluices.

DusiMamandur tank has only four sluices. They are divided into two sets. Each set consists of one low and one high level sluice. The difference of depth between two deep sluices is only three feet, but there are significant differences in water supply conditions of their respective *ayacuts*. Each set of sluices irrigates nine villages located in various reaches of them. According to our classification, three villages are available in each type of head, middle and tail reaches in each set of sluices.

On the basis of the above classification, nine villages from both the tanks were selected for detailed investigation. In doing so, care was taken to delete the villages irrigated by multiple sluices, where the conditions of water supply and management are more difficult to capture. The caste structure of the villages has also been taken into account. In some villages land ownership is diffused between two or more castes. In these villages the working of IIs are likely to reflect not only differences in water supply conditions but also intercaste tensions. Therefore, only villages with a single dominant land-owning caste were selected. The distribution of *ayacut* villages in the two tanks by sluice andreach, as well as the dominant land-owning castes is given in Tables 1.2 and 1.3. From this we haveThe specific

Table 1.2: Classifications of Sluices and the Villages Served by Kaveripakkam Tank.

Reach/Sluices	***Deep***				***Middle***			
	Village	***Total wells***	***Activeness of institutions***	***Caste***	***Village***	***Total wells***	***Activeness of institutions***	***Caste***
Head	**KATTALAI**	74 (57)	LA	S(NR)				
	Kaveripakkam	185 (31)	LA	S(NR)			-----	
	Cheri	192 (64)	A	S(NR)				
Middle	----------------------------				Kondapuram	162 (76)	LA	MC
					Maganipattu	129 (63)	A	MC
Tail	**ATHIPATTU**	148 (130)	D	S(MR)	**ALAPAKKAM**	76 (89)	LA	MC
	Sirukarumbur	238 (137)	D	S(MR)	Puthupattu	103 (107)	D	S(RY)
	Eralacheri	68 (38)	D	MC	TPP	260 (136)	LA	MC

(Contd...)

Elevated			
Village	***Total wells***	***Activeness of institutions***	***Caste***
KADAPERI	37 (137)	A	S(NR)
PANNIYUR	56 (192)	A	S(RY)

Kilveeranam	65 (175)	D	MC

Table 1.3: Classifications of Sluices and the Villages Served by DusiMamandur Tank

Reach/Sluices	*Mamandur sluices*				*Dusi sluices*			
	Village	*Total Wells*	*Activeness of institutions*	*Caste*	*Village*	*Total Wells*	*Activeness of institutions*	*Caste*
	Mamandur	133 (33)	LA	S(NR)	**VAGAI**	5 (19)	A	S(NR)
Head	**NARASAMANGALAM**	43 (96)	LA	S(NR)	Dusi	43 (24)	A	S(NR)
	Alingilpattu	15 (272)	D	S(NR)	Kuranganilmuttam	37 (51)	A	MC
	Valavandal	70 (73)	A	MC	Kanikiluppai	14 (31)	A	MC
Middle	Vadakalpakkam	39 (42)	A	MC	Kilnayakanpalayam	34 (33)	A	MC
	Girijapuram	39 (82)	A	MC	Menallur	75 (112)	A	MC
	SOTHIYAMPAKKAM	76 (139)	A	MC	Pallavaram	71 (43)	LA	S(MR)
Tail	Bagavanthapuram	61 (141)	A	S(NR)	Senianallur	31 (75)	A	MC
	Elacheri	123 (93)	D	MC	**PUNAITHANGAL**	37 (63)	LA	S(MR)

Note: Activeness: LA = Less active; A = Active; D = Defunct. Caste: S = Single Caste; MC = Multi Caste; NR = Nayakar; MR = Mudaliar;

SC = Scheduled Castes; RY = Reddy. PI = Pillai. The caste hierarchy is MR>PI>RY>NR>SC.

TPP = Thuraiperumpakkam. Figures in brackets indicate density of wells per 100 hectares of ayacut.

details on **water allocation** include: rules and regulations; PWD's role in this decision; formal and traditional water rights (in normal years, scarcity years and when water is insufficient from the beginning of a season), major changes, if any, in the entitlements in the past few decades, water allocation in the past, allocation procedures during scarcity period, their performance in actual practice, cropping decisions, role of the Irrigation Board and the PWD in allocation of water across villages, disputes and their settlement; *Mamulnama's* rules and procedures invoked; major changes in respect of rules/conventions on water release, allocation, irrigation practices, land use pattern, cropping pattern, extent of wells in the *ayacut* and how exactly they have influenced allocation rules; effect of changes in the rules and allocation in the last two to three decades, and experiences in this regard in the latest two supply years.

1.5.4 Selection of Sample Households for the Detailed Survey

The third stage of study was focused on the farmers experience of nine selected *ayacut* villages (five in KPT and four in DMT), followed by a detailed sample survey of 213 farmer households.

The village survey gave more details on the working of maintenance and water allocation, thenumber of functionaries in different categories and their remuneration. The sample survey of the *ayacutdars* was meant to get information on actual irrigation experience in 2016-17 and the next previous year when supply was available (2005-06); individual farmers involvement in maintenance of channels/management of tank water; value of their contribution to maintenance and payment to irrigation functionaries; problems regarding access to water and how they were resolved; details regarding availability and use of well water; the prevalence of purchase and sale of well water; and the nature and extent of cropping patterns, and yields.

The selection of households in the sample villages was done as follows: The unit of selection was the "farmer" (household) and not "plot" possessed by him in the *ayacut*. This decision was taken considering that particulars regarding conditions of water supply, maintenance activities, cropping pattern and yields can be documented more accurately for the holdings than for a particular "plot", especially when the plots are scattered. Respondents have difficulty in remembering what happened in terms of water supply in particular plots; also, many of the questions (such as those relating to participation in maintenance and costs) relate to the farmer and not to a particular plot.

But conditions of water supply and other related characteristics, even participation in maintenance, are affected by the location of the plot and whether or not it has access to groundwater. To capture these variations and to get a clear detail of the variability in the conditions of tank water supply within each village the *ayacut* was demarcated into **head reach** and **tail reach** with the help of the survey maps available with the village administrative officer. A complete list of *ayacut pattadars*, the extent and location of plots held by them in the *ayacut*, and the extent of well irrigation was obtained from the village land registers.

For the purpose of sampling, the *ayacut* farmers were broadly classified into well farmers and nonwell farmers. **Well farmers** are those on whose holdings according to the entries in the *adangal*, there was at least one well, and well owned by each farmer, irrigates more than 60 per cent of his total *ayacut* holdings. Others are nonwell farmers. The nonwell farmers were again divided into two. Head segment nonwell farmers and tail segment nonwell farmers. **Head segment nonwell farmers** are those who have more than 60 per cent of their total *ayacut* holdings in the head segment of the village. **Tail segment nonwell farmers** are those who have more than 60 per cent of their total *ayacut* holdings in the tail segment of the village.

In the case of well farmers, our main concern was to understand their dependence on tank water and attitude to tank related institutions/practices. Therefore, no further subdivision of this category by reach was made. After listing out the *ayacutpattadars* in each village into well farmers and nonwell head and tail farmers, eight farmers were randomly selected in each category. Altogether a maximum of 24 farmers were interviewed in each village. Where the number of well farmers was less than eight, all well farmers were interviewed. Hence information for third stage of study was derived from a detailed survey of a sample of 213 farmer households in the selected nine villages (for details of village level /HH level survey see photos in pp. 426-27).

1.5.5 Scope of the Household Survey, 2016-17

From each sample household the survey sought information on: *Patta* number, name, caste of present owner and tenant cultivator, household characteristics (hh composition, educational status, nature of occupation, sociopolitical status within and outside the village), land holdings, access to well irrigation, nature and extent of participation in maintenance in the latest two supply years, adequacy and reliability of tank supply, specific problems, efforts taken to resolve them, method of irrigation at field level, collective efforts for water distribution both in normal and scarcity supply years, opinion regarding the functioning of village, tank and system level IIs.

Information and data for the following aspects were collected: Crop wise details (on the variety and input used, area sown), the sources and conditions of water supply, nature of well water use (own/purchased) and the yields for the latest two supply years, payments to irrigation functionaries (basis of contribution, activities performed and actual payments given), details on water sale, area irrigated and amount received by water sale, characteristics of well water supply, purchase of well water (crop, area, rate and amount paid), dispute settlements, changes in water distribution rules due to specific conflicts over allocation of water.

To collect the above details, an intensive field survey from March to July 2017 was conducted. The cultivators of sample households were identified and contacted with the help of village assistants (VAO) and other knowledgeable persons in each village. However, some of the respondents were always 'busy' and reaching them was very difficult. By repeated visits and through persuasion by village friends, we eventually contacted them and also got satisfactory responses from them.

Interviews with respondents owning wells with enough water irrigating a relatively large *ayacut* took about four to five hours. As against this, interviews of a few tail reach respondents who did not sow anything in the *ayacut* for both the reference years (as a result of marginal holdings, their involvement with other jobs and the nonavailability of tank water due to encroachment of channels) took hardly an hour. Time for all other interviews ranged between these two extremes. Although in both the reference years, 2016-17 and 2005-06, *ayacutdars* received good tank water supply the gap between the two years is large. In two of the KPT sample villages, namely **Kattalai and Athipattu** served by the deepest and middle level sluices respectively did not get the tank supply throughout the latest supply year. The reason for no supply to the former village is due to breakdown of sluice, which was not repaired by the PWD till March 2018, and for the latter due to full encroachment of supply channel, at Kaveripakkam village, leading from the KPT.

With this backdrop, chapter 2 deals with the Palar Anicut System from which the sample tanks received its source of supply.

Notes

1. For an overview of the factors responsible for variations in nature of irrigation works and irrigation organisations see Vaidyanathan 1985.
2. The available literature is reviewed in Vaidyanathan 1985.
3. The experiments, their achievements and the problems encountered in different parts of the world have been documented in papers presented at an "International Conference on Irrigation Management Transfer" (IIMI 1994); for a brief review of the conference proceedings see Vaidyanathan 1994b.
4. Geertz 1967; Fernea 1970; Gray 1963; Downing and Gibson 1974; Coward 1980; Vaidyanathan 1985; Kelly 1980; Leach 1961.
5. Chambers 1977, 1988; Narayana 1982; Vaidyanathan and Janakarajan 1989; Sengupta 1980, 1982, 1984, 1991, 1993; Rajagopal 1991, Ramanathan and Ghose 1994; Sivasubramaniyan 1995, 2006, Vaidyanathan 2001.

Chapter - 2

The Palar Anicut System

2.1 Historical Background

The Palar Anicut system (PAS), from which the selected tanks (Kaveripakkam and Dusi-Mamandur) get the supply, is situated in the North Arcot district[1] of Tamil Nadu. (For the location of KPT and DMT in PAS see Figure 2.1). The inland districts of Vellore and Tiruvannamalai lies

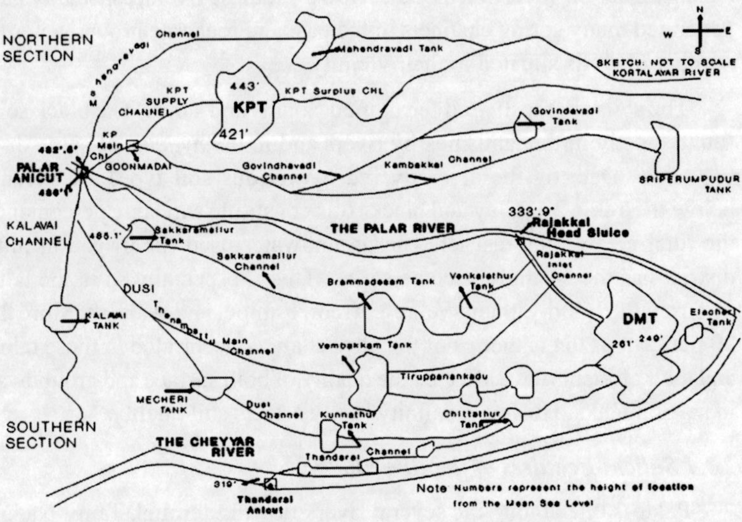

Figure 2.1: Location of KPT and DMT in the Palar Anicut System.

between 12 degrees 0 minutes and 13 degrees 15 minutes of the Northern latitude and 78 degrees 20 minutes and 79 degrees 50 minutes of the Eastern longitude. Its total area is about 12,268 sq. kms spread over 13 taluks in four Revenue divisions. It is bounded on the north by the State of Andhra Pradesh and on the south by the South Arcot district of Tamil Nadu. On the west are the district of Dharmapuri while on the east is Chengalpattu-MGR district of Tamil Nadu.

The districts have sizeable hilly and mountainous regions. The important mountain ranges are the Javadis and the Elagiri. The normal rainfall of the districts is 936.2 mm (Vellore) and 1040.0 mm (Tiruvannamalai) which is significantly more than the State's normal rainfall of 921.0 mm. The districts benefits from both the South-West and the North-East monsoons. About 45 per cent of the total normal rainfall of the districts occurs in the South-West monsoon period of June-September. This is followed by the North-East monsoon (October-December) which brings about 40 per cent of the total rainfall. The winter season (January-February) normally accounts for about four per cent of annual precipitation. The summer months (March-May) bring about 11 per cent, which helps the farmers to grow rainfed crops such as groundnut and pulses.

The districts have no perennial rivers. The rivers Palar, South Pennar, Cheyyar and Pennai which drain most parts of the districts remain dry during all but the monsoon months. However, these rivers, especially Palar, generated many spring channels in the past and recharge groundwater to a number of wells situated in their vicinity.

The soils are mostly red ferruginous loams and sands. The black soil is found mostly in the lands nearby rivers and in the *ayacuts* of some of the big tanks. Data on the area covered by various soil types are given in below the Table 2.1. Clay and black soils constitute a meagre percentage to the total area of the district. The groundwater used for irrigation in the district has increasingly become saline. This is especially so in the taluks of Tirupattur, Gudiyatham, Vellore, Arcot, Ranipet and Walajah. More than 70 per cent of the tanneries of the district are concentrated in these taluks, and their effluents have affected the quality of both surface and groundwater in several tanks. However, salinity rarely affects soil quality.

2.1.1 Salient Features of the River Palar

Palar is one among the several rivers flowing through Tamil Nadu. It rises at *Ambaijidurga*, about 32 kms to the east of *Nandidurg* in the Kolar district of Karnataka State. It travels a length of 80 kms in that State and

Table 2.1: Taluk wise Area by Soil Types in North Arcot District

Sl. No.	*Soil*	*Type Area in %*	*Predominantly Found in the Taluks*
1.	Clay	1.00	Tirupathur, Katpadi, Vaniyambadi, Ambur, Gudiyatham, Walaja, Arcot, Vellore, Arakkonam and Cheyyar.
2.	Clay loam	11.60	Tirupathur, Vellore, Vandavasi, Katpadi, Vaniyambadi, Arcot, Ambur, Cheyyar, Arakkonam, Walaja and Gudiyatham.
3.	Black soil	3.20	Tiruvannamalai, Chengam, Polur, Arani, Vandavasi, Arakkonam and Gudiyatham. Walaja, Arcot, Tirupathur,
4.	Red loamy	36.70	All taluks and Mostly in Polur.
5.	Sandy loam	47.50	All taluks and Predominantly in Chengam, Tiruvannamalai and Vandavasi.

Source: District Statistical Handbook 2016-17, Vellore and Tiruvannamalai districts.

then flows through Andhra Pradesh for a short distance of 34 kms before entering Tamil Nadu. Within Tamil Nadu it runs for 240 kms and finally joins the Bay of Bengal some 67 kms to the south of Chennai.

The Palar river has seven major tributaries joining its upstream of Palar Anicut: these are Mannar, Kallar, Malattar, Agaram, Koundinyanadi, Periyar and Poiney. The Cheyyar and Killiyar tributaries join the Palar below the Anicut. The total drainage area is 10,773 sq. kms, of which 9032 sq. kms are in Tamil Nadu and the rest in Andhra Pradesh and Karnataka[2]. The normal rainfall of Palar basin is 1023 mm per annum, of which 40 per cent is precipitated during the north-east monsoon.

2.1.2 Irrigation in the Palar Basin

Though several irrigation channels take off from its course, direct irrigation from Palar is negligible; most of the irrigation is under tanks. As per the latest available data (GOI 1993: 10) there are 2614 tanks in the Palar basin, commanding 1,35,715 hectares in Tamil Nadu of which 79% are rainfed non-system tanks (which are not connected to any river/anicut and are solely dependent on the supply available from their catchment run-off). At its origin in Karnataka, the Palar feeds a couple of large tanks, namely, Bethamangala and Ramasagar, which are mainly used for irrigation purposes in that State. These are the two significant tanks in the upper reaches of the Palar and there is no reservoir across the Palar. At the turn of the last century (1900s) throughout the Palar basin (within Tamil Nadu) there were 211 spring channels irrigating an ayacut of 4615 hectares, but most of them have dried up gradually due to depletion of groundwater table by large scale extraction of water since the 1960s. A few springs (as of the year 2000) are still in operation however, these springs are possible only when the Palar river gets its adequate flow at least for more than two weeks. Important among these are given in Table 2.2.

Table 2.2: Spring Channels Above and Below the Palar Anicut, 1950s.

Sl. No.	*Name of Spring Channel*	*Length from river bed (metre)*	*Tank fed/ direct*	*Discharge (cumecs)*	*Ayacut (ha.)*
1.	Arapakkam (17.6 kms above the anicut)	900	nil	6.4	97.2
2.	Vanni Vedu (98.40 kms above anicut)	900	direct	9.6	170.9
3.	Vanni Vedu (9.6 kms above the anicut)	900	nil	6.4	66.4
4.	Gudimallur (6.4 kms above the anicut)	900	Gudimallur tank	9.6	171.6
5.	Puduppadi (3.2 kms above the anicut)	600	Puduppadi tank	8.0	145.3
6.	Tirumalaichery	1800	nil	1.2	149.3
7.	Kalam Padi	1050	nil	4.0	295.8
8.	Sakkaramallur-Mudalipuram (0.80 km below the anicut)	1650	direct	9.6	104.1
9.	Sakkaramallur kammapuram (1.2 kms below anicut)	1800	nil	8.0	139.1
10.	Kalambadi	390	nil	1.2	225.1

Source: Report of the Palar Anicut System, Office of the E.E., Vellore division, Vellore district, Undated.

2.1.3 Main Features of Irrigation in the Command before the Construction of the Palar Anicut

Prior to the construction of the Palar Anicut, lands on either side of the river were irrigated by means of channels with open head, from a temporary *Korambus* (a diversion structure made up by wood and sand-bags) raised by the *ayacutdars* themselves across the river during every supply year. These *Korambus* (whose exact number is not known) were frequently washed away by floods and took much time and effort in repeated reconstruction. One can get some idea of the process from the *Mamulnamas* for KPT.

Water flowing in Palar was diverted to the KPT by a *Korambu* constructed, repaired or reconstructed as necessary by *ayacutdars* from 19 ayacut beneficiary villages mentioned in below the Table 2.3. During the flood season, whatever may be the number of men going to the river channel to divert water, two men from Alapakkam village were to remain

Table 2.3 KPT Villages for Appropriation of Water from the Palar

Sl. No.	*Name of Village*	*Number of Persons*
1.	Kaveripakkam	20
2.	Kondapuram	12
3.	Athipattu	4
4.	Villaham	4
5.	Sirukarumbur	4
6.	Eralacheri	4
7.	Perukarumbur Nanderi	2
8.	Kattalai	8
9.	Abiramacheri	4
10.	Cheri	15
11.	Maganipattu	6
12.	Thaluvanampattu	2
13	Thuraiperumpakkam	6
14	Sirasthavalapuram	2
15	Puthupattu	2
16	Kilveeranam	1
17	Panniyur	1
18	Gengatharanallur	1
19	Alapakkam	2
	Total	**100**

Note: During the 1882 Settlement and the 1913 Resettlement some small villages were clubbed with big villages. These are sl. nos. 4, 7, 9, 12 and 18.
Source: Mamulnamas of KPT (see Appendix 3.1).

at the tank and make Paddy *straw fire* to vigil at night on the tank bund to alert the people including the *Talaiyari* (village revenue functionary to assist the Village Administrative Officer), the *Karnam* (village accountant), and the *Maniyam* (village headman) and inform them about the flow measurements from the river channel. They took turns at eating and sleeping. It was compulsory that one of these two persons was always present to watch the water flow.

Whenever there was flow from the river, a *sluice guard* (head fisherman) was also to be present at the tank. His main function was to **protect the sluices from damage and to remove the silt deposited in and around it.** During the supply period if there was any problem in the sluice, only the *sluice guard* was authorized to undertake the needed repairs.

If supply continued even after it touched the surplus weir at a storage depth of 8.4 meter (28 feet), the construction of surplus embankment was to be taken up to a height of 0.9 meter. Ten of the villages (Kattalai,

Abiramacheri, Maganipattu, Thaluvanampattu, Thuraiperumpakkam, Sirasthavalapuram, Kilveeranam, Panniyur, Cheri and Gengatharanallur) had to construct the surplus weir according to the procedure detailed in the *Mamulnamas*. All these villages are located at the tail end of the distribution network and therefore stand to benefit more from increased storage in the tank. The remaining persons, allotted for each village as stated in Table 2.3, were to keep vigil at the diversion site to make sure that the *Korambu* built across the river is not breached as long as water flow continued in the river.

During the period when water flowed into the tank, one person was to inform the level of supply available to the inlet channel from the Palar. Another person was to inform the level of supply from two points: the *Singha Kinathalavu* and *Kachchi Nadhi* channel. Thus, three persons were required to inform the level of flow from their respective places to the officials stationed at Kaveripakkam village. These processes might be continued until water ceases to flow in the inlet channel. These persons also took note of the water level in KPT at *Bommai sluice* (Paracheri sluice). During the process of opening and closing of sluices the *sluice guard* might also be present. During the flood season the bunds of *Sumaithangi* and *Jadayeri* tanks (which are located adjacent to the KPT) should not be raised beyond its original capacity.

For DMT, we do not get any detail from the *Mamulnamas*, regarding the number of persons from each village required to participate in the diversion of flow, either from the Rajakkal channel or from the diversion spot at Palar Anicut. However, the general method adopted for getting supply to the tank is specified clearly.

According to the specification, when there was no supply from the Palar, two *Kondams* were to be built, one at Tiruppanankadu and the other at Vallakulam to divert the drainage and rain water flowing in the inlet (Thennampattu) channel to DMT. Further, along with the following five *madagus* (sluices) (two in Arasankuppam, one each in Settithangal, Olugampakkam, Vellakulam), the tank surpluses of Hariharapakkam, Cholapuram, Perumbulimedu, Sellaperumbulimedu, Sithathur and Alingilpattu villages and the field drainages also reach DMT through the foreshore area of Chithathur Odai. In the above said manner, if water reaches DMT at Full Tank Level (FTL), it should be feasible to irrigate the *ayacut* for 18 months.

The *Kondams* at Tiruppanankadu and Vallakulam were to be removed during the Tamil month *Aadi* (July), and the channel should not be obstructed by constructing *Kondam* till the Tamil month Thai (January). The *Kondam* could be built only after Thai. During the month of *Chithirai* (April), when rain water and the surpluses and drainages of Venkalathur village are available, the *Kondam* should be removed. Only if the water stopped itself as a result of poor rainfall or drainages the *Kondam* might be built.

2.1.4 Construction of the Anicut and Improvements in the Channels

After prolonged debate, the desirability of a permanent Anicut to replace the *Korambus* was investigated by the Superintendent of tank repairs in 1815, who recommended the construction of such an Anicut at the present site. This work was taken up in 1855 and completed in September 1857.

The Palar Anicut was built across the Palar at a point about 7 kms below the Arcot town. As constructed in 1857, it was 1.5 meter in height, 790 meters in length with a vertical fall having 22 under-sluices on the northern side and 10 on the southern side. There were three head-sluices namely the Kaveripakkam and Mahendravadi on the north bank and the Dusi-Sakkaramallur on the south bank. The Mahendravadi channel supplied the tank of that name. The Kaveripakkam main supply channel branched into two (viz., Kaveripakkam tank supply channel and Govindavadi channel, the latter supplying to Govindavadi tank). On the south side there were three channels taking off from the main Dusi-Sakkaramallur head-sluice, namely, the Sakkaramallur channel supplied by the under-sluices, the Dusi or Thennampattu channel, and the Kalavai channel. The other technical details are given in Table 2.4.

Table 2.4 Distribution of Head Sluices and the Main Channels at Palar Anicut before Improvements in 1883

Section	*Name of the head-sluice*	*Number and size of Vents*	*Main supply channels*
NORTH	Kaveripakkam Main	18; 1.5x1.5 metre	Kaveripakkam tank-supply channel and Govindavadi channel
	Mahendravadi	9; 1.5x1.5 metre	Mahendravadi channel
	Sand-Sluices	22; 1.5x1.5 metre	
SOUTH	Sakkaramallur	2; 1.5x1.5 metre	Sakkaramallur channel
	Dusi main	15; 1.5x1.5 metre	Dusi or Thennampattu channel and Kalavai channel
	Sand-Sluices	10; 1.5x1.5 metre	

Note: Sakkaramallur head-sluice was merged with the Dusi main head-sluice, a few years after its construction.

Source: The Palar Anicut System, Sl. no. 8, 1886, p. 4.

In the late 1880s, as part of an improvement programme, all the vents in the head-sluices, except in the Kaveripakkam main sluice, were modified to be made in line with the requirements of the supply to the *ayacut*. Originally, the Mahendravadi sluice was constructed to supply Mahendravadi tank and numerous other small tanks solely by means of this channel. Owing to the rocky nature of the soil in which the channel was excavated, it was never cut to the full length; hence the supply to the tanks commanded had to be supplemented from the KPT through the Ponnappanthangal sluice situated closer to the Mahendravadi channel. Consequently, in order to conserve water, six of the nine vents in the head-sluice were closed altogether. After modification the northern section has two head-sluices, viz., the Kaveripakkam head-sluice and the Mahendravadi head-sluice.

Initially only one head-sluice, namely Dusi Main, was formed in the southern section. The Sakkaramallur channel took off from this main channel. During the late 1870s a separate Sakkaramallur head-sluice was formed by adapting the original under-sluices which were meant to serve as sand-vents in the Anicut. This arrangement proved deficient and the channel could not draw much water because of silt formation in front of the sluice soon after its construction. As a result, the Sakkaramallur head-sluice was closed and the under-sluices restored to their original function. But a new head in the main Dusi channel was formed in the early 1880s to feed the Sakkaramallur channel. The construction of the new head resulted in considerable silting of the head of the Dusi channel and the banks of Sakkaramallur channel were also breached on several occasions. It was, therefore, decided to construct once again a separate head sluice supplying to the Sakkaramallur channel from a new cut in the left bank of the southern section. This was done in the late 1880s with three vents each with the dimensions of 1.8 mt length and 1.5 mt breadth.

In an attempt to check/mitigate the large accumulation of sand in the river bed being drawn into the channels, in the late 1880s the sills of the sluices on this side of the river were raised to a higher level than those on the north side. Below the table 2.5 provides the technical details of the main channels and the head-sluices in the Palar Anicut after the improvements.

Towards the end of the 19th century, the sluices as well as the shutters were further improved by fitting the screw-gearing shutters instead of the needle shutters as provided at the initial stage. In 1930 the sand vents, which had become blocked completely, were reopened and all vents were provided with screw gearing shutters. During 1960s all the main channels

Table 2.5: Technical Features of the PAS after Improvements

Name of the main channel	*Sill level (MSL) (Metre)*	*Full supply Depth (Metre)*	*No. of vents with size (Metre)*	*Theoretical discharge (cumecs)*	*Total length of channel*	*Bed width (Metre)*
NORTH SIDE						
Mahendravadi	144.9	1.05	3; 1.5 x 1.5	443	38 Kms	13.5
Kaveripakkam	144.9	1.05	18; 1.5 x 1.5	2761	16 Kms	64.8
Under-sluice	144.9	-	22; 1.5 x 1.2	-	-	-
Sand-vents	144.9	-	10; 1.5 x 1.98	-	-	-
Falling shutters	147.3	-	25; 3.6 x 3.6	-	-	-
SOUTH SIDE						
Sakkaramallur	145.5	.2	3; 1.8 x 1.5	500	32 Kms	12
Dusi-Kalavai (Thennampattu)	145.8	1.5	15; 1.5 x 0.9	1282	37 Kms	18.6
Sand sluice	144.9	-	20; 1.5 x 1.2	-	-	-

Note: Dash indicates no such detail could apply to these sluices/ vents; MSL = Mean sea level.
Source: A.F. Cox, Madras District Manuals, N.A. district, Vol. II, 1894, pp. 194-8; Improvements to Palar Anicut System below Ranipet, PWD, IRS report no. 3/61, Poondi, 1957, p. 5.

taking off from the Anicut head up to a certain distance were lined with concrete to prevent silt accumulation and facilitate free flow of water in the channels. It should also be noted that between 1900 and 1950, considerable sums were spent on repairing the damages and breaches which occurred at the Anicut wall as a result of heavy floods (as many as nine times during the first half of the last century).

The distribution network was gradually improved after the construction of the Anicut. The Kaveripakkam channel was much extended and new channels with masonry head works were excavated. Originally all the channels connected with the system were narrow and of irregular breadth. During the famine of 1876-77, substantial work to widen the channels and reduce the falls were started; it took nearly two decades to complete the construction. During this period the southern section of the system was improved and 21 tanks were given an assured supply by the construction of the Kukkundi branch of the Kalavai channel. Improvements to distributaries which were in a "deplorable" condition were also part of the programme. However, the progress on this seems to have been slower than in the case of main channels. It is reported that the southern channels were "badly designed and carelessly excavated at the first instance" (Vaidyanathan and Janakarajan 1989: 36).

Despite the improvements it would seem that the channels have not functioned at designed capacity basically due to the elevation of head-sluices as well as the accumulation of silt in front of the sluices. The deficiency is more marked in the southern channel. An estimate in 1953 showed that the actual drawal capacity of the northern channels was about 75 per cent of designed capacity, while in the southern channels it was only 52 per cent. This is one of the major reasons to believe that even during the normal supply years in the PAS, supplies from the southern channels are always inadequate to fill up the tanks fed by the system.

2.1.5 Effect of PAS Modernisation on the Availability of Supply

The construction of the Palar Anicut and the improvements in the channels carrying the water to tanks made a significant difference in several respects. The replacement of the temporary *Korambus* by a permanent anicut practically eliminated the substantial amount of labour and other resources which the *ayacutdars* of several tanks drawing from the Palar had to spend every year in order to construct the *Korambus* and the inlet channels and maintain them in good repair. Since the height of Palar Anicut was much higher and more stable than the earlier *Korambus* at that site, a larger proportion of river flow is likely to have been diverted at that site. Substantial improvements were also made in the main and branch channels carrying the water to the tanks (by way of realignment, widening and deepening, more gradual and better designed gradients and siding-wall, the lining of considerable lengths in the head of all the four main channels). All these not only made supply to the tanks already drawing water from the river more assured but also enabled more tanks to be connected to the system.

The *ayacut* served by the system increased substantially due to improvements made in the Anicut. Prior to the construction of the Anicut the *ayacut* of the system was estimated at less than 4,050 hectares. By 1883, the *ayacut* was estimated to have increased to about 21,500 hectares; and further to 30,400 hectares at the turn of the century. It further increased to 32,800 hectares in the first half of the century at which level the ayacut has stabilized since. The expansion of the *ayacut* reflects partly the increase in the total number of tanks fed by the system (from 201 during 1880s to 317 in 1950s) and partly by an increase in the *ayacut* served by newly connected tanks (see below the Table 2.6) reflecting more abundant water supply. But very little is known about the latter.

Table 2.6: Extension of Ayacut (hectares) under PAS between 1883 and 1953

Section	*No. of tanks*	*Ayacut (1883)*	*No. of tanks*	*Ayacut (1953)*	*Difference (5-3)*	*Column 6 as % of col. 3*
North	127	15,494	187	22,635	7,141	46.1
South	74	5,973	130	10,125	4,151	69.5
Total	201	21,467	317	32,760	11,293	52.6

Source: The Palar Anicut System, sl. no. 8, Calcutta, 1886, p. 7; and Improvements to Palar Anicut System below Ranipet, PWD, IRS report no. 3/61, Poondi, 1957, p. 5.

Although the *ayacut* was extended, the incremental *ayacut* was not evenly distributed over the two sections of the PAS. In absolute terms, over 60 per cent of the additional *ayacut* was under the northern channels and only 40 per cent was under the southern channels. However, since the *ayacut* originally served by the northern channel was much larger, the proportionate growth of *ayacut* was lower in the northern section (about 46 per cent increase over 1883 and 1953) compared to the southern section (where the increase was nearly 70 per cent).

2.2 Characteristics of Tank Irrigation Under the Palar Anicut System

There is little or no direct irrigation from the main or branch channels under the PAS. In almost every case the water is led into tanks in which it is stored. The total *ayacut* under PAS is 32,760 hectares, spread over 228 revenue villages, situated in the Walajah, Arcot and Arakkonam taluks in **Vellore district**, Cheyyar taluk in **Tiruvannamalai district**; and Kancheepuram andSriperumpudur in **Kancheepuram district**; and Tiruvallur in **Tiruvallur district** (Below the table 2.7).

Altogether the system serves 317 tanks spread over 228 villages. The northern Kaveripakkam channel includes a maximum of 108 villages, with an *ayacut* of 20,672 hectares, which constitutes about 63 per cent of the total *ayacut* of the present system. The next important is the southern Dusi-Kalavai channel which covers 92 villages with an *ayacut* of 8,058 hectares. This is roughly 25 per cent of the total *ayacut* under PAS. The remaining two channels namely, the northern Mahendravadi and the southern Sakkaramallur are serving a total of 28 villages with a combined *ayacut* of 4,030 hectares.

2.2.1 General Features of Tanks Served by the Four Main Channels

The aggregate storage capacity of the PAS tanks is around 275 mcm, which is about 0.0034 mcm per hectare of *ayacut*. The storage capacity per

Table 2.7 Command Area Details of PAS: Channel wise and Taluk wise

(Area in Hectares)

Sl. No.	*Name of taluk*	*Left Flank Channels*				*Right Flank Channels*				*Total no. of tanks*	*Area*	*Average ayacut per tank 12/11*
		Mahendravadi		*Kaveripakkam*		*Sakkaramallur*		*Dusi Thennampattu*				
		No.of tanks	*Area*	*No.of tanks*	*Area*	*No.of tanks*	*Area*	*No.of tanks*	*Area*			
1	*2*	*3*	*4*	*5*	*6*	*7*	*8*	*9*	*10*	*11*	*12*	*13*
1.	Walajah	11	903	2	159	-	-	-	-	13	1068	82
2.	Arcot	-	-	-	-	7	635	36	2219	43	2853	66
3.	Arakkonam	17	1060	81	9447	-	-	-	-	98	10507	107
4.	Cheyyar	-	-	-	-	16	1432	65	5839	81	7271	90
5.	Kancheepuram	-	-	50	6246	-	-	-	-	50	9246	125
6.	Sriperumpudur	-	-	29	4530	-	-	-	-	29	4530	156
7.	Tiruvallur	-	-	3	290	-	-	-	-	3	290	97
	Total Villages served	28	1963 (15)	165	20672 (108)	23	2067 (13)	101	8058 (92)	317	32760 (228)	103

Note: Figures in brackets indicate total number of villages served by each channel.

Source: National Water Management Project, India, PWD, Preliminary Report, March 1991.

hectare ranges from 0.0029 mcm to 0.0088 mcm. While there is no systematic relation between tank size and storage capacity per hectare, it seems that tanks with *ayacut* of 400 hectares and more have a significantly larger capacity per hectare of *ayacut* than those with *ayacut* of less than 200 hectares. The amount of water available in a year, however, depends not only on storage capacity but also on the number of fillings. Though the tanks are classified according to duration of supplies, the data are not readily available; nor are they up-to-date.

The volume of storage per unit of *ayacut* is relatively low in Kaveripakkam and Sakkaramallur channels; it is close to the average in Mahendravadi channel and relatively high in Dusi-Kalavai channel. On the average, the tanks under the northern channels have a storage capacity of 0.0026 mcm per hectare, which is almost half the level i.e., 0.0051 mcm per hectare in the tanks of southern channels (see Table 2.8). The larger storage capacity in the southern channels may be designed to store the

Table 2.8: General Features of the Main Channels under PAS.

Classification	*North*		*South*		*PAS as a whole*
	Mahen-dravadi Channel	*Kaveri-pakkam Channel*	*Sakkara - mallur Channel*	*Dusi-Kalavai Channel*	
Total number of villages	15	108	13	92	228
Total number of tanks	28	165	23	101	317
Storage capacity (mcm)	17.842	128.856	11.894	116.112	274.704
Registered ayacut (hectares)	1963	20672	2067	8058	32760
Storage capacity/ayacut (mcm/hectare)	0.0091	0.00623	0.00227	0.0058	0.0084
Storage capacity per ha of ayacut (metre)	0.2768	0.1895	0.1747	0.4394	0.2555
Catchment area (sq. kms)	5.83	293.08	45.34	177.80	522.05
Catchment area (hectares)	583	29308	4534	17780	52205
Free catchment per ha of ayacut (hectares)	0.30	1.42	2.19	2.21	1.59

Source: PWD, Palar Anicut office records; Vaidyanathan and Janakarajan 1989; GOI. National Water Management Project, Preliminary Report on PAS, 1991, pp 5-6; GOI. Revised estimates for the PAS, 27-10-1883. p.7.

larger run-off available from their local catchments. It may be seen from Table 2.8 that the free catchment of the southern channels is almost twice that of the northern channels. On storage capacity the PAS report has the following to say:

> In some years when the river has a large flood discharge, a much larger area could be irrigated than that provided for, the terminal tanks selected

in the system are of considerable size and are provided with weirs, over which the available water can be passed on to tanks lower down, of which there are a considerable number. No credit is taken for this additional supply, as it will only be available occasionally, and it will be utilised with great advantage (PAS, Serial no. 8: 1886: 2).

It follows that a large tank like Dusi-Mamandur located at the tail end of the system was not designed to be fed adequately from the Anicut during every supply year. These tanks can get adequate supply only if the river gets heavy floods provided these are effectively diverted at the Anicut.

The frequency distribution of PAS tanks by size of *ayacut* and the average storage capacity per hectare of *ayacut* in each sub-class are given in Table 2.9. Majority of tanks are small. About 35 per cent of the tanks served by the system have an *ayacut* of less than 40 hectares and account for merely six per cent of the area irrigated by the system; another 26 per cent have an *ayacut* of 40-80 hectares and account for 15 per cent of area. Only eight tanks serve more than 400 hectares that irrigates 25 per cent of the total ayacut of the system. Both KPT (capacity 32.91 mcm and *ayacut* 2370 hectares) and DMT (capacity 50.95 mcm and *ayacut* 1676 hectares) are in the last category.

Table 2.9: Frequency Distribution of PAS Tanks by Size of Ayacut (hectares) and Storage Capacity (Mcm/Hectare)

Size of ayacut	*No. of tanks*	*% to total*	*Ayacut*	*% to total*	*Storage capacity*
< 40	82	31	1577	5	0.0034
40 – 80	80	30	4855	16	0.0031
80 – 120	31	12	3064	10	0.0033
120 – 200	39	14	6006	20	0.0033
200 – 280	16	6	3595	12	0.0029
280 – 400	11	4	3267	11	0.0035
400 – 600	6	2	2900	9	0.0038
> 600	3	1	5142	17	0.0088
Total	268	100	30369	100	0.0044

Source: Vaidyanathan and Janakarajan 1989: 16.

2.2.2 Distribution of PAS Tanks by Size of Ayacut and Storage Capacity

The area of *ayacut* served is more important than the number of tanks. From this viewpoint small tanks (with ayacut <80 ha), which form 60% of all tanks, account for barely a fifth of the system's total *ayacut*. Tanks with *ayacut* of more than 200 ha are only 11% of the total which serve about 45% of *ayacut*. More than one third of total *ayacut* is served by medium size tanks (80 - 200 ha) which account for 28% of total tanks. The size distribution of PAS tanks by channels is given in below the Table 2.10. Tanks served by the northern channels (especially, Kaveripakkam channel) have lower storage capacity per hectare of *ayacut* compared to the southern channels. The larger storage capacity (>0.0029 mcm/hectare) of *ayacut* in all size-classes of tanks within the channel is seen only in Dusi-Thennampattu channel.

Channel wise frequency distribution of PAS tanks by size of *ayacut* and storage capacity per hectare of *ayacut* is given in below the Table 2.11. Almost all the tanks served by Mahendravadi channel are small sized (*ayacut*< 200 hectares). Only one tank serves an *ayacut* of about 280 hectares. Except three tanks, the storage capacity per hectare is less than 0.0055 mcm/hectare.

In the Kaveripakkam channel the average size of tank is somewhat large; more than half the tanks have *ayacuts* below 80 ha, and they account for about a sixth of the total ayacut under this channel. However, there are as many as ten tanks serving more than 280 ha each (and two, more than 600 ha each). They account for more than 30% of the total *ayacut* (Below the Table 2.12).

The average storage capacity per hectare of *ayacut* under Kaveripakkam channel is considerably less than in Mahendravadi channel. Nearly 77 % of the tanks under the former have a storage capacity less than 0.0031 mcm/hectare as against 59 % in the latter. Only five per cent of the tanks have a storage capacity more than 0.0050 mcm/ hectare (Sivasubramaniyan 2017: 55).

Under Sakkaramallur channel also small tanks predominate: about three-fourths of tanks possess a storage capacity below 0.0040 mcm/ ha. The smaller sized (below 120 ha) tanks have a relatively more capacity per ha than the larger tanks (Below the Table 2.13).

The average size of tank, and the size-wise distribution in Dusi-Thennampattu channel (Below the Table 2.14) is not materially different from others, but the average capacity per ha of *ayacut* is generally higher than in others. About one-fourth of the tanks account for a storage of more than 0.0040 mcm/ ha.

Table 2.10: Distribution of PAS Tanks by Size of Ayacut under Different Channels, and Capacity per Hectare of Ayacut

(Capacity: mcm/ha. Area in hectare)

Size of ayacut (Hectares)	*<------------------ NORTHERN SECTION ------------------>*									
	MAHENDRAVADI CHANNEL					*KAVERIPAKKAM CHANNEL*				
	Tanks		*Ayacut*		*Capacity*	*Tanks*		*Ayacut*		*Capacity*
	no.	*%*	*area*	*%*		*no.*	*%*	*area*	*%*	
< 40	9	32.1	137.6	6.8	0.0023	48	30.0	950.6	4.9	0.0031
40 - 80	9	32.1	534.6	26.3	0.0022	39	24.4	2299.9	11.9	0.0021
80 - 120	5	17.9	516.4	25.4	0.0023	22	13.8	2102.0	10.9	0.0023
120 - 200	4	14.3	612.7	30.2	0.0056	27	16.9	4197.9	21.8	0.0021
200 - 280	1	3.6	230.3	11.3	0.0053	14	8.8	3362.6	17.4	0.0013
280 - 400	0	0.0	0.0	0.0	0.0	6	3.8	1913.4	9.9	0.0016
400 - 600	0	0.0	0.0	0.0	0.0	2	1.3	1013.0	5.2	0.0026
< 600	0	0.0	0.0	0.0	0.0	2	1.3	3457.7	17.9	0.0055
Total	28	100	2031.6	100	0.0036	160	100	19297.0	100	0.0026

(Contd...)

Size of ayacut (Hectares)	*<--------------- SOUTHERN SECTION --------------->*										*PAS AS A WHOLE*			
	SAKKARAMALLUR CHANNEL					*DUSI-THENNAMPATTU CHANNEL*								
	Tanks		*Ayacut*		*Capacity*	*Tanks*		*Ayacut*		*Capacity*	*Tanks*		Ayacut	
	no.	*%*	*area*	*%*		*no.*	*%*	*area*	*%*		*no.*	*%*	area	%
< 40	7	33.3	134.8	6.4	0.0025	42	43.8	853.9	9.7	0.0055	106	34.8	2077	6.4
40 - 80	8	38.1	468.2	22.4	0.0040	23	24.0	1391.7	15.8	0.0029	79	25.9	4694	14.6
80 - 120	1	4.8	85.4	4.1	0.0052	14	14.6	1402.3	15.9	0.0029	42	13.8	4106	12.7
120 - 200	3	14.3	470.7	22.5	0.0025	10	10.4	1556.1	17.6	0.0043	44	14.4	6837	21.2
200 - 280	0	0.0	0.0	0.0	0.0	5	5.2	1119.0	12.7	0.0039	20	6.6	4712	14.6
280 - 400	0	0.0	0.0	0.0	0.0	0	0.0	0.0	0.0	0.0	6	2.0	1913	5.9
400 - 600	2	9.5	934.4	44.6	0.0018	1	1.0	515.6	5.8	0.0058	5	1.6	2463	7.6
< 600	0	0.0	0.0	0.0	0.0	1	1.0	1989.9	22.5	0.0104	3	1.0	5448	16.9
Total	21	100	2093.5	100	0.0026	96	100	8828.4	100	0.0054	305	100	32251	100

Source: Tabulated from Water Resources Consolidation Project, Tamil Nadu, Palar Anicut System, Project Report, August 1993.

Table 2.11: Frequency Distribution by Storage Capacity (mcm/hectare) of PAS Tanks under Mahendravadi Channel

Size of ayacut (hectares)	*Number of tanks with storage capacity (mcm)/hectare*									
	<.0014	*0.0014 to 0.0028*	*0.0028 to 0.0042*	*0.0042 to 0.0057*	*0.0057 to 0.0085*	*0.0085 to 0.0142*	*0.0142 to 0.0283*	*0.0283 to 0.1416*	*>.1416*	*Total*
< 40	2	3	1	0	0	3	0	0	0	9
40 - 80	4	2	2	1	0	0	0	0	0	9
80 - 120	2	1	2	0	0	0	0	0	0	5
120 - 200	1	1	0	1	0	1	0	0	0	4
200 - 280	0	0	0	1	0	0	0	0	0	1
280 - 400	0	0	0	0	0	0	0	0	0	0
400 - 600	0	0	0	0	0	0	0	0	0	0
< 600	0	0	0	0	0	0	0	0	0	0
Total	9	7	5	3	0	4	0	0	0	28
% to total	32.1	25.0	17.9	10.7	0.0	14.3	0.0	0.0	0.0	100

Table 2.12: Frequency Distribution by Storage Capacity (mcm/hectare) of PAS Tanks under Kaveripakkam Channel

Size of ayacut (hectares)	Number of tanks with storage capacity (mcm)/hectare									
	<.0014	0.0014 to 0.0028	0.0028 to 0.0042	0.0042 to 0.0057	0.0057 to 0.0085	0.0085 to 0.0142	0.0142 to 0.0283	0.0283 to 0.1416	>.1416	Total
< 40	17	12	7	5	4	1	2	0	0	48
40 - 80	13	19	3	3	1	0	0	0	0	39
80 - 120	5	10	5	2	0	0	0	0	0	22
120 - 200	8	12	7	0	0	0	0	0	0	27
200 - 280	10	3	1	0	0	0	0	0	0	14
280 - 400	4	1	1	0	0	0	0	0	0	6
400 - 600	0	1	1	0	0	0	0	0	0	2
< 600	0	1	0	0	1	0	0	0	0	2
Total	57	59	25	10	6	1	2	0	0	160
% to total	35.6	36.9	15.6	6.3	3.8	0.6	1.3	0.0	0.0	100

Source: Water Resources Consolidation Project, Tamil Nadu, Palar Anicut System, Project Report, August 1993.

Table 2.13: Frequency Distribution by Storage Capacity (mcm/hectare) of PAS Tanks under Sakkaramallur Channel

Size of ayacut (hectares)	Number of tanks with storage capacity (mcm)/hectare									
	<.0014	0.0014 to 0.0028	0.0028 to 0.0042	0.0042 to 0.0057	0.0057 to 0.0085	0.0085 to 0.0142	0.0142 to 0.0283	0.0283 to 0.1416	>.1416	Total
< 40	1	2	2	1	0	1	0	0	0	7
40 - 80	3	1	2	0	1	0	1	0	0	8
80 - 120	0	0	0	1	0	0	0	0	0	1
120 - 200	0	2	1	0	0	0	0	0	0	3
200 - 280	0	0	0	0	0	0	0	0	0	0
280 - 400	0	0	0	0	0	0	0	0	0	0
400 - 600	1	1	0	0	0	0	0	0	0	2
< 600	0	0	0	0	0	0	0	0	0	0
Total	5	6	5	2	1	1	1	0	0	21
% to total	23.8	28.6	23.8	9.5	4.8	4.8	4.8	0.0	0.0	100

Table 2.14: Frequency Distribution by Storage Capacity (mcm/hectare) of PAS Tanks under Dusi-Kalavai (Thennampattu) Channel

Size of ayacut (hectares)	Number of tanks with storage capacity (mcm)/hectare									
	<.0014	0.0014 to 0.0028	0.0028 to 0.0042	0.0042 to 0.0057	0.0057 to 0.0085	0.0085 to 0.0142	0.0142 to 0.0283	0.0283 to 0.1416	>.1416	Total
< 40	5	12	9	8	3	2	1	2	0	42
40 - 80	2	11	6	3	1	0	0	0	0	23
80 - 120	3	4	3	3	1	0	0	0	0	14
120 - 200	0	1	5	3	1	0	0	0	0	10
200 - 280	1	1	2	0	0	1	0	0	0	5
280 - 400	0	0	0	0	0	0	0	0	0	0
400 - 600	0	0	0	0	1	0	0	0	0	1
< 600	0	0	0	0	0	1	0	0	0	1
Total	11	29	25	17	7	4	1	2	0	96
% to total	11.5	30.2	26.0	17.7	7.3	4.2	1.0	2.1	0.0	100

Source: Water Resources Consolidation Project, Tamil Nadu, Palar Anicut System, Project Report, August 1993.

On the whole it is evident that about 86% of the total tanks in the three channels, namely, Mahendravadi, Sakkaramallur and Dusi-Kalavai have a storage capacity of less than 0.0057 mcm/ ha of *ayacut*. Out of 305 tanks in all the 4 channels, no tank has a capacity more than 0.028 mcm/ ha of *ayacut* (Sivasubramaniyan 2017: 55).

The regression results indicate that there is a significant positive relationship between the size of the *ayacut* and the storage capacity of tanks. The consistency increases if one moves from smaller tank group to larger one (see Table 2.15).

Table 2.15: Regression Results: Storage Capacity on the Size of Ayacut of Tanks under the Pas

Dependent Variable (Size of ayacut - ha)	*No. of tanks*	*Constant*	*Storage capacity (mcm)*	R^2
All tanks	305	62.57	1.36 (1.32)*	0.81
1-80	184	29.15	0.76 (5.03)*	0.13
81-200	87	107.65	0.62 (4.63)*	0.21
201-400	26	255.56	-0.017 (7.70)	0.0003
400+	8	438.02	1.08 (5.72)*	0.88

Note: Figures in brackets represent the standard errors of coefficients.
* Denotes-'t' values significant at one per cent level.

2.3 Conditions of Water Supply in Palar Anicut System

2.3.1 Water Realisation and Discharge in the PAS during 1906-2016

The total inflow at the Anicut (water realization) and the total quantity diverted through the four main channels and the amount which surplussed over the Anicut from 1906 to 2016 are given in below the Table 2.16. It is seen that the Anicut receives on the average 188.18 mcm a year, of which 74.47 mcm (barely 40%) is drawn through the channels for irrigation, and the balance of 113.70 mcm surplussed over the Anicut.

Table 2.16: Water Realization and its Discharge through the Four Main Channels in PAS, 1906-2016

Sl. No.	Year	PAS total	Surplus total	Four Channels total	Channels 5 years average	North Arcot rainfall (mm) *	<-- Column 5 --> as % of	
		<----- mcm ----->					Column 3	275 #
1	2	Realisation	4	Discharge	6	Rainfall	8	9
1.	1906	57	0	57	117	969	100	21
2.	1907	12	0	12	108	844	100	4
3.	1908	24	0	24	138	719	100	9
4.	1909	427	291	136	136	1118	32	50
5.	1910	860	505	355	115	1124	41	129
6.	1911	11	0	11	63	934	100	4
7.	1912	836	670	166	130	999	20	60
8.	1913	11	0	11	126	775	100	4
9.	1914	37	3	34	124	905	93	12
10.	1915	199	104	95	128	1107	48	35
11.	1916	819	474	345	116	1105	42	125
12.	1917	152	6	146	50	894	96	53
13.	1918	0	0	0	27	784	0	0
14.	1919	94	37	57	27	1134	60	21
15.	1920	32	0	32	19	953	100	12
16.	1921	13	0	13	15	1019	100	5
17.	1922	54	19	36	13	1041	66	13
18.	1923	0	0	0	9	641	0	0
19.	1924	14	0	14	15	849	100	5
20.	1925	10	0	10	12	983	100	4
21.	1926	6	0	6	35	836	100	2
22.	1927	13	0	13	62	820	100	5
23.	1928	30	0	30	74	957	100	11
24.	1929	1	0	1	81	836	100	0
25.	1930	638	513	125	84	1406	20	45
26.	1931	177	34	143	110	1131	81	52
27.	1932	79	6	73	86	916	92	27
28.	1933	63	0	63	83	831	100	23
29.	1934	15	0	15	75	941	100	6
30.	1935	567	311	256	130	1073	45	93
31.	1936	21	0	21	89	863	100	8
32.	1937	265	203	62	90	1185	23	22
33.	1938	18	0	18	82	779	100	7
34.	1939	373	81	293	107	971	78	106
35.	1940	62	9	53	65	1039	85	19
36.	1941	22	0	22	56	856	100	8
37.	1942	22	0	22	83	773	100	8
38.	1943	1431	1286	145	90	1356	10	53
39.	1944	455	370	85	65	1150	19	31
40.	1945	8	1	7	66	764	88	3
41.	1946	1335	1180	155	65	1387	12	57
42.	1947	59	0	59	34	766	100	21

(Contd...)

43.	1948	17	0	17	24	841	100	6
44.	1949	92	0	92	66	893	100	34
45.	1950	0	0	0	61	529	0	0
46.	1951	0	0	0	61	732	0	0
47.	1952	12	0	12	187	820	100	4
48.	1953	478	251	227	190	940	47	82
49.	1954	162	94	69	167	1034	42	25
50.	1955	0	0	0	158	991	0	0
51.	1956	818	189	628	158	1161	77	228
52.	1957	25	0	25	44	823	100	9
53.	1958	113	0	113	72	797	100	41
54.	1959	24	0	24	72	732	100	9
55.	1960	0	0	0	127	1183	0	0
56.	1961	83	21	61	129	1035	74	22
57.	1962	164	0	164	138	825	100	60
58.	1963	125	16	110	113	881	88	40
59.	1964	593	291	302	92	997	51	110
60.	1965	7	0	7	49	976	100	3
61.	1966	992	887	105	60	1409	11	38
62.	1967	44	0	44	48	900	100	16
63.	1968	0	0	0	65	606	0	0
64.	1969	163	75	88	73	1217	54	32
65.	1970	64	0	64	85	1176	100	23
66.	1971	43	0	43	105	1028	100	15
67.	1972	460	332	128	111	1056	28	47
68.	1973	43	0	43	101	894	100	16
69.	1974	150	2	148	106	902	99	54
70.	1975	275	111	165	126	980	60	60
71.	1976	127	56	72	93	1167	56	26
72.	1977	476	397	79	124	1284	17	29
73.	1978	73	6	67	108	933	92	24
74.	1979	630	383	247	104	1014	39	90
75.	1980	0	0	0	62	608	0	0
76.	1981	454	225	228	66	1065	50	83
77.	1982	0	0	0	21	693	0	0
78.	1983	46	0	46	21	1365	100	17
79.	1984	38	0	38	25	819	100	14
80.	1985	20	0	20	18	1233	100	7
81.	1986	0	0	0	14	709	0	0
82.	1987	0	0	0	66	998	0	0
83.	1988	69	0	69	67	927	100	25
84.	1989	0	0	0	54	946	0	0
85.	1990	0	0	0	54	699	0	0
86.	1991	1663	1401	262	54	1153	16	95
87.	1992	35	31	4	58	803	10	1
88.	1993	3	0	3	67	1093	100	1
89.	1994	0	0	0	116	1186	100	0
90.	1995	0	0	0	116	920	100	0
91.	1996	1478	1194	283	118	1602	19	103
92.	1997	47	0	47	61	1020	100	17

(*Contd...*)

93.	1998	609	361	248	81	980	41	90
94.	1999	0	0	0	31	650	0	0
95.	2000	11	0	11	31	814	100	4
96.	2001	0	0	0	75	770	0	0
97.	2002	146	0	146	75	624	100	53
98.	2003	0	0	0	57	1356	0	0
99.	2004	0	0	0	77	970	0	0
100.	2005	295	66	229	77	1297	78	83
101.	2006	0	0	0	52	899	0	0
102.	2007	56	0	56	52	1246	100	20
103.	2008	100	0	100	41	875	100	36
104.	2009	0	0	0	21	887	0	0
105.	2010	103	0	103	21	1165	100	38
106.	2011	0	0	0	3	1000	0	0
107.	2012	0	0	0	3	983	0	0
108.	2013	0	0	0		800	0	0
109.	2014	0	0	0		812	0	0
110.	2015	140	127	13		1087	9	5
111.	2016	0	0	0		0	0	0
	Avg.	188	114	74			40	27

Note: # Aggregate capacity of all PAS tanks is 275.71 mcm. NA = Not Available.

* North Arcot Rainfall since 1990-91 is given based on the actual average rainfall of Vellore and Tiruvannamalai districts, since both districts formed the North Arcot previously.

Source: PWD, Palar Anicut Office Records. Season and Crop Reports, Govt. of Tamil Nadu (Various Issues).

The volume of water realisation is liable to wide year-to-year fluctuations (See figure 2.2). Thus, no water was received at the Anicut in 24 years out of 111 years for which data are available; of the remaining

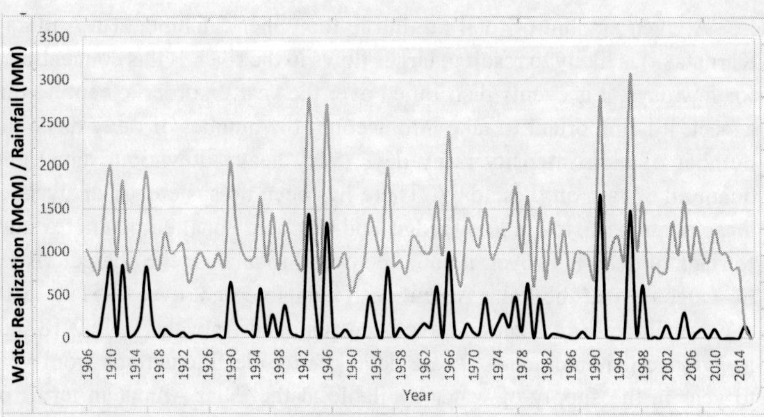

Note: 111 years – from 1906 to 2016: Water Realization MCM/RainfaLL MM.

Source: PWD Palar Anicut Office Records, 2016.

Figure 2.2: Water Realisation with Rainfall 1906-2016.

87 years, the yield was less than 70 mcm in 50 years, 71 to 137 mcm in 15 years, 138 to 166 mc3 in 9 years, 167 to 276 mcm in 7 years and exceeded 276 mcm in 6 years.

In 14 years the realization of flow exceeded 566 mcm. There is statistically a significant trend over last 80-90 years in the volume of water received by the Anicut (see Table 2.17). The rainfall in combined North Arcot district alone explains at best 38% of (the variation in dependent variable) flow realised by the Anicut. Year to year variations are due to variations in level and time pattern of rainfall across years. There is a significant positive association between volume of water realized (that explains 47%) at the Anicut and quantum of water diverted through the four main channels (Table 2.17). Clearly the pattern of rainfall in the catchment of river is far more important.

Table 2.17 Results of Regression of Realisation and Discharge on Select Variables, 1906-2016

Dependent variable	*Constant*	*Rainfall (RF)*	*Realisation*	R^2
Realisation	-965.794	1.214 (0.172)*	-	0.38
Discharge	48.654	-	0.198 (0.023)*	0.47

Note: Figures in brackets represent the standard errors of coefficients.
* denotes-'t' values significant at one per cent level.
Realization: column 3 in Table 2.16. Discharge: Column 5 in Table 2.16.

2.3.2 Characteristics of Water Flow at the Palar Anicut

A given amount of total rainfall in the upper catchment (in Andhra / Karnataka) is likely to result in larger flows in the river. If it is concentrated for few days, it is evenly distributed over the year. In order to capture this aspect, it is important to take into account the number of rainy days, the number of high intensity rainy days (very heavy downpour days) and quantum of rain on these days. There has been no systematic analysis of these relations up to our knowledge, and we could not attempt it here either for lack of data. However, it could be useful to analyse the characteristics of water flow at the Palar Anicut with available data over 1951 to 2016 (Years with zero realization have been removed from the table 2.18, but in average it is included). There have been wide fluctuations from year to year in the supply of water available at the Palar Anicut in terms of timing of first receipt of water at the Anicut, the earliest date on which the Anicut first flow was 26 July in 1964, while in 1971 it did not get its first flow got the even till December 16. Taking the 66-year period as a whole,

Table 2.18: Characteristics of Water Flow at Palar Anicut, 1951-2016

Year	Date of first receipt	No. of flow days	No. of over-flow days	-> Rainfall at <-		Total realization	Total discharge	Surplus over the Anicut and under vents
				Palar Anicut	N. Arcot District			
				----> mms <----		---------> mcm <---------		
1952	Nov. 12	11	-	na	820	12	12	0
1953	Oct. 3	62	9	na	940	478.1	226.6	251.5
1954	Oct. 16	49	4	na	1034	162.2	68.6	93.5
1956	Sep. 29	73	22	na	1161	817.6	628.1	189.5
1957	Oct. 24	17	-	na	823	24.7	24.7	0
1958	Sep. 27	32	-	na	797	112.8	112.8	0
1959	Oct. 29	19	-	na	732	23.6	23.6	0
1961	Sep. 21	29	1	na	1035	82.6	61.1	21.5
1962	Oct. 2	61	-	na	825	164	164	0
1963	Sep. 1	35	5	1144	881	125.1	109.5	15.5
1964	July 26	90	31	891	997	592.7	301.6	291.1
1965	Dec. 4	8	-	883	976	7.3	7.3	0
1966	Oct. 10	55	7	1289	1409	991.7	105.1	886.6
1967	Sep. 25	36	-	1064	900	43.7	43.7	0
1969	Oct. 22	90	9	1125	1217	163	87.6	75.4
1970	July 31	23	-	967	1176	64.2	64.2	0
1971	Dec. 16	22	-	861	1028	42.6	42.6	0
1972	Oct. 7	62	6	875	1056	460.5	128.2	332.3
1973	Oct. 22	43	-	811	894	42.7	42.7	0
1974	Sep. 27	55	2	883	902	149.7	147.7	2.1
1975	Oct. 26	70	6	784	980	275.4	164.6	110.8
1976	Aug. 20	30	4	923	1167	127.2	71.5	55.7
1977	Oct. 24	67	5	1364	1284	475.9	79.3	396.7
1978	Sep. 27	60	3	976	933	73	67.2	5.8
1979	Sep. 30	88	3	751	1014	630.2	247.1	383.1
1981	Sep. 12	70	2	1050	1065	453.6	228.2	225.5
1983	Sep. 11	15	-	971	1365	46	46	0
1984	Oct. 8	18	-	934	819	37.7	37.7	0
1985	Nov. 12	7	-	1067	1233	20.5	20.5	0
1988	Sep. 2	46	-	876	927	69.1	69.1	0
Total 1951-1990		**1343**	**119**			**6769**	**3433**	**3336**
Average		**34**	**3.0**	**930**	**964**	**169**	**86**	**83**
1991	Sep. 28	65	17	1334	1153	1662.8	261.5	1401.3
1992	Oct. 7	13	1	851	803	34.5	3.6	30.9
1993	Dec. 2	4	-	1117	1093	2.9	2.9	0
1994	Nov. 16	1	-	1071	1186	0.1	0.1	0
1995	Oct. 22	1	-	1749	920	0.4	0.4	0
1996	Oct. 5	58	13	1428	1602	1477.7	283.4	1194.3
1997	Sep. 19	11	-	1901	1020	46.9	46.9	0
1998	Sep. 24	63	8	1111	980	608.6	248	360.6
2000	Nov. 17	6	-	1155	814	11.3	11.3	0
2002	Oct. 12	35	-	709	624	145.5	145.5	0
2005	Oct. 16	41	3	1548	1297	295.1	229	66.1
2007	Nov. 28	8	-	1180	1246	56	56	0
2008	Nov. 13	14	-	1000	875	99.7	99.7	0
2010	Sep. 27	19	-	935	1165	103.2	103.2	0
2015	Nov. 15	25	5	1451	1087	139.8	13.1	126.6
Total (1991-2015)		364	47			4685	1505	3180
Average		15	**1.9**	1092	992	187	60	127
Total (1951-2015)		**1707**	**166**			**11454**	**4937**	**6516**
Average		**26**	**2.6**	**1007**	**981**	**176**	**76**	**100**

Note: Dash indicates nil; na = not available.

'0' realization years are removed for clarity. Average includes all years.

Source: PWD, Palar Anicut Office Records.

there was no flow at all in the river in 21 years. The Anicut got a steady supply mostly in September and October. For 15 years, the flow started in September, and in 17 years, it started in October. The months of September and October are the most important months in terms of water realisation and the number of days when flow is received at the Anicut. Overall, these two months supplied a major portion of water to the total quantum received at the Anicut.

Between 1951 and 2016 (excluding no flow years) river flow lasted less than 30 days in 21 years; between 31 and 60 days in 12 years; and more than 60 days in 12 years. In two years (1964 and 1969) the river flow occurred for a maximum of 90 days. Water overflowed the Anicut only in 22 years (out of 45 years in which there was some flow), the duration of overflow varying from one day in two years (1961 and 1992) to a maximum of 31 days in 1964.

Table 2.18 shows that there exists a relationship between the number of days of flow and total realisation. Whenever there was a flow in the Anicut for 50 days or more in a year, the total realisation was more than 141.6 mcm (except in 1978 when the number of days flow registered was 60, but the realisation was only 73 mcm). Apparently, in all the years (except 1954 when the registered flow days was less than 50) the total realisation was below 141.6 mcm. Further, in the years 2002 and 2005, although the number of days of flow was 35 and 41 respectively, the quantum of water realization exceeded more than 145 mcm. This indicates that in recent decades heavy down pour in less number of days has been increasing compared to previous decades. However, the last 26 years data (1991-2016) indicate that on an average both the number of flow days and number of overflow days decreased considerably than the previous period (1964-1990). The average number of days of flow declined from 34 to 15 between these two periods and the overflow days from 3.0 to 1.9 in the same period.

2.3.3 Surplus Flows in the PAS

As already evident from Table 2.16, the amount of water realised at the Anicut exceeded the amount drawn for irrigating the PAS *ayacut* in 43 out of 116 years. The maximum amount surplussed over the crest of the Anicut was 1401 mcm during 1991, it exceeded 600 mcm in 5 other years (see below the Figure 2.3). On an average, water overflows in one out of every two years when water is available; but in some decades, the degree of surplus was more than others. During 1930s and 1970s the water surplussed in 7 out of 10 years, while in 1920s, 1980s and 2000s water

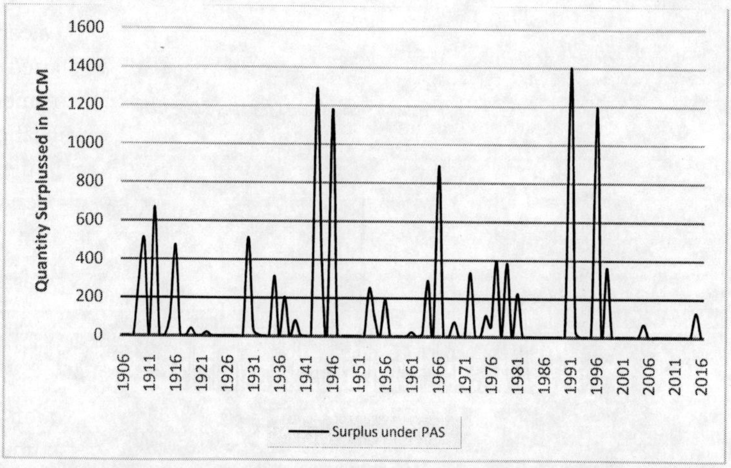

Source: PWD Palar Anicut Office Records, 2016.

Figure 2.3: Surplus Flows under the PAS – 111 Years from 1906 to 2016.

overflowed the Anicut only once in 10 years. Interestingly, it seems that most surplus years (30 out of 43) were also the same years in which there was above normal amount of rainfall in North Arcot district. This clearly indicates that when there is excess water available (or less demand due to rainfall), the surplus was also seen comparatively higher.

The quantum of supply diverted for irrigation averaged around 40% of the flow at the Anicut in years when water was available. It fluctuated between 10% and 100% of the total flow. The entire available flow was diverted in 43 out of 87 years; while in another 22, less than 50% was diverted (see below the Figures 2.4 and 2.5). There is no significant trend statistically in the volume of water drawn for irrigation from the Anicut. However, there are relatively dry years during which (as in 1920s, 1980s and 2000s) the frequency of below-average drawal is very high even as in the decades of relatively high rainfall. As in 1910s, 1960s, 1970s and 1990s) the frequency of above-average drawal was relatively higher. It seems that in most cases below-average realisation years are also years of below-average utilisation in absolute terms but not relative to realisation. Thus in 56 out of 62 years when the realisation was below 200 mcm, more than 80% of flow was diverted; while in 24 years when the flow exceeded 255 mcm, the proportion diverted was less than 30% in 11 years and less than 80% in 13 years (see below the Figure 2.6). Hence, it is clear, that the rainfall in each location plays a major role in deciding the realisation and water diversion.

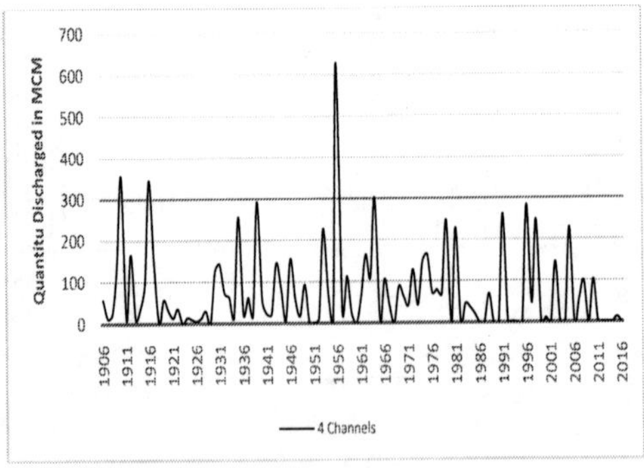

Source: PWD Palar Anicut Office Records, 2016.

Figure 2.4: PAS Water Discharge through Four Main Channels: 1906-2016.

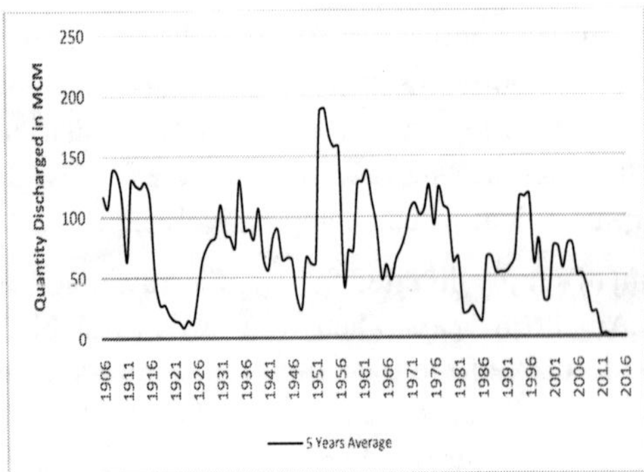

Source: PWD Palar Anicut Office Records, 2016.

Figure 2.5: The PAS Water Discharge through Four Main Channels: 1906-2016.

The variations in quantum of water drawn by the PAS depend on several factors such as volumes of water realised at the Anicut, drawal capacity of the channels, the extent of siltation in the channels and so on. It is to be expected that greater the volume realised at the Anicut, the larger is the quantum drawn by the channels. For any given total realization, the more evenly it is distributed within the season, the greater is likely to be the

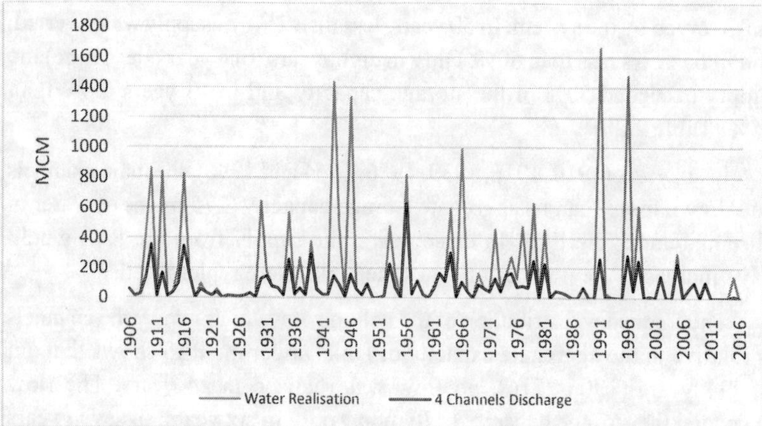

Source: PWD Palar Anicut Office Records, 2016.

Figure 2.6: PAS Water Realization with Water Discharge through 4 Main Channels: 1906-2016

volume, which can be drawn into the channels. On the other hand, when the flow exceeds the capacity of the channels, the water overflows the Anicut.

The regression results indicate that the quantity of water realised at the Anicut bears a significant positive relationship to the rainfall of undivided North Arcot district (Table 2.19). The analysis also indicates a positive relationship of the volume of water discharged and number of overflow days and realization. However, the relationship between discharge and flow days seems very weak. The same is true in the case of discharge to realisation on days of overflow.

The volume of water diverted through the four main channels of the PAS works on an average of about 27% of aggregate storage capacity of

Table 2.19: Determinants of Discharge Relative to Realisation in PAS, 1951-2016.

Dependent variable	***Constant***	***Rainfall (RF)***	***Flow days***	***Overflow days***	***Realisation***	***R^2***
Realisation	-559.3566	0.751129 (0.1742)*	-	-	-	0.23
Discharge	75.508098	-	1.309875 (1.2557)	7.80305 (3.9842)	0.0946 (0.0741)	0.54
Ratio of discharge to realization	0.4574925	-	0.002908 (0.00317)	0.00508 (0.0010)	-0.00039 (0.00015)**	0.27

Note: Figures in brackets represent the standard errors of coefficients; * and ** denote 't' values significant at 1% and 5% levels respectively.

tanks served by the system. In 50 years, less than 25% of supply was diverted; and in 65 years less than 50%. Only in eight years (one year out of nine) the supply exceeded 50% of the storage capacity and in 13 years more than 75% (Table 2.20).

In six years (1910, 1916, 1939, 1956, 1964 and 1996), the main channels had drawn more than its aggregate storage capacity (275 mcm) of water in all tanks under the PAS. In these years, the supply from the PAS would have increased the number of times the tanks were replenished.

The frequency distribution of the volume of water drawn by the channels is relative to the aggregate capacity of tanks under the PAS shows that out of 111 years (1906-2016), no flow was realised in 24 years. The flow exceeded the storage capacity (276 mcm) only in six years; in seven years it was in the range of 76-99%; 51 to 75% of flow was diverted in nine years, 26 to 50% in 15 years and less than 25% in a maximum of 50 (about half of the supply) years. Table 2.20 shows that taking no flow of 24 years

Table 2.20: Frequency Distribution of Utilisation as Percentage of Capacity of Tanks in the PAS: 1906-2016.

Range of flow in %	*No. of years*	*Volume of Water Drawn (mcm)*					
		0	*1 to 70*	*71 to 137*	*138 to 166*	*167 to 276*	>276
No flow	24	24					
1 – 25	50		50				
26 – 50	15			15			
51 – 75	9				9		
76 – 99	7					7	
100	6						6
Total	111	24	50	15	9	7	6

Source: Calculated from Table 2.16.

Note: Capacity of all PAS tanks combined is **275.71** million cubic metre (mcm).

and years when the flow was 1 to 70 mcm range years (which constitute about two thirds of the total years), the quantity realised at the anicut was very low compared to the storage capacity.

2.3.4 Extent of Water Availability from Catchment Run-Off and the PAS

Data available on rainfall and water drawn by the four main channels of the Palar Anicut (see below the Table 2.21) indicate that in all channels, supply from the PAS is more than 50% to the total capacity of tanks (except in Dusi-Thennampattu where it gets only less than 12%). Likewise in all channels, supply from the catchment is more than 50% of the total capacity

Table 2.21: Extent of Water Availability from Catchment Run-off and the PAS.

Sl. No	Details	*Northern Channels*		*Southern Channels*		*PAS as a Whole*
		Mahen-dravadi	*Kaveri-pakkam*	*Sakkara-mallur*	*Dusi-Then-nampattu*	
1.	Free catchment (sq.km) (for the channel as a whole)	5.8	293.0	45.3	177.7	521.8
2.	Yield from rainfall (mcm)	5.5	277.3	42.9	168.2	494.0
3.	Rainfall yield at 35% (mcm)	1.9	97.1	15.0	58.9	172.9
4.	% To total capacity of tanks	11	75	126	51	63
5.	Supply from PAS (mcm)	10.56	68.22	7.50	13.79	100.08
6.	% To total capacity of tanks	59	53	63	12	36
7.	Total availability (mcm) (3)+(5)	12.5	165.3	22.5	72.7	273.0
8.	% To total capacity of tanks (7) as % of (9)	70	128	189	63	99
9.	Total capacity of tanks (mcm)	17.84	128.86	11.89	116.11	274.71
10.	Registered ayacut (Ha)	1962.36	20663.29	2065.97	8054.63	32746.26
11.	Average depth (m) (7)/(10)X0.10000	0.0006	0.0008	0.0011	0.0009	0.0008

Note: The normal rainfall in North Arcot District (Vellore and Tiruvannamalai) is 988 mm. According to official sources, 1 mm rainfall in 1 sq.km catchment provides 0.00097 mcm.

Source: The Palar Anicut System, Selections from the Records of the Government of India, PWD, No. 8, Calcutta, 1886, p. 13; and Palar Anicut office records.

of tanks (except in Mahendravadi where it gets only 11%). Catchment supply relative to the capacity of tanks is the highest in Sakkaramallur channel (126%) followed by Kaveripakkam (75%).

Available supply from both the PAS and the catchment is about two thirds of the total capacity of tanks for Mahendravadi and Dusi-Thennampattu channels, but it is over 100% in Kaveripakkam and Sakkaramallur channels. On an average, tanks fed by these two channels get more than one filling in a year, while in the others the supply is inadequate for one filling.

In the Dusi-Thennampattu channel Dusi Mamandur tank gets additional supply from two non-PAS sources, viz., the Rajakkal supply and the Thandarai supply. The former source is assured whenever the Palar gets heavy floods. The latter source hardly provides supply to DMT even from the beginning (1975) of the construction of the Thandarai Anicut. However, it is still considered as one of the sources of supply to DMT.

2.3.5 Water Drawn and Variability in Water Supply across Channels

Data on the channel wise distribution of water drawn by the PAS are available only since 1964. The data summarised in Table 2.22, show that in 18 years out of 53 years during 1964 to 2016 there was no flow in the river and none of the channels got any water. In the remaining 35 years the average annual drawal was 106.2 mcm. The major quantity (51%) of water was drawn by the Kaveripakkam main channel; 21% by Dusi-Thennampattu; followed by 15% by Mahendravadi and 13% by Sakkaramallur.

Relative to the ayacut under each channel, Kaveripakkam main (0.038 metre) and Dusi-Thennampattu (0.036 metre) draws the maximum mean amount followed by Sakkaramallur (0.014 metre) and Mahendravadi (0.012 metre per unit area). This is lower than the average for 1964-1990 (except in Kaveripakkam main channel), largely because of the relatively high incidence of 'zero' flows (for 38% of the years) during the 1964 to 2016 period compared to the 26% 'zero flows' for the earlier period 1964-1990. The regression results for the period 1964-1979, that is, excluding the years (1980-89) on the supply drawn by the four main channels indicate that the quantum of irrigation supply was on an increasing trend in all the four channels.

Below the Table 2.22 shows that on an average, taking the period 1964 to 2016, the PAS supply amounts to 39% of total storage capacity of tanks served by the system. The proportion is much higher in the northern bank (15.7%) than in the south (9.4%). But within south, while the amount drawn represents 121% of the capacity of the tanks under Sakkaramallur channel, it supplies barely 19% of the capacity of the tanks under Dusi-Thennampattu. The supplies to northern channels are relatively more stable (observed from the CV of annual drawal) than in south with maximum variability in Dusi-Thennampattu during the period of 1964-1990. However, this pattern differs considerably taking the whole period of 1964-2016, where the CV of southern channels is more stable than northern channels. Dusi-Thennampattu, in both the periods, gets relatively smaller amount and the supplies are more volatile. Further, it should be noted that the volume of water drawn by each channel is not proportionate to the capacity of tanks served by the system. The implication is that the incidence of scarcity is unevenly distributed, and falls more heavily on the Dusi-Thennampattu than others.

Table 2.22: Volume of Water Drawn (MCM) through the Four Main Channels: 1964-2016

Year	*Northern Channels*		*Southern Channels*		*PAS as a whole*
	Mahen-dravadi	*Kaveri-pakkam*	*Sakkara-mallur*	*Dusi-Thennampattu*	
1964	30.22	187.65	22.43	61.31	301.61
65	1.10	5.83	0.08	0.25	7.28
66	16.09	52.02	12.77	24.21	105.10
67	4.42	36.93	0.88	1.47	43.70
69	8.55	63.61	6.49	8.92	87.57
1970	5.49	54.23	1.53	2.92	64.17
71	5.64	33.67	0.99	2.29	42.59
72	11.84	92.24	8.98	15.15	128.21
73	4.62	37.75	0.08	0.25	42.71
74	13.31	107.73	10.25	16.37	147.66
75	16.37	111.50	15.75	20.96	164.57
76	6.17	42.14	7.87	15.32	71.51
77	8.50	61.51	7.36	1.90	79.27
78	8.16	46.73	3.60	8.69	67.18
79	21.38	145.91	26.28	53.55	247.13
1981	25.26	155.99	15.86	31.07	228.18
83	4.96	35.91	1.98	3.12	45.96
84	5.35	31.89	0.37	0.11	37.72
85	2.69	14.70	1.19	1.90	20.48
88	11.24	46.62	5.24	5.98	69.07
Total	211.4	1364.6	150.0	275.8	2001.7
Average*	**10.6**	**68.2**	**7.5**	**13.8**	**100.1**
(% to total)	(11)	(68)	(7)	(14)	(100)
SD	781	4940	770	1743	8019
CV (SD/Average)	74	72	103	126	80
Ayacut (ha)	1962.4	20663.3	2066.0	8054.6	32746.3
Storage Capacity (mcm)	**17.8**	**128.9**	**11.9**	**116.1**	**274.7**
Avg/ water drawn/ Ayacut (metre)	**0.019**	**0.030**	**0.028**	**0.058**	**0.033**
Avg/ water drawn to Storage capacity (%)	**59**	**53**	**63**	**12**	**36**
1991	44.33	78.90	47.56	90.72	261.51
92	0.53	1.09	0.63	1.36	3.62

(*Contd...*)

93	0.50	0.78	0.57	1.00	2.85
96	63.18	60.14	68.57	91.48	283.37
97	4.96	15.15	7.23	19.60	46.94
98	56.10	69.26	53.75	68.91	248.01
2000	2.23	3.12	2.78	3.17	11.31
2	27.92	39.52	31.16	46.87	145.48
5	56.10	65.15	55.95	54.65	229.02
7	11.21	14.74	10.92	19.14	56.02
8	22.39	26.21	21.25	29.86	99.72
2010	25.20	32.58	22.62	22.84	103.25
15	2.79	3.75	2.73	3.87	13.14
2016	0	0	0	0	0
Total	528.8	1775.0	475.7	729.3	3505.9
Average*	**16.0**	**53.8**	**14.4**	**22.1**	**106.2**
(% to total)	(15)	(51)	(13)	(21)	100
SD	17	46	18	26	90
CV (SD/Average)	106	85	126	119	85
Ayacut (ha)	1962	20663	2066	8055	32746
Storage Capacity (mcm)	**17.8**	**128.9**	**11.9**	**116.1**	**274.7**
Avg/ water drawn/ Ayacut (metre)	**0.012**	**0.038**	**0.014**	**0.036**	**0.031**
Avg/ water drawn to Storage capacity (%)	**90**	**42**	**121**	**19**	**39**

Note: * Refers excluding '0' years. 1 hecta-metre = 10,000 cubic metres.
No water drawn years are removed from the table for clarity in presentation.

Source: PWD, Palar anicut office records.

This may be highlighted by a comparison of the relative positions of different channels in the last 53 years (Table 2.23). The PAS as a whole is estimated to have supplied on the average 66.1 mcm per annum (average includes no supply years also) through the channels taking off from the Anicut (see below the Table 2.22). The average during the periods of 1964-1980 and 1981-1998 was considerably higher (as 106.68 mcm and 113.43 mcm respectively) than in the period of 1999-2016 (93.99 mcm). In fact, during 1999-2016 there was no supply in 60% of the years. However, the supplies were available in all years from 1964 to 1980 and in three years it was more than 500 mcm. The low level of supplies during 1999-2016 was largely a reflection of unusually high proportion of years with less than average rainfall.

Table 2.23 Volume of Water (MCM) Drawn at Palar Anicut through the Four Main Channels during 1964-2016.

Channels	*Average water drawn for all years (mcm)*			*Increase/ Decrease in % over 1981-98 to 1999-2016*	*Average water drawn/storage capacity (mcm/mcm)*			*Increase/ Decrease in % over 1981-98 to 1999-2016*	*Avg. water supply by PAS per hectare of ayacut (mcm)*			*Increase/ Decrease in % over 1981-98 to 1999-2016*
	1964 to 1980	*1981 to 1998*	*1999 to 2016*		*1964 to 1980*	*1981 to 1998*	*1999 to 2016*		*1964 to 1980*	*1981 to 1998*	*1999 to 2016*	
Mahendravadi	10.79	19.92	21.12	6	0.61	1.12	1.19	6	0.0055	0.0102	0.0108	6
Kaveripakkam	71.96	46.40	26.44	-43	0.56	0.36	0.21	-42	0.0035	0.0022	0.0013	-41
Sakkaramallur	8.36	18.45	21.06	14	0.70	1.55	1.77	14	0.0040	0.0089	0.0102	15
Dusi-Thennampattu	15.57	28.66	25.77	-10	0.13	0.25	0.22	-12	0.0019	0.0036	0.0032	-11
PAS as a whole	106.68	113.43	93.99	-17	0.39	0.41	0.34	-17	0.0033	0.0035	0.0029	-17

Source: PWD, Palar Anicut Office Records.

There were considerable variations across channels: Kaveripakkam main channel drew the highest volume and the Sakkaramallur, the lowest in all the three periods (Table 2.23). More importantly, on an average, there has been a positive trend in all the channels (except Kaveripakkam where it has been decreasing) of supplies drawn gradually increasing from 1964-1980 period to the 1999-2016 period.

Overall, compared to the Northern channels, the Southern channels draws lesser quantum of supplies in all the periods. This variation is because at some extent the intake points of Dusi-Thennampattu channel are at higher altitude than northern channels. Further, southern channels are liable to silt accumulation at the head. This means that they draw less water at times of low flow in the river than the left bank channels. However, it must be noted that these disadvantages of the right bank channels in relation to the PAS supplies are adjusted to some degree by the larger extent of free catchment per hectare of ayacut and to a higher degree of utilisation of sub-surface water.

The average water supply per ha of ayacut is the highest in Mahendravadi, followed by Sakkaramallur and Kaveripakkam and the lowest in Dusi-Thennampattu. Again, reduction in flow years is the strongest in Dusi-Thennampattu compared to Sakkaramallur. Further, the variation in supply is lower in Mahendravadi and Sakkaramallur channels than in Kaveripakkam and Dusi-Thennampattu (see Figures 2.7 and 2.8). It is the highest in Kaveripakkam followed by Dusi-Thennampattu.

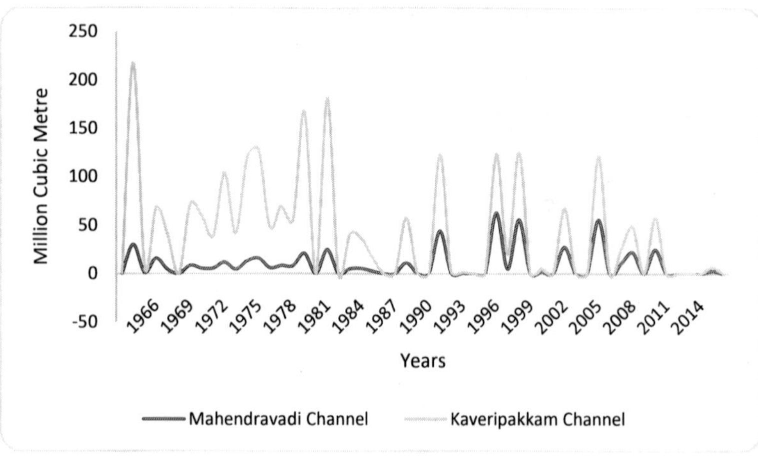

Source: Derived from Table 2.22.

Figure 2.7: Water Drawn under Northern Channels: 1964-2016.

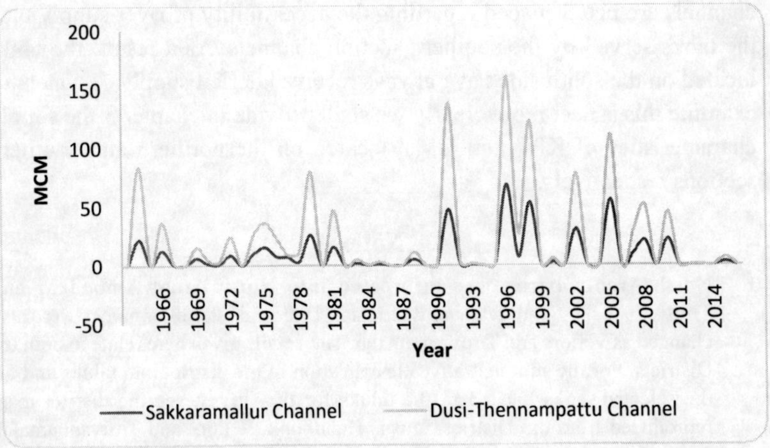

Source: Derived from Table 2.22.

Figure 2.8: Water Drawn under Southern Channels: 1964-2016.

The tanks under Dusi-Thennampattu channel have a larger capacity per hectare of the ayacut - which is more than that of Kaveripakkam channel - the water drawn by Dusi-Thennampattu channel is less than 25% of capacity.

With regard to average water drawn from 1964-2016 the variations of water drawal was much wider between the 1981-1998 and 1999-2016. In the system as a whole, the water drawn during the latter period was 17% lower than that of the 1981-1998 period. Among the four channels, the decrease in variation is found as high as 41% in Kaveripakkam to as low as 10% in Dusi-Thennampattu channel. Overall, the incidence of supply drawn was much lower in southern channels than in northern channels.

2.4 Summary

This chapter discussed the evolution of the PAS and the main features of irrigation in the command before and after the construction of the PAS. The improvements of the system were made in several phases over a period of a century and included structural improvements at the Anicut as well as in the main/branch distributaries leading from the head-sluice and the extension of ayacut. The detailed analysis of the availability of river supply to the Palar Anicut and its allocation to different tanks through the four main channels shows that the Palar Anicut did not receive adequate supply in most of the years and even if considerable floods occurred, the water was neither effectively diverted nor equitably distributed through the channels. The chapter indicates that the tanks served by the northern section channels are better placed regarding the accessibility of river supply than the tanks served by the southern section channels. As a result, the tanks located on the south side have always received deficit supply of water. To examine this aspect more clearly we shall provide in chapter 3 the supply characteristics of KPT and DMT located on the northern and southern sections respectively.

Notes

1. North Arcot district was bifurcated into North Arcot Ambedkar and Tiruvannamalai Sambuvarayar districts in 1989 and later their names were also changed as Vellore and Tiruvannamalai. The details given here relate to both the Districts. For the administrative classification of the district into taluks and for the selected tanks located in the taluks for the survey, see the district map, reproduced from the District Census Handbook, Vellore and Tiruvannamalai, 2016-17.
2. See GO 1617 I, PWD, June 19, 1931.

Chapter - 3

The Tanks of Kaveripakkam and Dusi-Mamandur

This chapter presents the physical and technical features of Kaveripakkam Tank (KPT) and Dusi-Mamandur Tank (DMT). The physical features include location, sources of supply, the inlet channels, the associated structures of the tanks such as bunds, tank bed, sluices and regulatory structures; water distribution network such as main, branch and field channels, and the *ayacut*. All technical details pertaining to the physical structures of the tanks are also given. The characteristics of water supply in the tanks are discussed in the context of the last five decades. The development of wells and agricultural operations in the *ayacut,* in terms of cropping pattern and cropping intensity are discussed towards the end of the chapter.

The KPT, situated in the head reach of the Kaveripakkam main supply channel with a storage capacity of 41.74 mcm, has an *ayacut* of nearly 2430 hectares spread over 14 villages. The DMT (capacity 50.95 mcm) is situated at the tail end of the Thennampattu channel on the right bank of the Palar and serves an *ayacut* of 1660 hectares spread over 18 villages. Other technical details are given in below the Table 3.1.

Both the tanks are of ancient origin: KPT is reportedly more than a thousand years old, dating back, according to one account, to the 9th century AD. DMT is even older, dating to the 6th Century AD. Before the construction of the Palar Anicut, KPT drew its supply from a *Korambu* - a temporary earthen embankment to divert water by using woods and sand-bags -

Table 3.1: Characteristics of Kaveripakkam and Dusi-Mamandur Tanks

Name of tank	KPT	DMT
Number of villages served	14	18
Number of sluices	10	4
Number of fillings	1	1
Number of supply sources	1	3
Number of surplus weirs	3	2
Depth of tank (metre)	9.18	9.09
Original storage capacity (mcm)	41.74	50.95
FTL storage available for irrigation by months	14	18
Storage capacity (mcm) in 1981	32.92	Na
Registered ayacut (hectares) in 1983 (UDR)	2370	1676
Storage capacity / hectare of ayacut (mcm/hectares)	Ori-0.**018**. Present-0.**014**	Ori-0.**435**
Water Spread area (m.sq. metre)	16.05	14.04
Free catchment area (sq.kms)	31.08	30.56
Combined catchment area (sq. kms)	126.91	300.44
Free catchment per hectare of ayacut (hectare)	0.013	0.018
Greatest width of water spread area (kms)	3.2	4.8
Length of bund (kms)	8.32	3.52
Maximum flood discharge through the surplus weirs (cumecs)	858.40	281.56
Length of main channels under all sluices (kms)	23.40	36.12
Length of branch channels under all sluices (kms)	67.19	23.19

Note: Na = Not available. UDR = Up-dated settlement registers. FTL = Full tank level. Storage capacity in 1981 in KPT is taken from the report of the silt survey conducted by Irrigation Research Station, Poondi, 1982. Ori = Original.

Source: PWD, Memoirs of Kaveripakkam and Dusi-Mamandur tanks.

constructed across the Palar roughly at the site of the present day Anicut. DMT's main supply seems to have been the Rajakkal channel fed by diverting river flow at a point some 21 kms below the Anicut and tapping the overflow and drainage of upstream tanks in the southern bank of the Palar.

This was changed significantly with the construction of the Anicut. The KPT benefited much more than DMT from the increased volume of water drawn from the river mainly because it was situated at the head of the system. It is in fact the first major tank on the Kaveripakkam channel. Though the channel does not draw the full extent of designed capacity, the ratio of actual drawal to designed capacity (as noted in Chapter 2) is considerably higher in Kaveripakkam channel than in the southern channels.

The construction of the Anicut presumably increased the volume diverted at the Anicut and the improvement of distribution channels were intended to ensure larger, more assured supply to the tail end in the south as well. However, given the poor performance of the southern channel, the extent to which DMT, located at the tail end of the chain, benefited from this is questionable. With the construction of the Anicut, the flows available for diversion into the Rajakkal channel were significantly reduced and became more uncertain. The construction of the Thandarai Anicut across Cheyyar also did not contribute much to DMT which again happens to be at the tail end of the channel. However, to some extent these disadvantages are offset by DMT's much larger catchment below the Anicut and the larger capacity relative to the *ayacut*.

3.1 Present Position of Inlet Channels

The inlet channels feeding both the tanks are quite lengthy. KPT is fed by the Kaveripakkam main channel, which takes off from the left side of Palar Anicut. Some five kms from the take off point, at a place known as Goonmadai, the main supply channel bifurcates one branch leads to KPT. Between Goonmadai and KPT there is only one tank, Kadaperi, with an *ayacut* of 148 hectares. Originally this supply channel was very tortuous, but substantial improvements were made, especially during the 19th century, with adequate provision for escapes to regulate the volume of water admitted into the tanks. Table 3.2 presents the hydraulic details of the channel.

Table 3.2: Hydraulic Details of KPT Supply Channel

Length of supply channel	9 kms
Bed width of channel	25.80 metre
Full supply depth	1.5 metre
Bed fall	0.75 metre per 1.6 km
Velocity of flow	0.8 metre per second
Maximum carrying capacity	1406 cumecs (1882)
Supply available at this rate per day	4.29 mcm
Present capacity of KPT	32.91 mcm (1981)
Number of days required to fill the tank at full supply from the channel	32.91/4.29 = 7.67 days

Source: Compiled from the PWD data, Section office, Kaveripakkam.

DMT is currently fed by three supply channels: The Thennampattu channel taking off from the Palar Anicut, the Rajakkal channel from Palar downstream of the Anicut and the Thandarai channel from Cheyyar. In addition, it derives the benefit of surplus and drainage flows from upstream tanks under the Sakkaramallur and the Dusi channels. DMT, being at the tail end of all the three channels, is not as favourably placed as KPT. It gets hardly any supply from Thandarai Anicut channel. Rajakkal remains an important source. In terms of water per unit of *ayacut*, these disadvantages are compensated to some degree by the fact that it benefits from the local rainfall in its combined catchment and from the larger storage capacity per hectare of *ayacut* compared to KPT.

The *ayacutdars* at the tail end of the southern channels have been pressing the government from the beginning of the last century for rectifying defects in the intake weir for Rajakkal, cleaning silt accumulation, raising the sill level at the head of the channel, and construction of the Thandarai Anicut. The suggested improvements in the Rajakkal channel were not fully taken up but improvements were made partially, because of the inadequacy of funds and variable flow at the weir site. A new Thandarai Anicut was eventually constructed across the Cheyyar river in 1975, but the amount of water available from this anicut has been relatively small.

In the case of Thandarai anicut, there is an additional factor restricting benefits to DMT. It feeds 11 tanks before reaching DMT. Each of these upstream tanks has a separate branch channel from the main Thandarai channel. DMT can get the supply only if the Chithathur tank (the last of the upstream chain) surpluses. Chithathur tank being small (an *ayacut* of 128 hectares) its *ayacutdars* are afraid of breaches if continuous supply is allowed in the channel and therefore seek out the officials at the Anicut to close the supply at the head sluice whenever their tank is filled. As a result, it is said that no water surpluses from the Chithathur tank, and DMT has got practically no supply from the tank since Thandarai Anicut was built. The DMT ayacutdars have been pressing the government for a separate supply channel from the Thandarai channel bypassing Chithathur tank. But the government has not acted on this request in the face of objections by the upper reachers against improvements of the channel to protect the tail enders.

3.1.1 Conditions of Inlet Channels

The Kaveripakkam channel is now lined up to a length of 10,270 feet. However, on both sides of lining a lot of sand has accumulated. Beyond the lined portion, the main channel is heavily silted to a depth of around three feet of its entire length up to Goonmadai. Near Goonmadai the siltation is more than four feet deep.

At the bifurcation point of Goonmadai, the KPT supply channel is silted to a depth of about three feet up to an obstruction weir, some three kms downstream, meant basically to lessen the bed fall and to stop silt accumulation further down. After this point, no silt is seen inside the channel, but thick bush growth is noticeable. In some places the bunds of the supply channel are partly cut and damaged. While authorised direct irrigation from this channel is limited, a large number of wells initially fitted mostly with oil engines in the 1980s and now converted into electric pump sets operate on both sides of the channel. It is reported by the ryots that these wells supply water throughout the year, most of them supplying for more than 12 hours in the summer. Sugarcane and paddy are the main crops cultivated here.

DMT has now three supply channels (including the two main channels supplying water from upstream tanks). The lengthiest channel is Dusi-Thennampattu 36 kms) followed by Thandarai channel (21 kms) from their take off points. The Rajakkal channel (13 kms) is a direct supply source to DMT. Table 3.3 presents the technical details of the inlet channels of DMT.

All the inlet channels of DMT are affected by heavy silt, encroachment and weed growth. In most places the breadth and depth of channels have been reduced leading to significant reduction in the volume of water they can carry during the flood season as well as in the normal supply periods. The main supply channel of Rajakkal - into which Thennampattu and Sakkaramallur channels confluence before reaching DMT - is in a deplorable condition. Apart from defective location of the take off point the entire length of this channel is affected by intensive bush growth. No silt clearance in this channel has been done by either the PWD or the *ayacutdars*. The only work done by the *ayacutdars* is to reconstruct the *Korambus* across the river to divert supply during the flood periods.

The responsibility of the maintenance of the inlet channels is vested with the PWD, but little has been done for many years. Some damage in the bunds of channels has been repaired but the work done falls short of needs. The neglect of inlet channels is mainly due to the following reasons:

(1) Inadequate funds allotted for maintenance of channels;

(2) Growth of population leading to encroachment in and around the channels, with encroachers cultivating in the bed of inlet channels in some places;

(3) Increasing fragmentation of land ownership among *ayacutdars* as reflected in the vast increase in the number of *ayacutdars* making

Table 3.3: Technical Details of the Inlet Channels of DMT.

THENNAMPATTU CHANNEL	
Number of tanks fed	36
Total ayacut (direct + indirect) hectares	3077
Supply available from this channel to DMT This channel joins with Rajakkal supply channel after 7 kms from the Rajakkal head sluice.	329 cumecs
RAJAKKAL CHANNEL	
Number of vents at the head sluice	6
Size of vents (metre)	1.5 x 1.2
Sill of head sluice	EL 100.18
Full supply level	EL 101.68
Maximum water level	EL 102.13
Channel bed width	1.35 metre
Bed fall	0.03 / 240
Slope	1.5:1
Designed full supply discharge	577 cumecs
Total number of days required to fill-DMT at the full supply discharge	4.75 days
THANDARAI CHANNEL	
Bed width (metre)	8.1x 1.8
Full supply depth (metre)	1.5
Bed fall	1 in 1584 meters
Side slopes	1 to 1
Discharge	265 cumecs
Length of channel	21 kms
Velocity	0.86 met/sec
Bed level at start	95.70 metre
Number of tanks served	16
Total ayacut fed by the channel	3080 hectares

Note: EL = Earth level.
Source: PWD Records, Section office, Iyenkarkulam.

it more difficult to mobilize them in cooperative efforts to improve channels; and

(4) More frequent dry years, and years of below average supply in the last decade (eg. 1980s) leading to neglect of maintenance of channels.

All these reduce the volume of water, which the supply channels carry compared to the designed levels. The effects are likely to be more severe in the tail end where DMT is situated than in the head reach like KPT.

3.2 The Tanks and Associated Structures

3.2.1 The Bund

In KPT water is stored behind an earthen bund about 8.3 kms long. Though the KPT bund is in good condition some improvements need to be made in the following aspects: Throughout the length of the bund, prickly pears grow densely, making it very difficult to pass through the bund. They should be removed at least twice a year. In most places the big granite stones placed to strengthen the bund are not held securely in their position and the packed granite stones adjacent to them are in danger of falling and need to be strengthened.

Compared to KPT, the bund of the DMT is both shorter and free of any tree or even grass cover. DMT has a larger storage capacity, bound by 1.6 kms of earthen bund and 2.4 kms of natural hillocks. During the late 1970s villagers discovered a leakage at the site of the Dusi deep-sluice and it increased as the tank got filled up. As a result, the bund was immediately strengthened by constructing a cement and granite masonry structure throughout the length of its outside surface.

3.2.2 Siltation and Encroachment in the Tank Bed

Silt accumulation is a long-standing and significant problem in KPT. The problem was understood even a century ago. A government order of 1900 made the following observations:

Originally the capacity of KPT was sufficiently large to supply water for about 14 months irrigation. But of the late years, the tank has become so much silted up, as to be incapable of holding much quantity of water as it formerly used to do. Besides, the channel which was excavated during the famine of 1875-76, near Moolai sluice, Lion sluice, and the Fox sluice has also now become so much silted up.... one main cause for silt collection is the need for repairing the breaches in the supply channel from the Palar river (GO 832 I, PWD, 14-03-1900).

Another document of the same year noted:

> The removal of silt from KPT would cost about Rs.1 lakh, this facilitates the tank to enhance the capacity sufficient to irrigate some more 80 acres. Its disadvantages are that it would be little remunerative, and that the road from the foot to top of the bund being buried by silt hence the transport of silt would be very difficult, and even if it is done, then it requires a heavier permanent maintenance charge, hence at these considerations it is not fruitful to do so (GO 470 I, PWD, 04-10-1900).

The economic viability of investments in desilting (especially where one considers the problem of disposing of the removed silt) remains a serious question.

Some 100 years ago the inlet channel breached two kms above its merging point to the tank. As a result of this breach and natural formation of new course of the channel (see figure 3.1) a huge accumulation of silt is deposited in and around the sluices. Presently the silt level is more than 5 meters at the deepest sluice in KPT.

However, the sill of this sluice is not closed by silt because, in and around 15 meters from the sluice, the silt is excavated and prevented by constructing silt obstruction wall. The silt accumulation has considerably reduced the volume of tank storage. A survey conducted in KPT by the Poondi Research Station in 1978 estimated the storage capacity of the tank as 81.50 mcm, which is 20 per cent less than the estimated original capacity of 41.74 mcm. Another survey conducted in 1981 showed a further reduction to 81.28 mcm. These estimates, approximate as they are, do point to a reduction in the storage capacity of KPT.

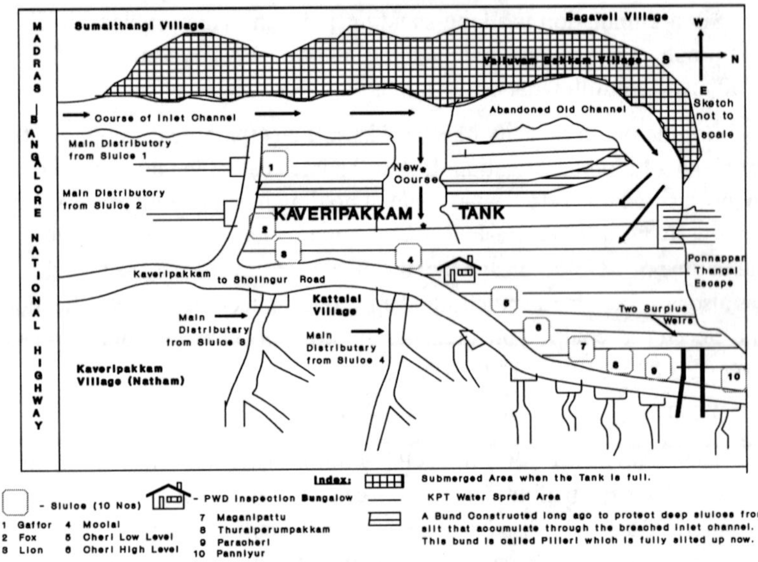

Figure 3.1: A View of Kaveripakkam Tank

There has been no systematic silt survey in DMT. According to the delegates of the Irrigation Board (IB), existed till 1990s, and other informed local people, this tank is also silted up to ten feet. This would suggest that siltation is considerably less than in KPT. One cannot be sure of the extent of siltation and the loss of storage capacity without a proper survey. However, given that DMT is at the tail end of a long chain of tanks, it is reasonable to expect the extent of siltation to be considerably lower than in KPT.

Foreshore area cultivation is not a new phenomenon. Even 130 years ago, the government granted permission to the local *ayacutdars* to raise dry crops in the tank foreshore area during non-supply (dry) years. This process continues still. The *Mamulnamas* (for details of KPT and DMT see Appendix 3.1) suggests that 200 years ago the tank foreshore area was also permitted to be used for cultivation to the surrounding *ayacut* villagers.

The KPT *Mamulnamas* has this to say on foreshore area cultivation:

> The free catchment area of KPT is spread over in 8 surrounding villages, namely, Kaveripakkam, Pilleri, Sumaithangi, Ponnappanthangal, Musiri, Edayanthangal, Bagaveli, and Valluvampakkam. Whenever KPT is filled immediately the waters spread over all these 8 village lands. Hence these villagers have a *mamul* right to crop the submerged land after depletion of water in the foreshore lands. According to this right, each village can install ettram to irrigate the surrounding lands. If they need more water they can dig shallow wells for irrigation to raise nanjai crop, garden crop and also punjai crops like cholam, cumbu and varagu (KPT *Mamulnamas*-documented in 1815).

Hence, it is clear that tank foreshore cultivation is a permitted and long-standing practice, which is still in vogue. A few government orders of the present century also indicate that the government granted permission to cultivate the foreshore area when there was no water in it.

Presently, in a year when there is no supply in the tank, there is intensive cultivation in the tank bed by the KPT villagers. Since the silt is fertile and the yield from the crop raised in the tank bed are good the ryots are reluctant to give up the tank bed cultivation. Since the tank has not filled frequently in recent decades the cultivators do not bother about the problem of crops being submerged as the tank fills up (see photos in p. 423). At the same time increasing population pressure on land has induced the people to have some pockets of land in the tank for self-cultivation even without any *patta*. In DMT tank bed cultivation is not as wide as in KPT. The reason is that DMT bed is rocky and not favourable for cultivation.

3.2.3 Sluices and Regulatory Structures

Sluice is a regulatory device to draw water from the tank through its *sill* and supplying it to the *ayacut*. Being large tanks KPT and DMT have several sluices, ten in KPT and four in DMT. On an average each sluice serves about 240 hectares in KPT and 400 hectares in DMT. The size of vents, their sill level and the *ayacut* served by each sluice vary a great deal.

Table 3.4 indicates that in KPT the three deepest sluices irrigate about 35 per cent of the tank's *ayacut*; the middle level sluices irrigate 60 per cent, and the elevated sluices cover barely five per cent. While the size of vents (measured by the area of cross section) is not strictly correlated with the sill level, it is apparent that in general within each category of sluices, higher the elevation larger is the vent size relative to *ayacut*.

Table 3.4: Sluice Particulars of Kaveripakkam Tank.

Name of sluice	*Vent size (sq. metre)*	*Sill level (metre)*	*Depth at FTL (metre)*	*Ayacut (hectares)*	*Cross section area of vent (sq. metre)*	*Ayacut / Vent size (hectares)*
Moolai	DS 0.149 x 0.065	126.34	9.27	317.0	0.0818	360
Cheri LL	DS 0.121 x 0.11	127.80	7.80	88.3	0.1310	63
Lion	DS 0.214 x 0.455	128.34	7.26	449.0	0.9931	42
Maganipattu	MS 0.158 x 0.009	128.70	6.90	604.0	0.1226	244
Thuraiperum-pakkam	MS 0.153 x 0.084	128.94	6.60		0.1078	
Cheri HL	MS 0.121 x 0.084	129.57	6.03	92.5	0.0929	99
Fox	MS 0.372 x 0.316	130.09	5.52	679.9	1.164	51
Paracheri	ES 0.149 x 0.084	132.03	3.57	16.5	0.1050	15
Panniyur	ES 0.149 x 0.121	132.58	30.3	66.5	0.1747	35
Gaffor	ES 0.093 x 0.093	132.82	2.79	27.1	0.0929	27

Note: DS =Deep sluice; MS =Middle sluice; ES =Elevated sluice. LL =Low level; HL =High level.

Source: PWD, Section office, Kaveripakkam.

Of the four sluices in DMT, two are low level sluices and the remaining two are high-level sluices. However, these are not separated from one another. A combination of one high level sluice and one low level sluice (1 and 2) irrigate nine villages; and the other set of sluices (3 and 4) also irrigate nine villages. Table 3.5 presents the sill level of sluices, its depth at FTL, and the ayacut of DMT. In this tank also, the data indicate that the elevated sluices have a larger vent size.

Table 3.5: Sluice Particulars of Dusi-Mamandur Tank

Name of Sluice	*Vent Size (sq. metre)*	*Sill level (metre)*	*Depth at FTL (metre)*	*Ayacut (hectares)*	*Cross section area of vent (sq. metre)*		*Ayacut / Vent size (hectares)*
Dusi HL	0.093 x 0.269	78.33	5.34	758	0.260	0.548	129
Dusi LL	0.121 x 0.242	75.39	8.28		0.288		
Mamandur LL	0.149 x 0.186	74.58	9.09	918	0.279	0.557	153
Mamandur HL	0.009 x 0.279	78.39	5.25		0.279		

Note: HL = High level; LL = Low level.

Source: PWD, Memoir of DMT, Section office, Iyenkarkulam.

In both the tanks all the sluices at present are fitted with tough wooden shutters with screw gears. They are generally in good condition. However, all the sluices fitted with shutters are not tightly closed. In order to avoid siltation, all the shutters in the sluices are left with a gap of 2 to 3 inches to allow water always to pass along with the mud. This cannot be stopped even during the rainy season or even if water is not necessary to the *ayacut*.

The maintenance of sluices and regulatory structures is vested with the PWD. In both the tanks most of the sluices are in good working condition. Whenever higher rank officials make a formal visit to the tanks for inspection, all the sluices are put in good working order. In KPT during 1988 the shutter in Cheri low level sluice was broken. As a result, a large quantity of water was wasted through it. The repair of the sluice took a couple of months. The PWD gives preference to repair other damages, whenever necessary, within the resources available to them. Likewise, in the supply year 2015-16, in KPT when it was at full storage, three of the sluices (Mulai, Thuraiperumpakkam and Panniyur) have got stuck up and supply stopped automatically. Consequent to people's request and effort, only the Panniyur sluice was repaired and the other two did not get repaired. Since the leakages in the deepest Mulai sluice was enlarged, the sluice was altogether closed and it was not yet repaired even till April 2018. So, the Kattalai villagers did not cultivate by using tank water in the two supply years viz. 2015-16 and 2017-18.

3.2.4 The Surplus Weirs of the Tanks

One of the important arrangements in all tanks is the surplus weir to carry the flows in excess of Full Tank Level (FTL) to downstream tanks. KPT has three surplus weirs, and DMT has two. The type of weirs differs between the tanks. In KPT all the three weirs are provided with tough wooden shutters with screw gearing arrangements. In DMT only open vents are provided. There are no shutters because the weirs are built at the maximum FTL point. When the flow exceeds FTL, the surplus flows out automatically. In the case of KPT the weirs are built three feet below FTL. This means that the tank will reach its FTL only if all the shutters in the surplus weirs are in position. If all the shutters in the surplus weirs are kept open the maximum level would be three feet below FTL. The flexibility in KPT is built into the design in order to regulate the amount of water used by KPT and ensure supply to lower down tanks. As a rule, the surplus weirs of KPT are not to be closed fully till December 15 but must be put one foot below the FTL in order to facilitate supply to downstream tanks. In contrast, DMT being the tail end tank, has no downstream tank. Hence there is no need to place the weirs below FTL like KPT. Nor is there any need to regulate the flow at FTL.

Further in KPT the Ponnappanthangal escape is used not only for emergency purpose but also for supply purpose: whenever KPT reached FTL, and if water is required for Mahendravadi tank, the Ponnappanthangal regulator must be opened to supply water to the Mahendravadi tank. This actually is one of the rules regarding water regulation in KPT which is still observed. An account of the regulatory arrangements existing in KPT and DMT is given in Table 3.6.

Table 3.6: Surplus Weir Arrangements in KPT and DMT

Tank	*Name of Weir*	*Size of shutters (metre)*	*No. of shutters*	*Total discharge (cu.mecs)*
KPT	{ Weir no. 1	2.4 x 1.05	30 }	
	{ Weir no. 2	2.4 x 1.05	27 }	24,310
	{ Ponnappanthangal	1.5 x 1.60	9 }	
		Size of vents	Length of weir	Total discharge (cu.secs)
DMT	{ Weir no. 1	0.6 x 0.90	114.6 }	
	{ Weir no. 2	045 x 0.57	71.1 }	7,974

Source: PWD records, Kaveripakkam and Iyenkarkulam.

3.2.5 Sluice Elevation

Each sluice is situated at different sill levels and command different parts of the *ayacut*. Naturally, the deepest sluice is in a better position in terms of both the volume of water received and duration of access to irrigation from the tank compared to elevated sluices. The latter provides a smaller volume for a shorter period. Below the table 3.7 indicates that at FTL, water depth at the deepest sluice is nearly 9.3 metre and the sluice with the highest elevation has only 2.8 metre level. When the tank level falls to 6.5 metre, the highest sluice has practically no water to draw. The deepest sluices can continue to draw water even if the level drops to as low as 2.0 to 2.5 metre.

3.3 Distribution Network

3.3.1 Main and Branch Channels

The main channels taking off from the tank sluices are lengthy. The combined length of all the main channels is 23 kms in KPT, whereas it is about 36 kms in DMT. In the case of branch channels, the combined length in KPT (67.19 kms) is much more than that of DMT (23.19 kms).

Table 3.7: Water Level in Various Sluices Compared to the Deepest Sluice in KPT

Name of sluice and elevation		*Ayacut (Ha) (1983)*	*Sill level (Mt)*	*Water level in different sluices compared to the deepest sluice (metre)*									
1. Moolai	**DS**	317.0	126.3	9.3	6.8	6.5	5.9	4.1	3.5	2.9	2.7	2.3	1.8
2. Cheri Low Level	**DS**	88.3	127.8	7.8	5.3	5.1	4.5	2.6	2.1	1.4	1.2	0.8	0.3
3. Lion	**DS**	449.0	128.3	7.3	4.8	4.5	4.0	2.0	1.5	0.9	0.7	0.3	-
4. Maganipattu	**MS**		128.7	6.9	4.4	4.2	3.6	1.7	1.2	0.5	0.3	-	-
	}	604.0											
5. Thuraiperumpakkam	**MS**		128.9	6.7	4.2	3.9	3.4	1.4	0.9	0.3	0.2	-	-
6. Cheri High Level	**MS**	92.5	129.6	6.0	3.5	3.3	2.8	0.8	0.3	-	-	-	-
7. Fox	**MS**	679.9	130.1	5.5	3.0	2.8	2.3	0.3	-	-	-	-	-
8. Paracheri	**ES**	16.5	132.0	3.6	1.1	0.8	0.3	-	-	-	-	-	-
9. Panniyur	**ES**	66.5	132.6	3.0	0.5	0.3	-	-	-	-	-	-	-
10. Gaffor	**ES**	27.1	132.8	2.8	0.3	-	-	-	-	-	-	-	-

Note: Dash indicates nil supply; DS = Deep sluice; MS = Middle sluice; ES = Elevated sluice.
Source: Calculated from the sill levels of various sluices.

The responsibility of maintenance of all the main and branch channels is vested with the *ayacutdars* in each village. But the system works only when enough tank supply is available. If the tank storage reaches less than FTL, most of the tail end farmers in both the tanks abstain from maintenance. During the decade preceding the initial survey (2005 to 2015), the tanks did not reach FTL in any year. Therefore, most of the channels were not maintained properly at the time of the initial survey (2016). The same situation was also reported in the decade 1980s.

In the late 1980s the government included both the tanks under the tank modernisation scheme, which, among other things, envisages lining all the main and branch channels. An estimated amount of Rs. 598.1 lakhs for KPT and Rs. 137.5 lakhs for DMT was sanctioned. In the case of DMT, the scheme also provides for lining of both its inlet channels (the Rajakkal and Dusi-Thennampattu). This scheme was taken up for execution in both the tanks in 1990 with financial assistance of the World Bank. The basic survey and measurement of all the main/branch channels have been completed and work on lining was started in DMT during 1993.

Likewise, during the early 2010s also both the tanks have got benefited in getting World Bank assistance through Irrigated Agriculture Modernization and Water Bodies Restoration and Management (IAMWARM) Project. About Rs. 10 million was utilised by KPT mainly to construct a retaining wall all along the bund of about five kms. This helped the bund position on a safer side. In DMT, the bund and inner walls of the tank was modernized to strengthen the bund. Wherever required minor repairs along the inlet channels of both the tanks were also benefitted by this scheme.

3.3.2 Field Channels

Field channels get their supply from either the branch channels or main distributaries to supply water finally to plots located in different parts of the *ayacut*. In both the tanks at the time of the survey (2016) it was reported that a major portion of field channels were silted up considerably. However, when supply is available from the tank, these channels are again formed by the *ayacutdars*. Increasing fragmentation of holdings, changes in land ownership pattern and the gradual increase in the use of well water in the *ayacut* have all added to the problems in sharing the responsibility for maintaining the field channels. Informants from all the riparian villages of both the tanks reported that the encroachment in field channel is substantial. These problems are recurrent due to the fact that there has not been any 'channel survey' conducted after the resettlement (1913) and, even when the breadth and length of channels are reduced due to encroachment, no authority was concerned about it and had not bothered to evict the encroachers although specific rules are still in force.

3.4 The Ayacut

According to the updated revenue settlement of 1983 the total *ayacut* of KPT is 2370 hectares and of DMT is 1676 hectares. Tables 3.8 and 3.9 give the below village-wise distribution of the *ayacut*. In the case of DMT, there was little change in the extent of the *ayacut* reported in the 1882 and 1913 settlements. In KPT there was a major change between 1882 and 1913; and the 1913 ayacut is less than the figure given in 1882. The difference is largely on account of the exclusion of 179 hectares of *poramboke* (uncultivated government land) area of Kaveripakkam from the *ayacut* in the 1913 settlement.

Estimates of the *ayacut* prior to 1882 vary - especially in the case of KPT: An 1856 report (Smith 1856) mentioned 3117 hectares as irrigated; the settlement of 1873-74 gave the figure of 2298 hectares in 23 villages; and the PWD list of irrigation sources and component works in 1900 reported 2417 hectares in 15 villages (see below the Table 3.10). These differences need to be properly verified and accounted for. But it seems likely that they are mainly on account of inclusion/ exclusion of *ayacut* of small tanks in the villages covered by KPT and getting supplies from KPT. The general impression one gets from these figures is that the construction of the Anicut and improvement of channels did not add significantly to the KPT *ayacut*.

Table 3.8: KPT Ayacut (Ha) during the Settlements

Sl. no.	*Village*	*Settlement Years*		
		1882	*1913*	*1983*
1.	Kadaperi	26.89	27.06	27.09
2.	Kaveripakkam	766.30	587.64	590.64
3.	Kondapuram	215.86	212.91	213.68
4.	Sirukarumbur	177.78	174.03	173.57
5.	Athipattu	115.43	115.06	113.70
6.	Eralacheri	185.20	179.38	180.15
7.	Gengatharanallur	43.55	*	*
8.	Cheri	296.79	298.04	298.03
9.	Kattalai	130.32	130.10	129.87
10.	Thuraiperumpakkam	200.08	190.91	190.61
11.	Maganipattu	210.94	205.73	204.00
12.	Alapakkam	103.48	84.45	85.62
13.	Puthupattu	53.62	96.59	96.15
14.	Panniyur	26.30	29.55	29.22
15.	Kilveeranam	37.25	37.25	37.25
	Total	2589.79	2368.70	2369.59

Note: * During the resettlement sl. no. 7 was merged with Puthupattu village.

Source: Settlement (1882), Resettlement (1913) and Updated settlement (1983) registers.

Table 3.9: DMT Ayacut (Ha) during the Settlements

Sl. no.	*Village*	*Reach*	*Settlement years*		
			1882	*1913*	*1983*
Dusi sluices 1 and 2					
1.	Vagai	HR	26.25	26.04	26.19
2.	Dusi	HR	175.19	176.15	180.66
3.	Kurankanilmuttam	HR	57.44	73.13	72.58
4.	Kanikiluppai	MR	45.74	45.01	44.92
5.	Kilnayakanpalayam	MR	102.23	102.23	102.68
6.	Menallur	MR	66.29	68.08	66.81
7.	Pallavaram	TR	162.40	164.56	163.44
8.	Senianallur	TR	41.62	40.49	41.55
9.	Punaithangal	TR	59.94	59.98	59.16
	Total		737.11	755.68	757.99
Mamandur sluices 3 and 4					
10.	Mamandur	HR	408.09	400.83	400.98
11.	Narasamangalam	HR	44.09	44.55	44.68
12.	Alingilpattu	HR	6.70	5.91	5.52
13.	Valavandal	MR	96.49	95.67	95.50
14.	Vadakalpakkam	MR	96.57	93.22	93.22
15.	Girijapuram	MR	46.45	48.01	47.70
16.	Sothiyampakkam	TR	57.39	56.01	54.52
17.	Bagavanthapuram	TR	42.72	43.53	43.17
18.	Elacheri	TR	131.59	132.36	132.40
	Total		930.10	920.10	917.69
	DMT as a whole		1667.21	1675.77	1675.68

Note: HR = Head reach, MR = Middle reach, TR = Tail reach.

Source: Settlement (1882), Resettlement (1913) and Updated.
Settlement (1983) registers of the DMT villages.

Table 3.10: KPT Ayacut as Reported in Various Sources

Sl. No.	*Source*	*Ayacut (hectares)*
1.	Smith, 1856	3117
2.	North Arcot Settlement Report, 1873-74	2298
3.	North Arcot district Manual, Vol II, A.F. Cox, 1886	2378
4.	PWD flow map ayacut, 1886	2573
5.	PWD hydraulic particulars, 1886	2542
6.	Revised list of component works and irrigation sources, North Arcot district, 1900-01	2417
7.	Revenue Settlement, 1882	2590
8.	Revenue Settlement - Survey, 1892 } (Direct ayacut)	2175
	} (Indirect ayacut)	381
	Direct + Indirect Total	2556
9.	Revenue Resettlement, 1913	2369
10.	Revenue Updated Settlement, 1983	2370
11.	V.N. Kudva, North Arcot district Gazetteer, 1953	2622

Note: Information for sl. nos. 3, 4 and 5 is taken from Vaidyanathan and Janakarajan op. cit. 1989, p. 52. For sl. no. 6 village-wise information is also available.

The DMT *ayacut* at the time of the 1873-74 settlement was placed at 1277 hectares. The 1882 settlement recorded it as 1667 hectares. The increase reflects the extension of the *ayacut* falling under DMT. The construction of the Anicut and the subsequent improvements in the distribution system which could get DMT supplies for a single crop season when the tank reached FTL, were included as part of the *ayacut*. The inclusion of the tail end village of Elacheri in the DMT ayacut during the 1882 settlement is an example of this process. After 1882 the DMT *ayacut* more or less stabilised at 1660-1676 hectares (Table 3.11).

Table 3.11: DMT Ayacut as Reported in Various Sources.

Sl. No.	*Source*	*Ayacut(hectares)*
1.	North Arcot Settlement Report, 1873-74	1277
2.	North Arcot district Manual, Vol II, A.F. Cox, 1886	1233
3.	PWD flow map ayacut, 1886	1667
4.	PWD hydraulic particulars, 1886	1667
5.	Revised list of component works and irrigationsources, North Arcot district, 1900-01	1739
6.	Revenue Settlement survey, 1912	1667
7.	Revenue Settlement, 1882	1667
8.	Revenue Resettlement, 1913	1676
9.	Revenue Updated Settlement, 1983	1676

Note: Information for sl. nos. 2, 3 and 4 is referred from Vaidyanathan and Janakarajan, 1989, p. 52. For sl. no. 5 village-wise information is also available.

The changes in the *ayacut* are already evident from Tables 3.8 and 3.9. Except for Kaveripakkam village there have been only marginal changes in the village-wise *ayacut* of KPT. The reduction in Kaveripakkam village is due to exclusion of the foreshore land (179 ha) consequent on their being given *Pattas* as dry land. This happened during the resettlement (1913) as a result of the *ayacutdars'* objection to the inclusion in the earlier settlement of foreshore land (single crop) in the *ayacut* where no crop could be raised both when the tank got filled and when it is dry. Further only dry crops can be raised in the tank foreshore area.

The decrease in double crop *ayacut* was spread over many villages. This also happened during 1913 in response to objections by the *ayacutdars* in many villages about the non-availability of supplies for second crop in certain segments (see also Table 3.13).

The KPT feeds its *ayacut* through 10 sluices and DMT through four sluices. The major part of the KPT *ayacut* (1377 hectares or 58% of the total) is under middle level sluices. The three deepest sluices of KPT have an *ayacut* of 850 hectares (about 36% of the total). The elevated sluices account for barely five per cent of the overall *ayacut*. The villages served and the extent of *ayacut* irrigated in each village by each sluice are indicated below the Table 3.12. Some villages (Kaveripakkam, Cheri and Eralacheri) in the KPT command get water from more than one sluice (see also figure 3.2).

The arrangements of DMT are rather different. The *ayacut* villages fall into two distinct groups: a group of nine villages being served by the Dusi low level and high-level sluices with a total *ayacut* of 758 hectares, while the other nine villages are shared by the Mamandur low level and high-level sluices (total *ayacut* 918 hectares). Unlike in KPT where each sluice is operated independently of others, in DMT each pair of low and elevated sluices is operated in combination. Whenever the tank reaches FTL, only the elevated sluices are opened. The channels of these sluices join with the lower sluice at some distance from the tank and before the *ayacut* starts (see figure 3.3). There are numerous branch channels taking off from each of the two sluices to feed various *ayacut* villages.

3.4.1 Class Rating of the Villages in 1882 and 1913

The Revenue Department has classified tanks into categories and lands into classes. As KPT and DMT are capable of supplying water for more

Table 3.12: Sluice-Wise Ayacut Details of KPT.

(Ayacut in hectares)

Sluice	*Reach*	*Village*	*Reach*	*Ayacut (1983)*	*Total Ayacut*
Moolai	DS	Kattalai	HR	129.9 }	
		Cheri	MR	73.1 }	317.0
		Kaveripakkam	TR	114.0 }	
Cheri Low Level	DS	Cheri	HR	88.3	88.3
Lion	DS	Kaveripakkam	HR	318.1 }	
		Eralacheri	TR	130.9 }	449.0
Maganipattu and}		Cheri	HR	27.7 }	
Thuraiperum- }	MS	Maganipattu	MR	204.0 }	
pakkam }		Alapakkam	MR	85.6 }	604.0
		Puthupattu	TR	96.2 }	
		Thuraiperumpakkam	TR	190.6 }	
Cheri High Level	MS	Cheri	HR	92.5	92.5
Fox	MS	Kaveripakkam	HR	129.7 }	
		Kondapuram	MR	213.7 }	
		Eralacheri	TR	49.2 }	679.9
		Sirukarumbur	TR	173.6 }	
		Athipattu	TR	113.7 }	
Panniyur	ES	Cheri	HR	16.5	16.5
Panniyur	ES	Panniyur	HR	29.2 }	
		Kilveranam	TR	37.2 }	66.5
Gaffor	ES	Kadaperi	HR	27.1	27.1
Surplus kondam		Kaveripakkam		15.5	15.5
Catchment		Kaveripakkam		13.3	13.3
		Total		2369.6	2369.6

Note: DS = Deep sluice; MS = Middle sluice; ES = Elevated sluice.
HR = Head reach; MR = Middle reach; TR = Tail reach.

Source: Village Adangal registers (1989) and Updated settlement (1983) registers of KPT villages.

than ten months to their *ayacuts* they are included in class I category of tanks (Below the Table 3.13). While the entire *ayacut* is classified as wet land, there are significant differences within the *ayacut* in respect of the period of supply of water to the *ayacut* lands. For instance, lands irrigated by the high-level sluices though situated very near the tank, and those lying in tail end villages though commanded by low level sluices, are unlikely to get supplies for ten months and the supply is likely to vary widely. These lands were therefore placed in class II.

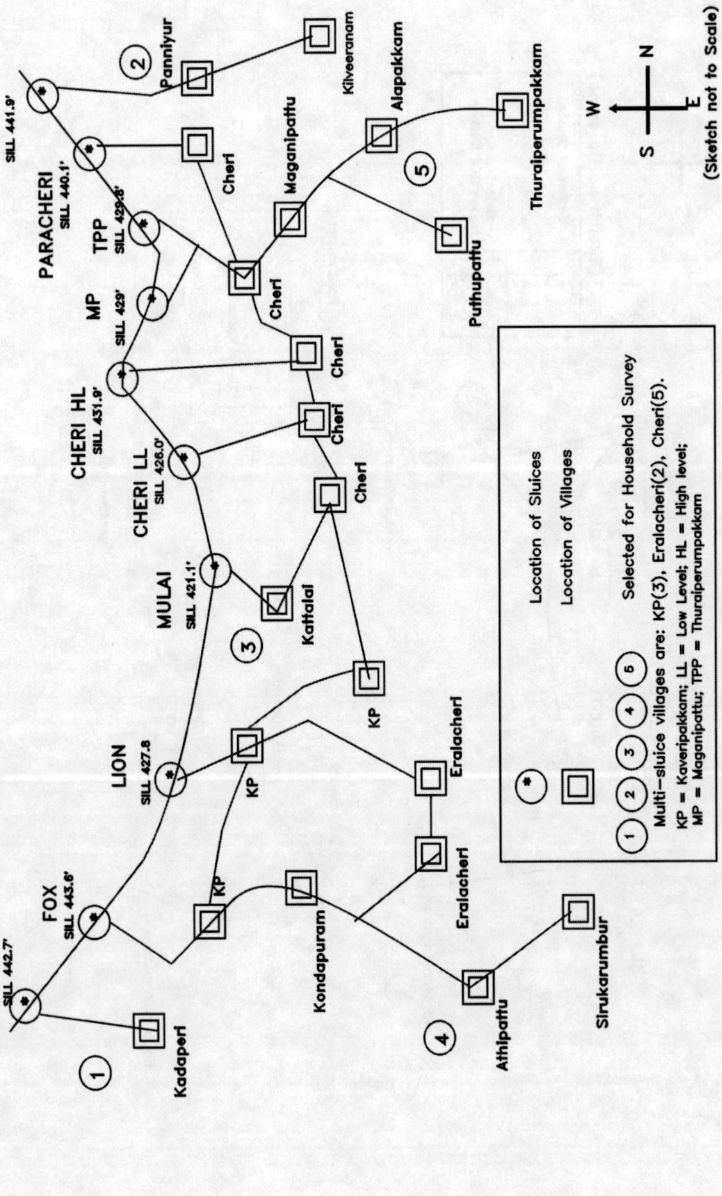

Figure 3.2: Kaveripakkam Tank: Location of Sluices & Villages.

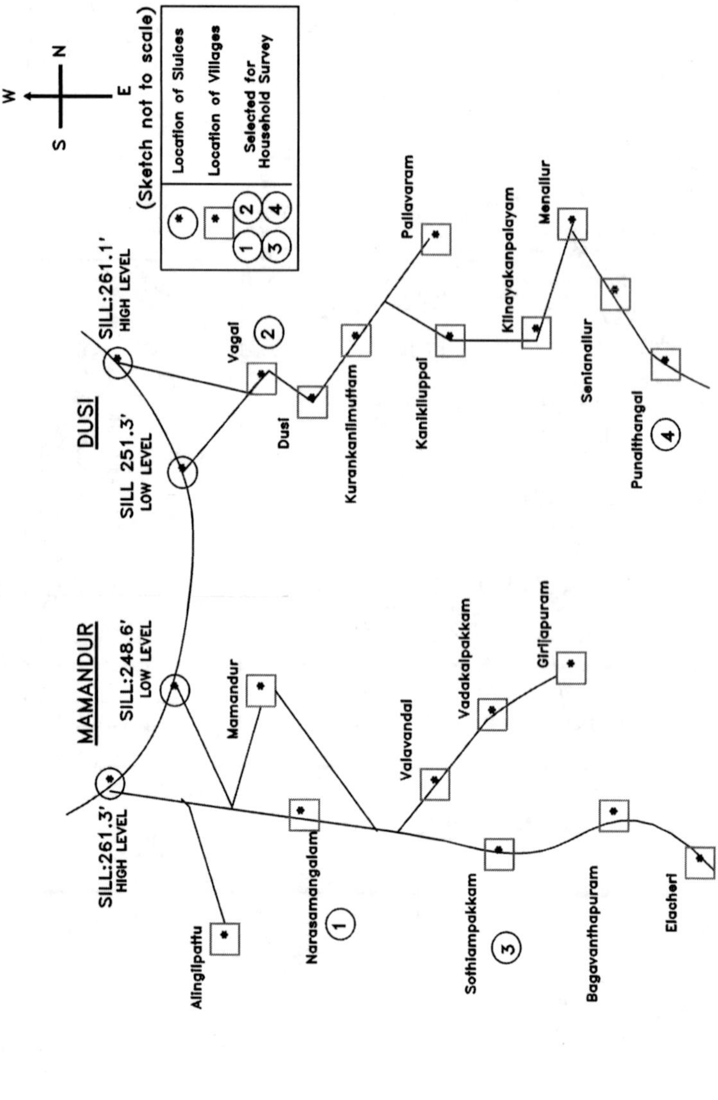

Figure 3.3: Dusi—Mamandur Tank: Location of Sluices & Villages.

Table 3.13: Villages Commanded by KPT and the Wet Class Assigned at the Time of Settlement and the Resettlement Periods.

Village	*Details of sluices and channels*	*Wet 1882*	*Class 1913*
Kaveripakkam	Fox, Lion and Moolai sluices	I	I
	Tank water spread	III	III
	Iruppanam Kondam	I	I
Kondapuram	Fox sluice	I	I
Athipattu	-----do----	I	I
	Palar river spring channel	I	II
Sirukarumbur	Fox sluice	I	I
	Kasakkal	II	II
	Sitheri	I	I
Eralacheri	Fox and Lion sluices	I	I
	Iruppanam Kondam	I	I
Kattalai	Moolai sluice	I	I
Cheri	Moolai, Cheri Low Level, Cheri High Level, and M. Pattu & T.P.P. sluices	I	I
	Paracheri Sluice	I	I
	Panniyur	I	II
	---do--- and Upparanthangal	II	II
Maganipattu	Maganipattu & T.P.P. Sluice	I	I
Thuraiperumpakkam	----------do-------------	I	I
	Munchivakkam Kondam	I	I
Alapakkam	Maganipattu and T.P.P. sluice	I	I
	Arakkaneri	I	II
	Perumal Eri	I	II
	Sathavanatham Eri & Sitheri	I	II
Panniyur	Panniyur sluice and Panniyur tank	II	II
Puthupattu	----do---- & Gengatharanallur tank	II	II
	Maganipattu & T.P.P. sluices	I	I
	Puthupattu Periya Eri & thangal	I	II
Kilveeranam	Sitheri thangal Eri & Periya Eri	II	II
	Surakolam Eri	III	III

Note: Class I = >10-month supply. Class II = 6 to 8-month supply. Class III = 3 to 4-month supply.

Source: Go no. 250 I, PWD, Government of Madras, 03-06-1914.

The rest of the *ayacut* has been classified as class I. In KPT a small area in one village was placed in class III. At the time of the original settlement, the KPT *ayacut* was further classified into 30 groups depending on the sluice from which water is derived. Of these, 23 were categorised as wet class I; five as class II and two as class III.

During 1913 resettlement, objections were raised about the classification of tanks in Puthupattu and Alapakkam villages. The ryots of these villages pleaded that they never received KPT supplies even when the tank was full because the *Kondam* across KPT surplus channel to feed these villages (constructed in 1895) was washed away. The plea was accepted and resulted in a substantial reclassification with the rating of six segments being changed from class I to class II.

For DMT data are available only in respect of rating given in 1913 (Table 3.14) when the *ayacut* in all the nine villages served by the Dusi sluices were categorized as class I wet land. In the case of villages served by the Mamandur sluices, lands in six villages were classified as class I and lands in the remaining three (all in the tail end) were classified as class II.

Table 3.14: Villages Commanded By DMT and the Wet Class Assigned in the Resettlement Ayacut (Ha)

Sl. No.	*Village*	*Reach*	*Registered ayacut, 1913*		*Total*	*Class assigned in 1913*
			Double crop	*Single crop*		
Dusi sluices 1 and 2						
1.	Vagai	HR	26.15	0.0	26.15	I
2.	Dusi	HR	174.74	1.39	176.14	I
3.	Kurankanilmuttam	HR	57.23	15.34	72.57	I
4.	Kanikiluppai	MR	44.22	0.75	44.97	I
5.	Kilnayakanpalayam	MR	70.40	32.07	102.47	I
6.	Menallur	MR	42.98	24.83	67.82	I
7.	Pallavaram	TR	102.35	62.49	164.83	I
8.	Senianallur	TR	*	*	*	*
9.	Punaithangal	TR	26.84	32.74	59.58	I
	Total		544.92	169.61	714.53	
Mamandur sluices 3 and 4						
10.	Mamandur	HR	403.04	0.0	403.04	I
11.	Narasamangalam	HR	26.99	17.52	44.51	I
12.	Alingilpattu	HR	0.00	5.91	5.91	I
13.	Valavandal	MR	95.72	0.06	95.78	I
14.	Vadakalpakkam	MR	60.64	32.46	93.09	I
15.	Girijapuram	MR	13.55	34.42	47.96	I
16.	Sothiyampakkam	TR	0.28	55.74	56.02	II
17.	Bagavanthapuram	TR	0.00	43.56	43.56	II
18.	Elacheri	TR	1.22	129.53	130.75	II
	Total		601.44	319.20	920.64	
	DMT as a whole		1146.36	488.81	1635.17	

Note: Settlement for sl. no. 8 was made only in 1964.
HR = Head reach; MR = Middle reach and TR = Tail reach.
Class I represents the tank supply available for more than 10 months.
Class II represents the tank supply available for about 6 to 8 months.
Source: Details taken from the consolidated figures given in the Resettlement (1913) registers of the villages irrigated by DMT.

In KPT between 1882 and 1983, the registered extent decreased from 2590 hectares to 2370 hectares, with a net reduction of 220 hectares. This reduction is mainly due to the exclusion of single crop land of about 162 hectares in Kaveripakkam village alone located at the tank foreshore area (Table 3.15). Between 1882 and 1983, the extent of area under DMT changed, as single crop wet has increased about 186 hectares with a corresponding (145 hectares) decline in land classified as double crop wet (see below the Table 3.16). This conversion in 1913 happened mainly in response to objections by the *ayacutdars* against 1882 settlement classification.

Table 3.15: Cropwise Ayacut of KPT Villages During Settlement And Updated Settlement

(Ayacut in hectares)

Sl. no.	Village	*Settlement years*		*Ayacut*			
				1882		*1983*	
		1882	*1983*	*Single crop*	*Double crop*	*Single crop*	*Double crop*
1.	Kadaperi	26.89	27.09	26.02	0.87	26.87	0.21
2.	Kaveripakkam	766.30	590.64	251.69	514.61	87.76	502.89
3.	Kondapuram	215.86	213.68	133.29	82.57	143.81	69.87
4.	Sirukarumbur	177.78	173.57	128.27	49.51	130.17	43.40
5.	Athipattu	115.43	113.70	28.29	87.14	30.67	83.04
6.	Eralacheri	185.20	180.15	162.86	22.34	158.70	21.46
7.	Gangatharanellur	43.55	*	43.55	0.0	*	*
8.	Cheri	296.79	298.03	148.54	148.24	159.07	138.96
9.	Kattalai	130.32	129.87	4.04	126.28	3.85	126.02
10.	Thuraiperumpakkam	200.08	190.61	182.92	17.15	173.77	16.84
11.	Maganipattu	210.94	204.00	139.03	71.91	136.96	67.04
12.	Alapakkam	103.48	85.62	96.21	7.28	78.46	7.15
13.	Puthupattu	53.62	96.15	27.89	25.73	70.44	25.71
14.	Panniyur	26.30	29.22	18.93	7.38	21.87	7.35
15.	Kilveeranam	37.25	37.25	25.82	11.43	26.00	11.24
	Total	2589.79	2369.59	1417.36	1172.44	1248.40	1121.18

Note: Sl . no. 7 was merged with Puthupattu village.

Source: Settlement and Updated settlement registers of KPT villages.

3.5 Characteristics of Water Supply

There are no records of the amount of water received by each of the tanks and its utilisation in different seasons. The only information available is the record of the depth of storage maintained by the PWD. In principle the volume of water in storage corresponding to a given depth can be calculated if the configurations of the storage had been accurately surveyed. Such a survey has been recently done only for KPT. In any case, storage

Table 3.16: Cropwise Ayacut of DMT Villages During Settlement and Updated Settlement (Ayacut in hectares)

(Ayacut in hectares)

Sl. no.	Village	Reach	Settlement years		Ayacut			
					1882		1983	
			1882	1983	Single crop	Double Crop	Single crop	Double Crop
Dusi sluices 1 and 2								
1.	Vagai	HR	26.25	26.19	0.0	26.25	0.0	26.19
2.	Dusi	HR	175.19	180.66	4.36	170.83	9.81	170.85
3.	Kurankanilmuttam	HR	57.44	72.58	1.69	55.75	17.76	54.82
4.	Kanikiluppai	MR	45.74	44.92	1.19	44.55	0.00	44.92
5.	Kilnayakanpalayam	MR	102.23	102.68	8.06	94.17	32.95	69.73
6.	Menallur	MR	66.29	66.81	34.87	31.42	26.32	40.49
7.	Pallavaram	TR	162.40	163.44	58.88	103.53	63.10	100.34
8.	Senianallur	TR	*	41.55	*	*	37.94	3.61
9.	Punaithangal	TR	59.94	59.16	27.69	32.25	33.30	25.86
	Total		695.49	757.99	136.74	558.75	221.18	536.81
Mamandur sluices 3 and 4								
10.	Mamandur	HR	408.09	400.98	0.00	408.09	0.0	400.98
11.	Narasamangalam	HR	44.09	44.68	0.00	44.09	17.54	27.14
12.	Alingilpattu	HR	6.70	5.52	0.06	6.64	5.52	0.0
13.	Valavandal	MR	96.49	95.50	0.00	96.49	0.24	95.26
14.	Vadakalpakkam	MR	96.57	93.22	35.79	60.79	33.91	59.31
15.	Girijapuram	MR	46.45	47.70	32.90	13.55	33.85	13.85
16.	Sothiyampakkam	TR	57.39	54.52	16.89	40.50	54.52	0.0
17.	Bagavanthapuram	TR	42.72	43.17	3.91	38.81	43.17	0.0
18.	Elacheri	TR	131.59	132.40	121.36	10.23	132.40	0.0
	Total		930.10	917.69	210.91	719.19	321.15	596.54
	DMT as a whole		1625.59	1675.68	347.65	1277.94	542.33	1133.35

Note: Settlement for sl. no. 8 was made only in 1964.
HR = Head reach; MR = Middle reach and TR = Tail reach.

Source: Settlement, and updated settlement registers of DMT villages.

tells nothing about the flows in and out. During the rainy season, the actual volume of water used is likely to be greater than the storage. But once the filling season is over (normally in December), there is hardly any inflow and utilisation is almost exclusively based on depletion of stored water.

While these limitations preclude a detailed analysis of supply and use in the two tanks, it is possible to examine certain related issues:

(1) the relation between the quantum of water available from the main source of supply (diversion from the Anicut and rainfall in the free catchment) and the maximum storage attained over a period of year.

(2) the level of storage at different times of the year and its adequacy for sustaining irrigated crops in different seasons.

(3) the extent of variability in storage within the seasons in particular years as well as in a particular season across years.

All these are of course only subject to the reliability of storage depth record, the high level of approximation of estimated yield from rainfall, and the fact that storage is not a reliable index of utilisation especially in the rainy season. As DMT is fed by three channels, located at the tail end of the diversion channels, and data on diversion are not available for all channels, it is particularly difficult to relate storage to supply source. In the case of KPT, there are basically two supply sources: the Kaveripakkam supply channel and rainfall in its free catchment. Being at the head of the channel, close to the Anicut, and free catchment being equal to the combined catchment between the Anicut and the tank, a close nexus is expected between the amount of water diverted into the channel and the rainfall in the catchment.

Table 3.17 presents the estimate made on the following assumptions: (1) About 50 per cent of the water diverted into Kaveripakkam main channel at the Anicut is diverted to the KPT supply channel; (2) the yield from the rainfall in the free catchment is estimated on the basis of annual rainfall recorded at the tank site, and (3) 35 per cent of total rainfall is available in the form of surface run off. The data suggest that:

Table 3.17: Sources of Supply (PAS + Catchment Run-off) to KPT

Year	*KP Main supply channel*	*KPT supply channel*	*Maximum storage (metre)*	*Rainfall KPT (mm)*	*Total rainfall*	*Run-off 35% of col 6*	*Total supply (3+7)*
	<------ mcm ------>				*<------------ mcm ------------>*		
1	*2*	*3*	*4*	*5*	*6*	*7*	*8*
1964	187.65	93.8	9.18	723	18.8	6.6	100.4
1965	5.83	2.9	6.39	1107	28.7	10.1	13.0
1966	52.02	26.0	9.18	1144	29.7	10.4	36.4
1967	36.93	18.5	7.02	1162	30.2	10.6	29.0
1968	0	0.0	0	392	10.2	3.6	3.6
1969	63.61	31.8	9.18	1146	29.8	10.4	42.2
1970	54.23	27.1	8.61	1192	31.0	10.8	38.0
1971	33.67	16.8	7.05	1008	26.2	9.2	26.0
1972	92.24	46.1	9.18	969	25.2	8.8	54.9
1973	37.75	18.9	7.44	771	20.0	7.0	25.9
1974	107.73	53.9	9.06	829	21.5	7.5	61.4
1975	111.50	55.7	9.12	864	22.4	7.9	63.6
1976	42.14	21.1	8.97	1206	31.3	11.0	32.0
1977	61.51	30.8	9.18	1228	31.9	11.2	41.9

(*Contd...*)

1978	46.73	23.4	8.97	955	24.8	8.7	32.0
1979	145.91	73.0	9.12	823	21.4	7.5	80.4
1980	0	0.0	0	627	16.3	5.7	5.7
1981	155.99	78.0	8.97	779	20.2	7.1	85.1
1982	0	0.0	5.1	810	21.0	7.4	7.4
1983	35.91	18.0	7.29	1148	29.8	10.4	28.4
1984	31.89	15.9	5.16	1332	34.6	12.1	28.1
1985	14.70	7.3	6.99	1179	30.6	10.7	18.1
1986	0	0.0	0	724	18.8	6.6	6.6
1987	0	0.0	5.16	1029	26.7	9.4	9.4
1988	46.62	23.3	8.82	1180	30.6	10.7	34.0
1989	0	0.0	0	836	21.7	7.6	7.6
1990	0	0.0	0	681	17.7	6.2	6.2
1991	78.90	39.4	9.27	929	24.1	8.4	47.9
1992	1.09	0.5	4.14	480	12.5	4.4	4.9
1993	0.78	0.4	9.27	642	16.7	5.8	6.2
1994	0	0.0	0	799	20.8	7.3	7.3
1995	0	0.0	0	1172	30.4	10.7	10.7
1996	60.14	30.1	9.27	1687	43.8	15.3	45.4
1997	15.15	7.6	7.26	770	20.0	7.0	14.6
1998	69.26	34.6	9.18	1176	30.5	10.7	45.3
1999	0	0.0	0	901	23.4	8.2	8.2
2000	3.12	1.6	4.5	1178	30.6	10.7	12.3
2001	0	0.0	0	1191	30.9	10.8	10.8
2002	39.52	19.8	8.94	1108	28.8	10.1	29.8
2003	0	0.0	0	774	20.1	7.0	7.0
2004	0	0.0	0	806	20.9	7.3	7.3
2005	65.15	32.6	8.94	1235	32.1	11.2	43.8
2006	0	0.0	0	981	25.5	8.9	8.9
2007	14.74	7.4	5.55	995	25.8	9.0	16.4
2008	26.21	13.1	5.76	1118	29.0	10.2	23.3
2009	0	0.0	0	757	19.7	6.9	6.9
2010	32.58	16.3	5.82	873	22.7	7.9	24.2
2011	0	0.0	0	798	20.7	7.3	7.3
2012	0	0.0	0	735	19.1	6.7	6.7
2013	0	0.0	0	850	22.1	7.7	7.7
2014	0	0.0	0	901	23.4	8.2	8.2
2015	3.75	1.9	9.18	1395	36.2	12.7	14.6
2016	0	0.0	6.3	1081	28.1	9.8	9.8

Note: Column 2: About 50 % of the total supply from the Kaveripakkam (KP) main channel is diverted to KPT supply channel.

Column 4: Maximum storage at the beginning of the supply season (Oct-Dec.)

Column 7: Excluding 65 % loss due to evaporation, percolation, encroachment in the catchment area, etc.

Conversion formula for columns 6 and 7:

1 mm rainfall = 0.02597 mcm is available from the local rainfall.

1 mm rainfall on (1 sq. mile) 2.59 sq. km = 0.00252 mcm

Free catchment area of KPT = 31 sq. kms.

Source: Water reading registers of KPT and PAS.

(1) Whenever the volume diverted into the KPT supply channel increased, the maximum storage in KPT also increased. However, discharge in excess of 28.32 mcm does not make much difference to maximum storage while it reaches FTL. Also, at very low level of flow (< 2.832 mcm) there is no clear relation between discharge in channel and storage at KPT (see figure 3.4).

(2) There is little association between the quantum of run-off from the free catchment of KPT and its storage level (figure 3.5).

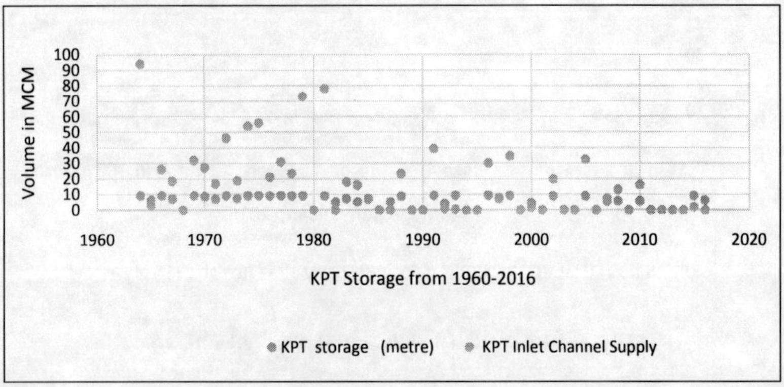

Fig. 3.4: KPT Storage with Inlet Chl. Supply.

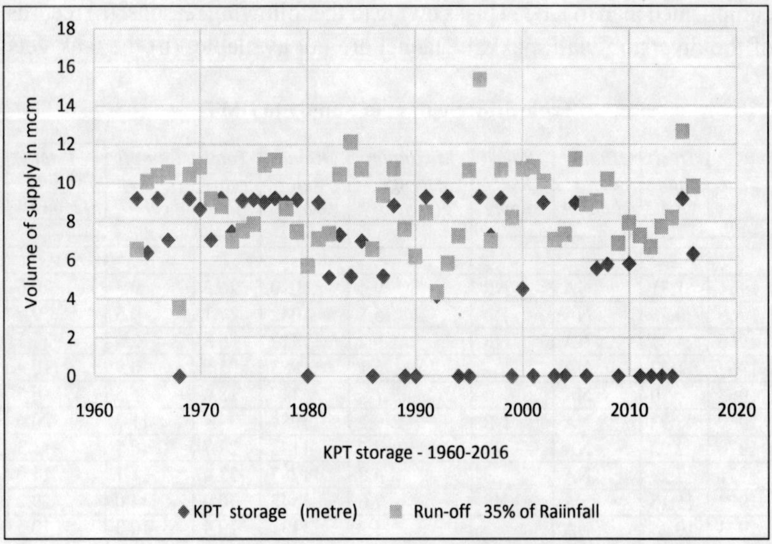

Fig. 3.5: KPT Storage with Run Off.

(3) There is a strong association between the sum of discharge into KPT level and run-off from the free catchment and the maximum storage in KPT (figure 3.6).

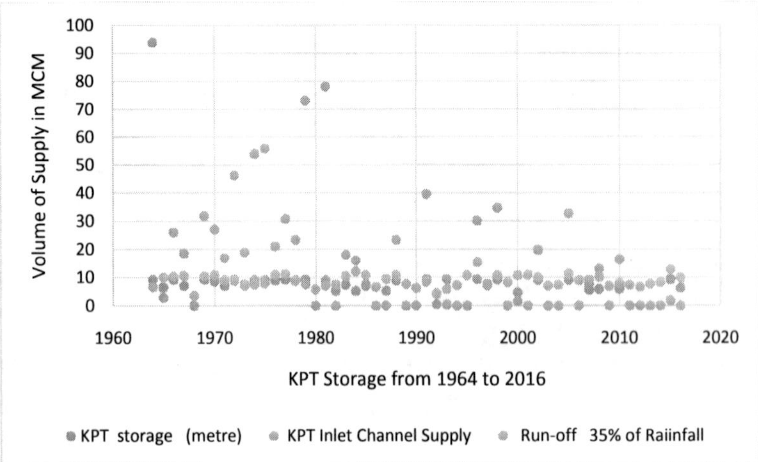

Fig. 3.6: KPT Storage with Inlet Chl. Supply + Run-off.

Table 3.18 gives the relevant source of supply data for DMT. Relating water available from its sources and its utilisation to DMT storage is more complicated than to KPT. This is owing to the following reasons: (a) records of the diversion via Rajakkal channel are not available; (b) the tank gets

Table 3.18: Sources of Supply to DMT

Year	*Dusi channel*	*Thandarai channel*	*Rajakkal supply*	*Maximum storage (metre)*	*Rainfall DMT*	*Total rainfall*	*Run-off 35% of col 7*	*Total supply (2+3+8)*
<-------------	*mcm*	------------->			*(mm)*	<----------	*mcm*	---------->
1	*2*	*3*	*4*	*5*	*6*	*7*	*8*	*9*
1964	61.31	NA	NR	9.4	1020	26.5	9.3	70.6
1965	0.25	NA	-	5.8	1042	27.1	9.5	9.7
1966	24.21	NA	NR	9.2	1417	36.8	12.9	37.1
1967	1.47	NA	-	6.1	983	25.5	8.9	10.4
1968	0	NA	-	-	717	18.6	6.5	6.5
1969	8.92	NA	NR	9.2	1287	33.4	11.7	20.6
1970	2.92	NA	NR	7.0	1143	29.7	10.4	13.3
1971	2.29	NA	-	4.8	1287	33.4	11.7	14.0
1972	15.15	NA	NR	7.0	1542	40.0	14.0	29.2
1973	0.25	NA	-	3.2	1131	29.4	10.3	10.5

(Contd...)

1974	16.37	NA	NR	5.1	887	23.0	8.1	24.4
1975	20.96	NA	NR	7.1	1432	37.2	13.0	34.0
1976	15.32	-	NR	9.1	1343	34.9	12.2	27.5
1977	1.90	953	NR	9.6	2019	52.4	18.4	973.2
1978	8.69	1229	NR	7.4	1614	41.9	14.7	1252.4
1979	53.55	719	NR	9.3	982	25.5	8.9	781.5
1980	0	-	-	3.5	895	23.2	8.1	8.1
1981	31.07	151	NR	4.4	1061	27.6	9.6	191.7
1982	0	-	-	3.9	688	17.9	6.3	6.3
1983	3.12	894	-	7.3	1177	30.6	10.7	907.8
1984	0.11	204	-	17.3	1006	26.1	9.1	213.3
1985	1.90	507	NR	5.2	1205	31.3	11.0	519.9
1986	0	50	-	-	535	13.9	4.9	54.9
1987	0	142	-	6.6	755	19.6	6.9	148.9
1988	5.98	-	-	2.1	732	19.0	6.7	12.6
1989	0	-	-	2.7	741	19.2	6.7	6.7
1990	0	-	-	-	1085	28.2	9.9	9.9
1991	90.72	77		8.79	1162	30.2	10.6	178.3
1992	1.36	-	-	4.11	700	18.2	6.4	7.7
1993	1.00	-	-	5.55	971	25.2	8.8	9.8
1994	0	55	-	5.19	1095	28.4	10.0	65.0
1995	0	135	-	8.31	1938	50.3	17.6	152.6
1996	91.48	560	NR	9.24	2545	66.1	23.1	674.6
1997	19.60	475	NR	7.59	1689	43.9	15.4	510.0
1998	68.91	610	NR	8.49	1572	40.8	14.3	693.2
1999	0	-	-	7.8	1082	28.1	9.8	9.8
2000	3.17	-	NR	2.4	1034	26.9	9.4	12.6
2001	0	310	-	3.6	1389	36.1	12.6	322.6
2002	46.87	-	NR	4.5	629	16.3	5.7	52.6
2003	0	79	-	3.1	1330	34.5	12.1	91.1
2004	0	-	-	3.2	1001	26.0	9.1	9.1
2005	54.65	235	NR	9.09	1531	39.8	13.9	303.6
2006	0	-	-	3.99	987	25.6	9.0	9.0
2007	19.14	61	NR	4.8	1039	27.0	9.4	89.6
2008	29.86	58	NR	6.21	1165	30.3	10.6	98.4
2009	0	-	-	3.6	837	21.7	7.6	7.6
2010	22.84	98	NR	7.41	1121	29.1	10.2	131.0
2011	0	-	-	6.9	930	24.2	8.5	8.5
2012	0	-	-	6.51	879	22.8	8.0	8.0
2013	0	-	-	4.41	992	25.8	9.0	9.0
2014	0	-	-	3.9	863	22.4	7.8	7.8
2015	3.87	670	NR	9.24	1596	41.4	14.5	688.4
2016	0	-	-	6.35	1181	30.7	10.7	10.7

Note: Thandarai anicut was constructed in 1975. NA = Not applicable.
NR = Data not recorded; Dash indicates no supply/storage.
Column 5 indicates storage at the beginning of the supply season.
Figures in column 8: excluding 65 % loss due to evaporation, Percolation, encroachment in the catchment area, and so on.
Conversion formula for columns 7 and 8:
1 mm rainfall = 0.02597 mcm is available from the local rainfall.
1 mm rainfall on (1 sq. mile) 2.59 sq. km = 0.00252 mcm
Free catchment area of DMT = 30.6 sq. kms.
Source: Water reading registers of DMT and PAS.

supply from the surpluses of tanks fed by Sakkaramallur channel; (c) the extent of combined catchment between the Anicut and the tank is several times larger than its free catchment; and (d) it is situated at the tail end of the channel. One would, therefore, expect the relation between yield from ultimate supply sources and storage to be much fuzzy in DMT. And indeed, it seems to be so.

As may be seen from figure 3.7, there is no strong association between the volume drawn by the Dusi-Thennampattu and the Thandarai channels and the maximum storage reached in DMT. On the other hand there is a strong positive association between run-off from the free catchment and maximum storage reached at DMT (figure 3.8) as well as between the sum

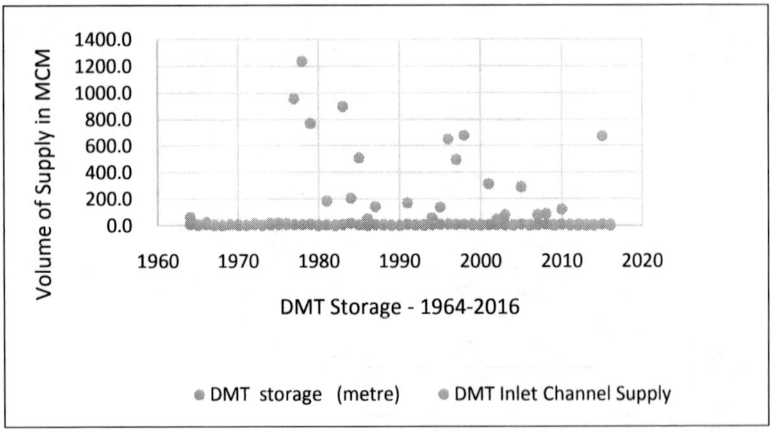

Fig. 3.7: DMT Storage with Inlet Chls. Supply.

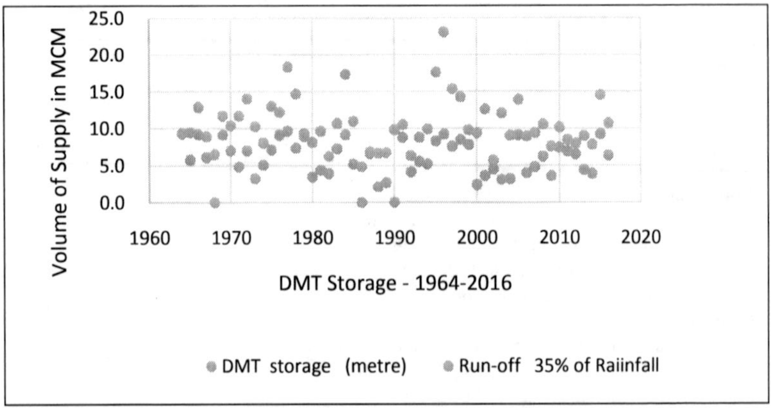

Fig. 3.8: DMT Storage with with Run-off.

of volume diverted by the supply channels at the source and supply from the free catchment run-off and the DMT storage (figure 3.9). This means that the free catchment run-off is the more important determinant of supplies to DMT while in KPT it depends more on the volume diverted from the Anicut.

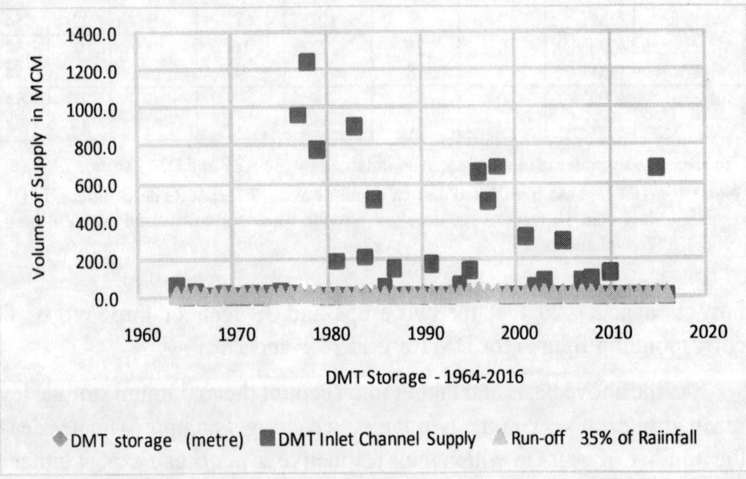

Figure 3.9: DMT Storage with Inlet CHLS. Supply + Run-off.

3.5.1 Storage Levels at the Tanks

Though the maximum depth of the tanks is more or less the same, KPT attained FTL in 14 out of the 54 years between 1964 and 2017, whereas DMT attained it only nine out of 54 years. In the case of KPT the maximum storage was between 20 and 30 feet in 16 years; and between 15 and 20 feet in 24 years (a level in which one crop irrigation is feasible for a few deep sluices head reach villages ayacut) and less than 15 feet no irrigation is considered feasible due to heavy siltation in the KPT bed. DMT attained maximum depth of 20-30 feet in 27 years, and less than 15 feet in 20 years (Below the table 3.19). In terms of full storage, KPT is clearly better placed than DMT.

However, DMT has a larger storage capacity per hectare of *ayacut* than KPT, and consequently if the storage is full it can irrigate for a period of 18 months compared to 14 months in the case of KPT. According to knowledgeable farmers, a minimum depth of 21 feet is required in KPT in order to raise at least one irrigated crop (in the deep sluices head reach

Table 3.19: Frequency Distribution of Years According to Levels of Maximum Storage Attained in KPT and DMT, 1964-2017

Month of Max. supply Available		*Storage levels* <---------->						
		30'	*25' – 30*	*20' - 25'*	*15' - 20'*	*10' -15'*	*<10'*	*Total*
December:	KPT	14	8	8	5	4	15	54
	DMT	9	6	12	7	9	11	54
March:	KPT	0	26	12	16	0	0	54
	DMT	0	18	3	9	11	13	54
July:	KPT	0	0	2	52	0	0	54
	DMT	0	0	9	9	11	25	54

Source: Daily water reading registers maintained by the KPT and DMT section offices.

Note: KPT bed has been silted up to a depth of over 17 feet at its deep sluices. In DMT, less than 10 feet storage does not provide water for cultivation, hence it may be considered as dry.

farmers ayacut), 29 feet for two crops and 30 feet for three crops. The corresponding figures for DMT are 15', 23' and 28'.

On the above basis and taking into account the maximum storage level reached in each year in the two tanks, we can get an approximate idea of the number of years in which their respective *ayacuts* can expect either no irrigated crop, or one, two, or three irrigated crops (see Table 3.20).

Table 3.20: Crop Feasibility in KPT and DMT, 1961-2017

Crop	*1961 – 1990*		*1980s*		*1991 – 2017*		*2005-2015*	
season	*No. of Years*		*Percentage*		*No. of Years*		*Percentage*	
	KPT	*DMT*	*KPT*	*DMT*	*KPT*	*DMT*	*KPT*	*DMT*
1 Crop	3	6	20	40	4	10	20	40
2 Crops	7	8	20	10	2	5	0	30
3 Crops	14	12	20	20	9	9	20	30
Tank dry	6	4	40	30	**12**	3	**60**	0
Total	**30**	**30**	**100**	**100**	**27**	**27**	**100**	**100**

Source: Data Derived from Tables 3.21 and 3.22.

In KPT out of 30 years (1961-1990) the storage in 14 years was adequate to raise three crops; in 21 years it permitted for raising two crops; and in 24 years it was enough at least a single irrigated crop. The tank remained dry in six years. During the same period, the respective feasibility for three crops, two crops and one crop in DMT was for 12 years; 20 years; and 26 years. The tank remained dry in four out of the 30 years.

It is also observed from Table 3.20, during the latter period 1991 to 2017 more number of dry years is recorded in both tanks compared to the earlier period. During this period from 1991, KPT storage was adequate to raise three crops in 9 years; two crops in 11 years and at least one crop in 15 years. The corresponding years of crop feasibility for DMT was 9 years, 14 years and 24 years. Also, more number of tank dry years (12) is seen in KPT than DMT (3).

It would seem from Tables 3.21 and 3.22 that the period 1980s is different from the 1960s and the 1970s. KPT was dry only once during the 1960s and the 1970s and DMT was dry only once in the 1960s. But during the 1980s both the tanks remained dry for three years mainly because of lower rainfall than in the 1960s and 1970s. During the 1990s both tanks dry for two years each.

Even then, both the tanks attained 4 times of full tank level storage (in DMT in 3 years the tank had reached about to full level). The last two decades of 2000s and 2010s invariably seen more of dry years compared to the other periods in both tanks. A more reliable - though still inadequate - basis for assessing feasibility of irrigation in different seasons would be the depth of storage at the beginning of each crop season, and the average depth of storage during each season. Three crop seasons have been distinguished: October-February, March-June and July-September. Of these the first is the rainy season and the tanks get most of their supply during this period. Storage level may be a less reliable indicator of water availability during this season; but in the second and third seasons when rainfall is low, available storage from the tank is the main source of water supply. The level of water is a crucial determinant of cropping possibilities.

The relevant data - based on a collection of daily water level recordings at the two tanks - for the period 1966-2017 are presented below the Table 3.23. The picture in summary may be seen from below the Table 3.24. The following features are noteworthy. In 80% of the years both the tanks had storage to permit at least one irrigated crop; in three fourths of the years two or more irrigated crops could be raised. Storage by July which is used for cropping in the third season is available more frequently (23 out of 51) in KPT than in DMT (21 out of 51).

Table 3.21: Maximum Water Levels Reached in KPT During Selected Months, 1961-2017.

Water levels	*1961-70*			*1971-80*			*1981-90*			*1991-2000*			*2001-2010*			*2011-2017*		
	Oct	*Nov*	*Dec*	*Oct*	*Nov*	*Dec*	*Oct*	*Nov*	*Dec*	*Oct*	*Nov*	*Dec*	*Oct*	*Nov*	*Dec*	*Oct*	*Nov*	*Dec*
Full tank level	-	1	4	-	2	5	1	1	-	-	-	4	-	1	1	-	1	1
22'-29'	1	-	2	-	2	-	1	1	-	-	1	1	-	-	-	-	-	-
15'-21'	-	-	1	-	-	-	-	1	2	1	1	-	-	-	2	-	1	1
Tank dry	-	-	1	-	-	1	-	-	3	-	-	2	-	-	6	-	-	3

Note: 15' - 21' = one crop. 22' - 29' = two crops. >30' = three crops. Dash indicates nil.
Source: Water Reading Register, Public Works Department, Kaveripakkam section.

Table 3.22: Maximum Water Levels Reached in DMT During Selected Months, 1961-2017.

Water levels	*1961-70*			*1971-80*			*1981-90*			*1991-2000*			*2001-2010*			*2011-2017*		
	Oct	*Nov*	*Dec*	*Oct*	*Nov*	*Dec*	*Oct*	*Nov*	*Dec*	*Oct*	*Nov*	*Dec*	*Oct*	*Nov*	*Dec*	*Oct*	*Nov*	*Dec*
Full tank level	-	3	1	-	1	2	-	-	-	-	-	1	-	-	1	-	1	-
22'-29'	1	-	1	-	-	1	-	-	2	-	1	3	-	-	1	-	-	-
15'-21'	-	-	-	-	1	1	-	1	3	-	1	-	-	1	1	-	1	1
Tank dry	-	-	1	-	-	-	-	-	3	-	-	2	-	-	4	-	-	1

Note: 10'-15'= one crop. 16'-23'= two crops. >24'= three crops. Dash indicates nil.
Source: Water reading register, Public Works Department, Iyenkarkulam section office, Dusi.

Table 3.23: Availability of Water Supply in Different Crop Periods in KPT and DMT, 1966-2017.

Period	*No Crop*	*1 Crop*	*2 Crops*	*3 Crops*	*Total*
KPT					
1966-1979	1	0	5	8	14
1966-1989	3	2	8	11	24
1980-1989	2	2	3	3	10
1990-2017	12	4	2	9	27
DMT					
1966-1979	1	1	2	10	14
1966-1989	3	4	4	13	24
1980-1989	2	3	2	3	10
1990-2017	3	10	5	9	27

Note: If the tank supplied water adequately to the third crop, one can definitely say that the second crop also got adequate supply. Because, the effective storage in the tank must be decided within the month of December. After that the tanks would not get any supply. Three crops total does not include one crop and two crops total.

Source: Daily water reading registers of KPT and DMT.

Excluding the years in which the tanks were dry in all the three seasons, it is found that out of the 51 years KPT had stored supply available in 23 years DMT in 34. The number of years in which storage was available in the second and the third seasons was 28 and 20 in the case of DMT and 32 and 18 respectively in the KPT. In KPT the mean depth of storage (considering only years in which water was available in the tank during the particular season) in the second season is lower than in the first; and lower in the third than in the second. In DMT, however, the average depth of storage in the second season is higher than in the first, but in the third it is substantially lower than in both the first and the second (Table 3.24).

Table 3.24: Water Availability under KPT and DMT, 1966-2017.

Season	*No. of Dry years*	*No. of non dry years*	*mean depth*	*mean CV within the year*
KPT				
I	18	23	22	30.3
II	9	32	20	25.1
III	23	18	11	52.5
DMT				
I	7	34	16	30.4
II	13	28	19	29.8
III	21	20	12	22.2

Source: Derived from the Table 3.23.

Inferences about the feasibility of irrigated cropping based on level of tank storage alone are subject to qualifications for at least three reasons:

(1) The level of storage at the beginning of a season only indicates whether or not any tank water is available for irrigation in that season. In general, given that the depth of water level declines after October-December, the extent of the *ayacut* which can be irrigated by the tank in the third and first seasons will be smaller than in the second. The extent of area which can be irrigated in a season varies depending on the level of storage in that season.

(2) The feasibility of cropping depends also on the extent and distribution of rainfall in each season. This is especially important in the first and the third seasons.

(3) As groundwater exploitation is significant in the ayacut, the **extent** of cropping in the second and third seasons will be determined by the quantum of groundwater pumped. Considerable changes have occurred in this respect as a result of growing tendency for conjunctive use of surface and groundwater. The extent of such development differs significantly between KPT and DMT, and also within the *ayacut* of each of these tanks.

3.6 Growth of Wells in the Ayacuts Under KPT and DMT

According to farmers the narrowing and silting of inlet channels (especially in DMT) and the siltation of the tank bed (especially in KPT) have reduced the supply available from the tank considerably. Although tank supply is reduced, the *ayacut* has not fallen. If anything, the widespread use of HYVs and increased application of fertilizers and the cultivation of water intensive perennial crops like sugarcane and banana, all of which are more demanding in terms of quantum and quality of irrigation, have increased. The spread of conjunctive use of groundwater has facilitated this.

The rapid growth of supplemental irrigation from wells came in the wake of the introduction of diesel pumps and, more importantly, after the large scale rural electrification. These innovations substantially reduced the cost of lifting water even as the rise in agricultural prices and the easy availability of liberal institutional finance increased the pump-technology of wells and induced their development.

The wells in the *ayacut* of the two tanks have grown rapidly is clear from Tables 3.25 and below the 3.26. In KPT the number has more than doubled between 1913 and 1989; in DMT it has increased by some 80 per cent in the same period. While in KPT the increase is distributed across villages, in DMT it is concentrated in villages under the Mamandur sluice.

Table 3.25: Wells Under KPT Villages During 1913, 1989 and 2017.

Sl. no. Village	Ayacut Wells				% increase during 1913 & 1989	% increase during 1989 & 2017	Categories of Wells in 2017			Ayacut 2017 hectares	Well Density
	1913	1989	2017								
			Open	Bore			EP *	WO	Total EP+WO		
1 Kadaperi	10	13	22	15	30	185	35	2	37	27.1	137
2 Kaveripakkam	73	109	127	58	49	70	179	6	185	590.6	31
3 Kondapuram	36	69	53	109	92	135	135	27	162	213.7	76
4 Sirukarumbur	25	79	190	48	216	201	203	35	238	173.6	137
5 Athipattu	24	69	112	36	188	114	139	9	148	113.7	130
6 Eralacheri	9	46	53	15	411	48	56	12	68	180.2	38
7 Cheri	15	96	138	54	540	100	158	34	192	298.0	64
8 Kattalai	27	41	44	30	52	80	20	54	74	129.9	57
9 Thuraiperumpakkam	39	61	114	146	56	326	175	85	260	190.6	136
10 Maganipattu	25	66	85	44	164	95	104	25	129	204.0	63
11 Alapakkam	14	18	42	34	29	322	55	21	76	85.6	89
12 Puthupattu	21	29	55	48	38	255	67	36	103	96.1	107
13 Panniyur	6	17	31	25	183	229	44	12	56	29.2	192
14 Kilveeranam	8	21	44	21	163	210	51	14	65	37.2	175
Total	332	734	1110	683	121	144	1421	372	1793	2369.6	76

Note: SW = Sole wells. AW = Ayacut wells; EP = Electric pumpset, WO = Well only.
* indicates electric pump set includes all bore wells.
Well Density = Wells per 100 ha of ayacut.
Source: Re-settlement (1913), Adangal Registers (1989 and 2017) of KPT Revenue Villages.

On an average KPT has 76 wells and DMT 53 per 100 hectares of *ayacut*. The overall well density is not sufficient to assess the importance of groundwater since the yield per well varies. But little is known about this. Also there is great variation within the *ayacut*. Thus in KPT the well density ranges from 31 wells per 100 hectares to 192 per 100 hectares. In DMT the range (19 to 250) is even greater. These variations in part reflect the extent and reliability of tank supplies and in part the groundwater supply conditions.

Thus the two tail reach villages of KPT, Athipattu and Sirukarumbur, which are unfavourably placed in respect of tank supplies have relatively high well density. These villages also happen to be located close to the Palar which helps to recharge the wells effectively. This reliable groundwater source induced the farmers to dig more wells (see Table 3.25) to reduce dependence on tanks.

As Kadaperi, Panniyur and Kilveeranam are irrigated by the most elevated sluices of KPT, they always get inadequate tank supply. Even if the water reaches its FTL, they get the supply for only a crop's season. So, there is need to depend on sub-surface water for the crop production. This has induced a higher density of wells in these villages. Despite higher density, the quantum of supply available is reported to be low compared to the wells in the command of deeper sluice/head reach villages of the tanks. But no precise estimates are available.

Although villages Eralacheri, Alapakkam, Maganipattu, Puthupattu and Thuraiperumpakkam are situated in the tail end, the well density in these villages is low (ranges between 38 and 136 per 100 ha) due to very limited availability of sub-surface water. Even within this category, the tail end villages of Puthupattu and Thuraiperumpakkam have a well density of more than 100 well per 100 ha of ayacut. Among the head reach villages, Kaveripakkam has only 31 wells per 100 ha. Being located favourably to get the water supply from the three deepest sluices, its lands are assured of tank water even if there is a month's water supply available in the tank.

The reach wise analysis of DMT (below the Table 3.26) shows that the density of wells is gradually increasing from head to tail reaches in both the Dusi and Mamandur. However, overall well density is 65 wells per 100 ha in Mamandur sluices, whereas it is only 46 wells per 100 ha in the Dusi sluices. But we do not have any yardsticks to show whether the available well water is high or low in a village.

Table 3.26 indicates that in the Mamandur sluices two of the head reach villages (Narasamangalam and Alingilpattu) have a higher well density than Mamandur. This is mainly due to the locational factor which includes two aspects: (1) Mamandur village being located very close to the tank, access to tank supply is more assured and *ayacutdars* do not need to depend on wells. The villages Narasamangalam and Alingilpattu are situated at an elevated level and have to depend on water supply mostly from wells. (2) The tail reach villages are located far away from the tank and they get only a season's supply if the tank storage reaches between 25 and 30 feet. Even in this case the village Elacheri is able to get a season's supply only when the tank has surplussed. Hence, they have to depend much on well water for cultivation. This induces a higher well density compared to the head and middle reach villages.

In all the middle reach villages one can see a moderate well density, i.e., equal to the tank average. This is due to the fact that the middle reach

Table 3.26: Wells under DMT Command During 1913, 1989 and 2017

Sl. no.	Name of village and reach		Ayacut Wells					% increase during 1913 & 1989	% increase during 1989 & 2017	Categories of wells in 2017		Ayacut 2017 hectare	Well Density	Average for reac
			1913	1989	2017									
					Open	Bore	Total			EP	OE			
1	2	3	4	5	6	7	8	9	10	11	12	13	14	15
Dusi sluices 1 and 2														
1	Vagai	HR	3	3	5	0	5	0	67	5	0	26.2	19	
2	Dusi	HR	1	13	39	4	43	1200	231	43	0	180.6	24	
3	K.muttam.	HR	3	13	28	9	37	333	185	37	0	72.6	51	30
4	Kanikiluppai	MR	0	5	5	9	14	NA	180	14	0	44.9	31	
5	K.n.palayam.	MR	20	25	26	8	34	25	36	31	3	102.7	33	
6	Menallur	MR	0	17	71	4	75	NA	341	72	3	66.8	112	57
9	Pallavaram	TR	17	24	71	0	71	41	196	71	0	163.4	43	
7	Senianallur	TR	4	55	1	30	31	1275	-44	31	0	41.5	75	
8	Punaithangal	TR	0	2	37	0	37	NA	1750	37	0	59.1	63	53
	Total		48	157	283	64	347	227	121	341	6	758.0	46	
Mamandur sluices 3 and 4														
10	Mamandur	HR	60	109	124	9	133	82	22	120	13	401.1	33	
11	Narasamangalam	HR	14	35	43	0	43	150	23	43	0	44.7	96	
12	Alingilpattu	HR	2	12	15	0	15	500	25	15	0	5.5	272	42
13	Valavandal	MR	37	53	70	0	70	43	32	70	0	95.5	73	
14	Vadakalpakkam	MR	15	30	30	9	39	100	30	39	0	93.2	42	
15	Girijapuram	MR	8	22	39	0	39	175	77	39	0	47.7	82	63
16	Sothiyampakkam	TR	37	65	76	0	76	76	17	76	0	54.5	139	
17	Bagavanthapuram	TR	50	50	61	0	61	0	22	61	0	43.2	141	
18	Elacheri	TR	69	77	123	0	123	12	60	118	5	132.4	93	113
	Total		292	453	581	18	599	55	32	581	18	917.7	65	
DMT as a whole			340	610	864	82	946	79	55	922	24	1675.7	56	

Note: SW= Sole welss. AW = Ayacut wells. EP = Electric pumpset. OE = Oil engine. WO= Well only
HR = Head reach, MR = Middle reach, TR = Tail reach. NA - Not Applicable.
Source: Resettlement (1913), Adangal registers of 1989 and 2017 in the villages irrigated by DMT.

villages are situated below Mamandur village and formed like a chain structure. They can get good access to tank water if they put enough effort, and the necessity to depend on wells is less compared to the tail reach villages.

The above analysis shows the macro-level picture of the growth of wells under the tanks. Whether, and to what extent the well water is used for irrigation and what difference it makes to cropping patterns and cropping intensity in different parts of the *ayacut* will be explored in detail in chapter

7 with the help of farm level survey. In this context it is important to analyse broadly the cropping intensity and cropping pattern during the settlements in the villages under both the tanks.

3.7 Cropping Intensity and Cropping Pattern in KPT and DMT Villages

The outcome of irrigation supplies may be seen through the intensity and type of crops grown and the yield in a particular year. Though yield details in the *ayacut* are not possible without conducting a specific survey, the details of cropping intensity and cropping pattern are available from the village records. These details help to understand how effectively the available land resources are utilised by the farmers in the *ayacut* and whether any change occurred over time in the cropping intensity and cropping pattern in the villages of both the tanks.

3.7.1 Cropping Intensity

Cropping intensity (CI) is the ratio of the total cropped area to the land area (in percentage). The triple factors which affect the CI in the tank irrigated areas are: (i) the extent and frequency of the availability of seasonal rainfall; (ii) the availability of tank water supply and (iii) the extent of well water use as supplementary to tank water. If all these factors are favourable the extent of land cultivated may increase and thus induce a higher CI. Poor seasonal rainfall and/or uncertain tank supply leads to low cropping intensity.

In the *ayacut* under both the tanks the resettlement period (1913) was a "normal" year which to some extent induced a higher cropping intensity in both the tank ayacuts. However, in DMT the CI is comparatively more (131%) than in KPT (92%). This was due to the widespread adoption of broadcast sowing of paddy in the villages under DMT, while it was almost rare in the KPT *ayacut*. One could observe from below the Tables 3.27 and 3.28 a progressive fall in CI from the head to the tail reach villages. But in the villages under DMT no such variation is observed.

Recent decades have witnessed the use of advanced technology for crop production as well as the extraction of well water and a lot of changes in cropping pattern. Further, the year 1985 (this was used as an average of the five years from 1985 to 1990) was an "above normal" rainfall year in both the tank *ayacuts*, especially in KPT, as a result the CI has increased. This fact is that the tank water supply in KPT is more favourable than in DMT. The low availability of tank supply was the basic reason for low

Table 3.27: Cropping Intensity and Cropping Pattern in KPT Villages During, 1913, 1985-90 and 2016-17.

Area in hecteres

Name of village and reach of sluice	Area (Ha) %	Cultivable Area (NSA)	<--- 1913 --->						Well Density 1913	<--- 1985-90 --->							Well Density 1989	<--- 2012 to 2017 --->							Well Density 2016
			GCA	GCA/NSAX 100	Area under crops: Paddy	Banana	Oil-seeds	Others		GCA	GCA/NSAX 100	Area under crops: Paddy	Banana	Sugar-cane	Oil-seeds	Others		GCA	GCA/NSAX 100	Area under crops: Paddy	Banana	Sugar-cane	Oil-seeds	Others	
1	2	3	4	5	6	7	8	9	10	4	5	6	7	7A	8	9	10	4	5	6	7	7A	8	9	10
Kattalai DH	Area	130.8	141	108	132	7	0	2	8	195	149	125	64	2	0	4	13	112.7	86	60	34	1	0	18	57
	%				(93)	(5)		(2)				(64)	(33)	(1)		(2)				(53)	(30)	(1)	(0)	(16)	
Cheri DH	Area	394.7	411	104	338	2	38	32	2	560	142	448	56	28	0	28	13	297.4	75	214	42	10	25	7	64
	%				(82)	(0)	(9)	(8)				(80)	(10)	(5)		(5)				(72)	(14)	(3)	(8)	(2)	
Kaveripakkam DH	Area	751.6	941	125	630	24	61	226	5	1038	138	726	197	42	31	42	7	281.1	37	132	145	4	0	1	31
	%				(66)	(3)	(6)	(24)				(70)	(19)	(4)	(3)	(4)				(47)	(51)	(1)	(0)	(0)	
Kondapuram DM	Area	218.2	226	103	202	3	2	19	7	288	132	196	46	29	12	6	13	79.2	36	54	16	5	1	3	76
	%				(90)	(1)	(1)	(8)				(68)	(16)	(10)	(4)	(2)				(68)	(20)	(7)	(1)	(4)	
Maganipattu MM	Area	392.3	314	80	205	3	61	45	5	502	128	301	65	55	55	25	13	195.9	50	106	1	26	45	19	63
	%				(65)	(1)	(19)	(14)				(60)	(13)	(11)	(11)	(5)				(54)	(0)	(13)	(23)	(10)	
Alapakkam MM	Area	432.7	353	82	210	0	30	114	7	454	105	250	0	82	73	50	9	212.2	49	147	0	9	44	13	89
	%				(59)		(8)	(32)				(55)		(18)	(16)	(11)				(69)	(0)	(4)	(21)	(6)	
Sirukarumbur DT	Area	374.9	315	84	182	38	59	36	6	465	124	246	56	153	0	9	18	276.4	74	255	6	6	2	7	137
	%				(58)	(12)	(19)	(11)				(53)	(12)	(33)		(2)				(92)	(2)	(2)	(1)	(2)	
Athipattu DT	Area	332.1	308	93	175	15	70	49	8	428	129	214	39	129	17	30	25	300.9	91	215	36	10	21	18	130
	%				(57)	(5)	(23)	(16)				(50)	(9)	(30)	(4)	(7)				(71)	(12)	(3)	(7)	(6)	
Eralacheri DT	Area	307.1	188	61	152	0	13	23	2	289	94	211	0	6	43	29	10	64.9	21	53	2	2	4	4	38
	%				(81)	(0)	(7)	(12)				(73)		(2)	(15)	(10)				(81)	(3)	(3)	(6)	(6)	
T.p.pakkam MT	Area	290.4	249	86	200	0	17	32	8	325	112	254	7	16	33	16	13	143.3	49	81	11	17	27	7	136
	%				(80)		(7)	(13)				(78)	(2)	(5)	(10)	(5)				(56)	(8)	(12)	(19)	(5)	
Puthupattu MT	Area	438.9	315	72	166	0	52	96	9	479	109	283	0	38	110	48	12	294.5	67	205	0	8	66	15	107
	%				(53)		(16)	(31)				(59)		(8)	(23)	(10)				(70)	(0)	(3)	(23)	(5)	
Kadaperi EH	Area	318.5	336	106	241	1	24	70	15	364	114	236	55	25	18	29	19	70.9	22	24	6	23	6	12	137
	%				(71)	(0)	(7)	(21)				(65)	(15)	(7)	(5)	(8)				(34)	(8)	(32)	(9)	(17)	
Panniyur EH	Area	308.5	228	74	90	0	60	78	8	333	108	227	0	33	63	10	24	128.5	42	87	1	3	26	11	192
	%				(40)		(26)	(34)				(68)		(10)	(19)	(3)				(68)	(1)	(2)	(21)	(9)	
Kilveeranam ET	Area	185.8	183	98	101	0	13	69	9	181	97	135	0	0	27	18	23	125.2	67	79	13	2	22	10	175
	%				(55)		(7)	(38)				(75)			(15)	(10)				(63)	(10)	(2)	(17)	(8)	
Total	Area	4876.5	4507	92	3024	93	499	891	6	5901	121	3852	584.2	638.1	482	343.7	13	2583.0	53	1713	312	125	289	144	76
	%				(67)	(2)	(11)	(20)				(65)	(10)	(11)	(8)	(6)				(66)	(12)	(5)	(11)	(6)	

Note: Column 3 represents the total land area (excluding poramboke) in the village.

Column 9 includes cereals other than paddy, pulses, vegetables and topes. Column 10 represents number of open and bore wells in the village.

Figures in brackets indicate percentage to the total. North Arcot Rainfall (mms): 1913 =1025; 1985-1990 = 934; 2012-2017 = 936.

DH = Deep sluice head village, DM = Deep sluice middle village, DT = Deep sluice tail village.

MT=Middle sluice tail village, MM = Middle sluice middle village, EH = Elevated sluice head village, ET = Elevated sluice tail village.

The low cropping intensity (53%) during 2012 to 2017 is mainly due to recording of crops only for the firt two seasons (in III season no crop raised is indicated) in Tamil Nadu villages in recent years.

Source: Resettlement (1913). the G Returns 1985-90 and (2012-13 to 2016-17) of the villages irrigated by KPT.

Table 3.28: Cropping Intensity and Cropping Pattern in DMT Villages during, 1913, 1985 to 90 and 2012-13 and 2016 - 17

Area in hecteres

Name of village and reach of sluice		Area (Ha) %	Cultivable area (NSA) 1913	<— 1913 —>					Well Density 1913	<— 1985-90 —>						Well Density 1989	Cultivable area (NSA) 2012	<— 2012-13 to 2016-17 —>					Well Density 2017
				GCA	GCA/ NSAX 100	Area under crops: Paddy	Oil-seeds	Others		GCA	GCA/ NSAX 100	Area under crops: Paddy	Banana& S.cane	Oil-seeds	Others			GCA	GCA/ NSAX 100	Area under crops: Paddy	Oil-seeds	Others	
1		2	3	4	5	6	7	8	9	4	5	6	6A	7	8	9	10	4	5	6	7	8	9
Dusi sluices 1 and 2																							
Vagai	HR	Area	117.7	134	114	112	14	8	3	114	97	87	0	18	9	3	113.0	48	42	38	7	3	19
		%				(83)	(11)	(6)				(76)	(0)	(16)	(8)					(79)	(15)	(6)	
Dusi	HR	Area	330.6	503	152	456	25	22	0	352	107	281	1	52	18	4	291.0	105	36	74	23	9	24
		%				(91)	(5)	(4)				(80)	(0)	(15)	(5)					(70)	(22)	(8)	
K.muttam.	HR	Area	100.5	192	191	174	5	12	3	104	104	79	1	16	8	13	99.0	39	39	26	9	4	51
		%				(91)	(3)	(6)				(76)	(1)	(15)	(8)					(66)	(24)	(10)	
Kanikiluppai	MR	Area	45.3	90	199	66	23	1	0	32	70	28	0	3	1	11	45.0	15	33	14	1	0	31
		%				(74)	(25)	(1)				(87)	(0)	(10)	(3)					(94)	(6)	(0)	
K.n.palayam.	MR	Area	137.4	180	131	143	30	7	15	146	106	100	0	36	10	18	135.2	63	47	38	16	9	33
		%				(79)	(17)	(4)				(69)	(0)	(24)	(7)					(61)	(25)	(14)	
Menallur	MR	Area	200.3	228	114	94	109	25	0	203	101	85	1	104	13	8	198.0	82	41	28	38	16	112
		%				(41)	(48)	(11)				(42)	(1)	(51)	(6)					(35)	(46)	(19)	
Senianallur	TR	Area	44.7	*	*	*	*	*	0	50	111	46	0	2	1	12	33.2	25	75	24	1	0	19
		%										(92)	(1)	(5)	(2)					(97)	(3)	(0)	
Punaithangal	TR	Area	101.3	147	146	104	33	11	0	98	97	68	0	23	7	2	83.9	46	55	29	13	4	99
		%				(70)	(23)	(7)				(69)	(0)	(24)	(7)					(62)	(29)	(9)	
Pallavaram	TR	Area	213.6	254	119	213	14	27	8	256	120	191	1	49	15	11	204.3	61	30	40	13	8	120
		%				(84)	(5)	(11)				(75)	(0)	(19)	(6)					(66)	(21)	(13)	
Total		Area	1291.4	1729	134	1362	253	113	4	1355	105	964	6	303	82	12	1202.7	483	40	311	120	52	46
		%				(79)	(15)	(6)				(71)	(1)	(22)	(6)					(64)	(25)	(11)	

(Contd...)

Name of village and reach of sluice	Area (Ha) %	Cultivable area (NSA) 1913	1913 GCA	1913 GCA/NSAX 100	1913 Area under crops: Paddy	1913 Oil-seeds	1913 Others	Well Density 1913	1985-90 GCA	1985-90 GCA/NSAX 100	1985-90 Area under crops: Paddy	1985-90 Banana & Scane	1985-90 Oil-seeds	1985-90 Others	Well Density 1989	Cultivable area (NSA) 2012	2012-13 to 2016-17 GCA	2012-13 to 2016-17 GCA/NSAX 100	2012-13 to 2016-17 Area under crops: Paddy	2012-13 to 2016-17 Oil-seeds	2012-13 to 2016-17 Others	Well Density 2016
1	2	3	4	5	6	7	8	9	4	5	6	6A	7	8	9	10	4	5	6	7	8	9
Mamandur sluices 3 and 4																						
Mamandur HR	Area	475.9	773	162	715	42	16	13	734	154	527	1	173	32	23	453.0	330	73	252	64	14	33
	%				(93)	(5)	(2)				(72)	(0)	(24)	(4)					(77)	(19)	(4)	
Narasamangalam HR	Area	133.0	147	111	92	41	14	11	112	84	37	1	66	9	26	128.0	82	64	34	34	14	96
	%				(63)	(28)	(9)				(33)	(1)	(59)	(7)					(42)	(41)	(17)	
Alingilpattu HR	Area	85.7	70	82	23	19	28	2	103	120	20	1	66	16	14	89.9	39	43	15	13	10	272
	%				(32)	(28)	(40)				(19)	(1)	(65)	(15)					(40)	(34)	(26)	
Valavandal MR	Area	138.7	179	129	154	21	4	27	170	123	96	2	62	11	38	139.3	95	68	39	29	26	73
	%				(86)	(12)	(2)				(57)	(1)	(36)	(6)					(41)	(31)	(28)	
Vadakalpakkam MR	Area	164.5	158	96	103	34	21	9	146	89	78	1	58	9	18	156.9	71	45	38	19	13	42
	%				(66)	(21)	(13)				(53)	(1)	(40)	(6)					(54)	(27)	(19)	
Girijapuram MR	Area	92.8	106	114	57	28	20	9	93	100	44	0	38	11	24	93.8	45	48	19	19	7	82
	%				(54)	(27)	(19)				(47)	(0)	(41)	(12)					(41)	(43)	(16)	
Sothiyampakkam TR	Area	169.0	177	105	79	71	27	22	164	97	92	0	59	13	38	183.0	86	47	49	28	8	140
	%				(45)	(40)	(15)				(56)	(0)	(36)	(8)					(57)	(33)	(10)	
Bagavanthapuram TR	Area	88.3	106	121	39	52	16	57	90	102	35	0	50	4	57	89.6	58	64	27	25	6	141
	%				(37)	(49)	(14)				(39)	(0)	(56)	(5)					(46)	(43)	(11)	
Elacheri TR	Area	298.9	395	132	169	163	63	23	304	102	100	1	193	11	26	311.2	167	54	55	73	39	93
	%				(43)	(41)	(16)				(33)	(0)	(63)	(4)					(33)	(44)	(23)	
Total	Area	1646.7	2113	128	1432	471	209	18	1916	116	1029	7	764	116	28	1644.7	971	59	529	303	138	65
	%				(68)	(22)	(10)				(54)	(0)	(40)	(6)					(55)	(31)	(14)	
DMT as a whole	Area	2938.1	3841	131	2795	725	322	12	3271	111	1993	13	1067	198	21	2847.4	1454	51	840	423	191	56
	%				(73)	(19)	(8)				(61)	(0)	(33)	(6)					(58)	(29)	(13)	

Note: Column 3 and 10 represent the total land area (excluding poramboke) in the village.

Column 8 includes cereals other than paddy, pulses, vegetables, flower, tree crops, greens and tope s.

Column 9 represents the well density in terms of number of wells per 100 ha of ayacut in the villag e as a whole.

The low cropping intensity (51%) during 2012 to 2017 is mainly due to recording of crops only in fi rt two seasons (in III season no crop raised is indicated) in Tamil Nadu villages in recent years.

HR = Head reach, MR = Middle reach. TR = Tail reach

Source: Resettlement (1913); The G Returns 1985-90 and 2012-13 to 2016-17 of the villages irrigated by DMT.

CI (111 %) in DMT in the year 1985. Also, it should be mentioned that in the *ayacut* under KPT, the CI in the year 1985 was comparatively higher than that of the resettlement period. This increase is mainly due to technological changes (i.e., the use of short duration crop varieties/annual crops and the development of well irrigation). In DMT the CI in the corresponding period was less (111 %) than the resettlement period (131 %). This is attributed to the fact that the tank water supply and the seasonal rainfall in the *ayacut* under DMT in 1985 was unfavourable which pushed down the CI considerably. Across villages there is no significant pattern prevailing in CI in both the tanks.

Multiple regression results (Table 3.29) in the two periods, 1913 and 1985, indicate that neither the well density nor the percentage of irrigated area (except in KPT in 1913, where the coefficient for the proportion of irrigated area is significantly positive) had a significant association with cropping intensity.

Apart from these two periods of comparison of CI, an attempt was also made to find out the trends in CI across villages and between the tanks in the latest years of mid-2010s. For this purpose, five year period, from 2012-13 to 2016-17, data collection was made and the results are presented in Tables 3.27 and 3.28. Before analysing the data the following points are noteworthy to mention. 1. In both tanks, except the last two years (2015-16 and 2016-17) the remaining three years are affected by drought. 2. Unfortunately, in recent years, the government of Tamil Nadu informally suggested the village administrative officers (VAOs) to record the cropping pattern details only for the first two seasons in a Fasli year. Consequently, even if a farmer cultivated in three seasons in a year or an annual crop /

Table 3.29: Multiple Regression of Cropping Intensity on Well Density and % of Irrigated Area under KPT and DMT, 1913 and 1985

Tank, year	*Dependent variable*	*Constant*	*Well density*	*% of Irrigated area*	*R^2*
1	*2*	*3*	*4*	*5*	*6*
KPT					
1913	Cropping intensity	51.91	0.78 (1.22)	0.48 (0.18)*	0.41
1985	Cropping intensity	100.78	-0.16 (1.19)	0.13 (0.31)	0.02
DMT					
1913	Cropping intensity	77.01	0.42 (1.00)	0.65 (0.46)	0.12
1985	Cropping intensity	98.86	0.22 (0.48)	0.58 (0.49)	0.11

Note: Figures in bracket represents the standard error of coefficients.
* denotes-'t' value significant at 1 per cent level.

perennial crop is grown throughout the year, the VAO has recorded it only for two seasons and in the Adangal registers it is mentioned in the third crop as NIL cultivation.

Due to this serious problem in data authenticity and availability, in our data also it is clearly seen that CI is nowhere recorded even 100% in any of the 32 villages served by both the tanks. In KPT, only in three villages – two are served by deep sluices and the other one (Athipattu) is rich with groundwater potential – the CI is between 75% and 91%. In all the remaining villages it is less than 75%. The tank as a whole the average CI during this 2012-2017 is only 53%.

Whereas in DMT, under Dusi sluice villages, except Senianallur (75%) and Punaithangal (55%) all remaining villages achieved only less than 40% of CI and this is the average range under Dusi-sluice villages. Under Mamandur sluices, an improvement in CI is noticeable. In five of the villages the CI is between 50% and 75%. In the remaining 4 villages it is more or less equal to that of the sluice average of 51%.

As indicated, although CI is an important measure to know the extent of utilisation of land resources under tank irrigated area, the official data do not provide an authentic picture of what actually happening under cultivation in the tank ayacuts. To some extent this problem could be overcome by analysing the cropping pattern which is practiced by the ayacutdars. That is, those who cultivate annual / perennial crops are considered better placed than cultivation of seasonal crops like paddy. Further, an apt details of cropping pattern, cropping intensity and productivity will be seen in this study through micro level farmers' survey that is elaborated in chapter 7.

3.7.2 Cropping Pattern

Since the CI is determined by the duration of crops occupied in a given land, it is again important to know what type of crops, i.e., short/long duration crops, grown in that land in a particular period of time. Cropping pattern is a sequence of different crops grown in a year on any particular field. In the villages under the two tanks paddy is the principal crop. There are three main cropping seasons in the tank ayacut: *Sornavari* from May to August, *Samba* from September to January, and *Navarai* from February to May. The important crop is *Samba* (see figures 3.10 and 3.11). In the villages under KPT, nursery preparation is done by using well water, whereas in DMT, sowing is done mostly with the help of south-west monsoon rains and only in limited pockets nursery preparation is done by using well water.

However, the required water supply for transplanting paddy and its subsequent growing period depends mostly on the northeast monsoon rainfall and the supplies made available from the tanks. The available maximum tank water storage is determined in the months of November/December. In this season if the rainfall or the tank supply fails the yield of the crop would be reduced considerably.

Sowing of *Navarai* crop is mostly conditioned by the availability of tank water supply. If there is no supply available from the tank, the area irrigated by well water would also be reduced drastically. The *Sornavari* crop is raised by using well water alone. Generally, the available well water in the tank *ayacut* is a function of the available tank supply. Hence if there is no tank supply there is practically no paddy cultivation in the *Sornavari* season.

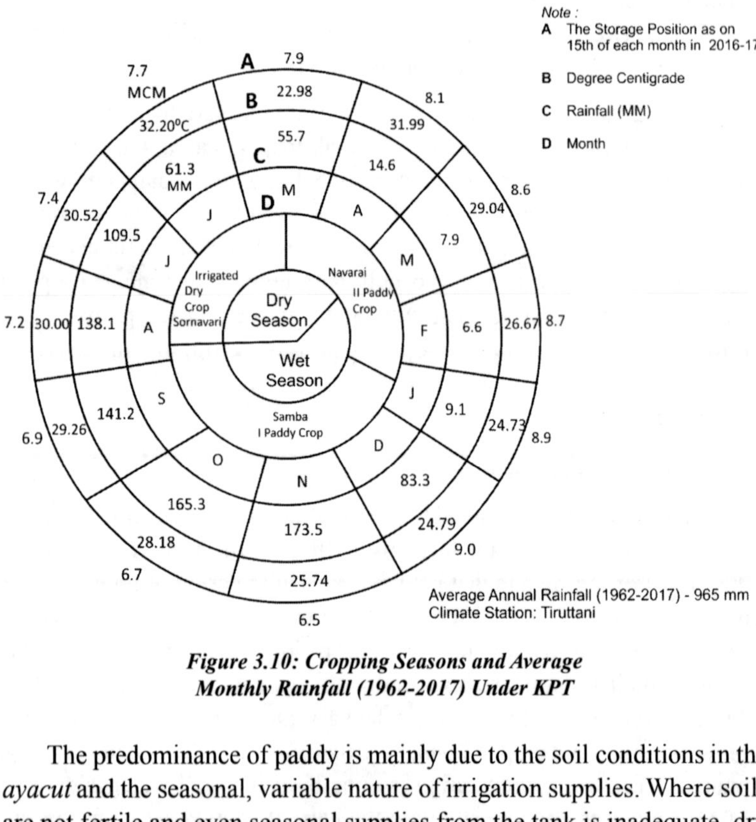

Figure 3.10: Cropping Seasons and Average Monthly Rainfall (1962-2017) Under KPT

The predominance of paddy is mainly due to the soil conditions in the *ayacut* and the seasonal, variable nature of irrigation supplies. Where soils are not fertile and even seasonal supplies from the tank is inadequate, dry irrigated crops like millets, pulses, and groundnut are raised. Where abundant

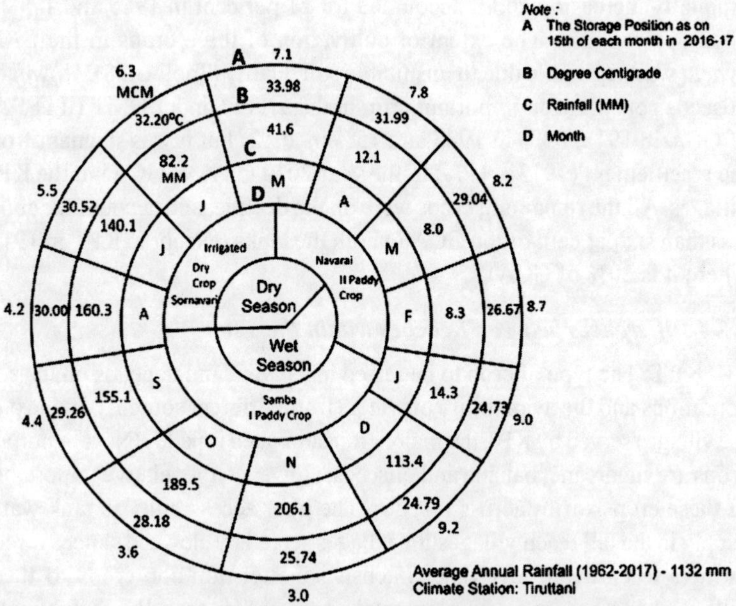

Figure 3.11: Cropping Seasons and Average Monthly Rainfall (1962-2017) Under DMT

and assured supplies of irrigation are available (mostly by using wells) water **intensive** annual crops like sugarcane and banana are grown. But their extent seems small. The cropping pattern in KPT and DMT may be analysed under two heads: (i) differences across years and between tanks and (b) differences between reaches within a tank.

3.7.3 Cropping Pattern: Differences across Years and between the Tanks

The available data on cropping pattern and cropping intensity relating to the resettlement (1913), in the year 1985 (it is an average of five years 1985-90) and also in the year 2012 (it is also the average of five years 2012-2016) in the villages served by KPT and DMT provide an interesting picture. In all the three periods, about two thirds of the gross cropped area were under paddy. However, the share of paddy has been steadily decreasing in DMT. In KPT, paddy extent is stabilized. It is 67% of GCA in 1913; 65% in 1985 and 66% in 2012. In DMT, a steady fall in area is observed: it is 73% of GCA in 1913; 61% in 1985 and 51% in 2012. This clearly points out that in DMT area under paddy has been gradually decreasing. The share of annual crops such as sugarcane and banana to the GCA in KPT was gradually increasing and it accounted for 21 per cent in 1985 and 17% in 2012 (Table 3.27). The extent of cultivation of these crops in the DMT *ayacut* was negligible due to unsuitable soil quality (Table 3.28). However, oilseeds constitute an important crop in the *ayacut* under DMT (it is 19% of GCA in 1913; 33% in 1985 and 15% in 2012) but it was stagnant from the resettlement (1913 – 11%) to the year 2012 (11% of GCA) in the KPT villages. All the remaining crops were on the decline, and account for about less than six per cent of the GCA in both the tanks (except in KPT in 1913, where it is 20% of GCA).

3.7.4 Differences between Reaches within the Tank

KPT: The type of crop to be raised in a given land depends on the soil conditions and the availability of water (from different sources). In most of the villages served by KPT the major crop cultivated is paddy. Next important crops are sugarcane, banana and oilseeds. However, the relative importance of these crops varies across reaches. The poor accessibility to tank water supply in the tail reach villages/the villages served by elevated sluices forced farmers to cultivate less water intensive/shorter duration crops. If these villages have access to supplemental source other than the tank supply, then one can expect the cultivation of annual crops such as sugarcane and banana. The head reach villages tend to raise usually water intensive/long duration crops due to locational advantage. The accessibility to tank water supply is not difficult to middle reach villages compared to the tail/ elevated sluice villages, but apparently is less compared to the head reach villages. Hence the middle reach villages can raise crops depending on both tank supply and well water. During the resettlement period (1913) no major change in crop cultivation (i.e., from paddy to sugarcane and banana) took place in most of the villages. Changes have occurred only in recent decades. Hence, it is important to analyse the variations in cropping pattern across villages in the 1980s and 2000s and we selected the year 1985 and 2012 - which were weather wise normal - for this purpose.

It may be observed from Table 3.27 that in KPT in the villages under elevated sluice (Panniyur and Kilveeranam) and in the tail reach villages (except Sirukarumbur and Athipattu), next to paddy, oilseeds constitute a major crop. Sugarcane and banana are insignificant to the GCA. The two tail reach villages -Sirukarumbur and Athipattu- are situated close to the Palar; hence the abundance of groundwater supply permits a more water intensive cropping pattern. In the same manner the elevated sluice village of Kadaperi is also having a supplemental source of the Palar river channel (*Athukalvai*) supply to a part of that village ayacut, which helped to raise groundwater table (here the well density is also high). This supplemental source induces farmers to raise annual crops to some extent. These possibilities (including the high well density) are not available to the other tail/elevated sluice villages such as Eralacheri.

In the three middle reach villages (Kondapuram, Maganipattu and Alapakkam), next to paddy, all the other three crops viz., sugarcane, banana and oilseeds are more or less equally distributed. In these villages, well density is equal to the tank average (13 wells / 100 ha in 1985 increased to 76 wells / 100 hectares in 2013). This means that these villages depend on both tank supply and well water to satisfy their requirements. The three head reach villages (Kaveripakkam, Cheri and Kattalai) are in a better position in obtaining tank water supply; hence next to paddy, banana and sugarcane occupy the highest share in GCA. Oilseeds are almost rare. In these villages, well density is also less than that of the tank average.

DMT: The pattern of cropping in the villages under DMT is different from that of KPT. In the former tank ayacut, next to paddy, oilseeds are the most important. Banana and sugarcane are almost insignificant. Crops other than paddy and oilseeds constitute a meagre proportion (six per cent) to the GCA (Table 3.28). In the lands irrigated between sluices, one can note differences in cropping pattern. In Dusi sluices paddy is mostly dominant in all the villages in both years 1985 and 2012, whereas in Mamandur sluices in most of the villages (eg. Narasamangalam, Alingilpattu, Bagavanthapuram and Elacheri) both paddy and oilseeds dominate. In all the remaining villages about one fourth of the GCA is devoted to oilseeds. Across reaches in both the sluices there is not much variation in cropping pattern.

The difference in the cropping pattern between the Dusi and Mamandur sluices is basically due to the extent of availability of wells in the villages. The density of wells in all the villages (including the tail reach) falling under Dusi sluices is below the tank average (12 wells per 100 hectares during 1985 and 46 wells / 100 ha in 2012), whereas it is more than the tank average (28 wells / 100 ha in 1985 and 65 wells / 100 ha in 2012) in the villages served by Mamandur sluices. This higher well density in Mamandur sluices induced the farmers to raise irrigated groundnut even in the *samba* crop season.

By and large, changes in cropping pattern are apparent in the KPT villages; but these changes are not significant in the DMT villages. The important point to be noted is the available tank water supply/ supplemented by wells is an inducing factor for the changes in cropping pattern as well as CI. Where the supply of water is abundant, the water intensive crops and the annual crops dominate. In the less water supply regions only dry/irrigated crops are grown predominantly. The soil and climatic conditions are also important for the changes in cropping pattern. In DMT where the soil condition is not suitable for sugarcane and banana cultivation, farmers are mostly raising paddy crop. Further, the broadcast sowing of paddy is possible in DMT *ayacut*, but not in KPT due to the climatic (combined with soil) conditions.

It is important to note that an analysis of annual water use,[1] based on crop pattern details in KPT and DMT villages during the years (1) 1913 and 1985 and (2) 1985 and 2015 indicates an increase of water use by about 18% in both the tanks in the first period; and during the second period, it increased to 13% over the first period of water use. During the first period, though tank water supply in KPT in 1985 was less than FTL (7.2 metre out of 9.3 metre), the proportion of water use in the year as a whole was slightly higher (KPT 17.63%; DMT 17. 53 %) than in DMT. In the latter, tank storage was about full level (8.4 metre out of 9.3 metre) in 1985.

In the second period, both tanks are at full level (9.3 metre). However, variations in tank water use across villages are noticed, especially in KPT, where four villages (Kilveeranam, Eralacheri, Athipattu and Sirukarumbur) altogether not possible to use the tank supply due to full encroachment of supply channels. In three other villages (Thuraiperumpakkam, Puthupattu, and Kondapuram) only partial irrigation by a portion of the ayacut received the KPT supply. However, some of these villages make use of their well irrigation where the groundwater source is rich. This shows that groundwater use in KPT villages contributed a higher proportion of total water use (both from the tank and wells) than in DMT villages. It may be noted that in KPT, across villages, increase in the proportion of water use was much higher in the tail end villages served by deep/middle sluices compared to head reach villages served by deep sluices. In DMT, the pattern is mixed in Dusi sluice villages. But in Mamandur sluice middle reach, tail end villages recorded a higher proportion of water use compared to head reach villages over 1913 and 1985. The same pattern is also applicable in the 1985 and 2015 water use comparison.

Summary

This chapter described the general features of the selected tanks of

KPT and DMT in a historical perspective. Both tanks are ancient and larger in size in the system but differ from each other in several respects. The supply available from the Palar Anicut is more assured in KPT than in DMT. Although DMT has three supply sources none of them is effective to fill the tank due to its tail end location.

The effect of PAS modernisation on KPT and DMT shows that KPT benefited both from the non-requirement of *Korambu* construction and from assured water supply. However, DMT did not benefit as much if at all, especially because of fall in supply through Rajakkal channel as a result of the construction of the Anicut. But it should be noted that DMT has a larger storage capacity than KPT which benefits to store more water during flood season. Present conditions of the inlet channels especially in DMT were not satisfactory due to improper and inadequate maintenance. The siltation and encroachment in the tank bed are a more severe problem in KPT than in DMT.

The discussion on water supply characteristics shows that the *ayacut* under KPT is in a better position than under DMT. This is mainly due to the locational factor. Apart from the poor availability of supply from the sources to DMT, the *ayacutdars* are in a disadvantageous position in their access to groundwater also. Access to groundwater is more conducive in KPT than in DMT, mainly due to rocky sub-surface and poor aquifer in the *ayacut* of DMT. Due to favourable water supply conditions from both the tanks and wells, the KPT villagers are gradually shifting the cropping pattern from paddy to banana and sugarcane; whereas this is not happening in DMT.

The cropping intensity has increased appreciably in both the tanks over time. In cropping pattern there has been a considerable shift from staple crop (paddy) to commercial crops (sugarcane and banana) in KPT, but hardly any change in DMT. The increase in CI and the changes in cropping pattern were mostly due to the development of wells in the *ayacuts*.

In order to get enough tank water supply to the *ayacut*, the *ayacutdars* under DMT have been working hard to get as much supply as possible from the sources to the tank, and distributing it in a fair manner. Hence, the functioning of IIs is better in DMT than in KPT. This aspect on the organisational structure and functioning of IIs in KPT and DMT is an important matter which will be discussed in chapter 4.

Note

1. For both the KPT and DMT villages, annual water use in 1913 and 1985 is calculated as follows: (1) All crops are categorised into three major groups, viz., paddy, banana and sugarcane and other. (2) The proportion of area cropped under each group is calculated by using A1 Registers. (3) The actual area for each group of crops is taken. (4) The total area cropped under each group of crops is calculated by multiplying the proportion of area under each group with the actual area in each season, i e., sl. no. 2X3. (5) Irrigation depth for each group of crops is taken into account, which is as follows: Irrigation depth for Kharif (June-September) season is: paddy, banana and sugarcane, and others = 0.64; 0.11; 0.11 metres per hectare respectively; Rabi (November-February) season is 0.83; 0.35 and 0.35 metres per hectare respectively; Summer (March-June) season is 0.64; 0.20 and 0.20 metres per hectare respectively. (6) Total water use is calculated by multiplying the total area with depth of irrigation, i e., sl. no. 4X5. (7) Annual water use is derived by adding the total water use for the three seasons and dividing it by the GCA.

Chapter - 4

Structure and Functioning of Irrigation Institutions

This chapter seeks to describe the form, structure and functions of the organisation of the three levels, viz., system, tank and village, and the inter-relations among them. In doing so, the chapter also provides, as a backdrop, a brief account of the position of irrigation management before the PAS was constructed, the changes which have occurred since, the current functioning of irrigation institutions (IIs) and the factors responsible for changes in their structure and working.

4.1 System Level Organisation

Before describing the structure and management of KPT and DMT organisations, it is useful to give a brief account of the main features of the management of Palar Anicut System as a whole of which these two tanks are a part. Nothing is known about the "system level" organisation prior to the construction of the PAS (1857). With the construction of the Anicut, a number of earlier systems which were fed by channels drawn from different *Korambus* (temporary earthen embankments to divert the river supply to inlet channels) were integrated into the system. From the time of construction of the Anicut, the maintenance of the Anicut, main and inlet channels of all major tanks, and of the tanks and associated structures has continued to be the direct responsibility of the PWD. The PWD is also responsible for regulation and distribution of water to different channels and tanks in accordance with certain principles (as detailed in chapter 6) governed by conventions and tradition. However, it is difficult to get a precise account of what these principles[1] are in the present context.

The PAS at present is controlled by 9 section offices located in three districts (see Table 4.1). Two of the districts, Vellore and Tiruvannamalai, are located on the left and right side of the PAS and the third of Kancheepuram district is situated at its tail end. Section offices of three, four and two are located respectively in each district, which have the overall responsibility for all the PWD tanks (system and non-system) in their jurisdiction. On the whole, the number of PAS tanks (317) is almost equal to the number of non-PAS tanks (332) controlled by the PWD in the three districts in which the system is situated. However, except in Kancheepuram district, the number of non-system tanks is much smaller than the PAS tanks.

Table 4.1: Tanks Controlled by Section Offices under the PAS, 2016-17

District/location of sectionoffice	*Name of sectionoffice*	*Number of tanks controlled*	
		PAS tanks	*Minor Irrigation + Non-system tanks*
Vellore dt.	1. Kalavai	37	17
Right side of Palar Anicut	2. Cheyyar	47*	14
System (PAS)	3. Dusi	33	13
	Total	117	44
Thiruvannamalai dt.	4. PA section	15	36
Left side of PAS	5. Kaveripakkam	36	4
	6. Panapakkam	37	3
	7. Takkolam	25	5
	Total	113	48
Kancheepuram district	8. Kancheepuram	57	104
Tail end of PAS	9. Sriperumpudur	30	136
	Total	87	240
	Overall	317	332

Note: * refers that this section controls apart from PAS tanks, Thandarai Anicut plus 16 of its tanks and Cheyyar Anicut plus seven of its tanks.

Source: Data collected from the section offices, 2017.

The top management of the PWD in the region where the PAS is located consists of a Chief Engineer, three Superintending Engineers and three Executive Engineers. The workers are the lower level functionaries who perform the day-to-day management work at the ground level. Table 4.2 shows the below that at the time of India's Independence in 1947 there were 22 officials managing the system with 105 workers. But changes

Table 4.2: Structure of System Level Organisation (PAS)

Personnel	*Category*	*Allotment in 1950s*	*1992 level*	*2017 level*
Chief Engineer Superintending Engineer Executive Engineer	Policy makers I	1 3 3	1 3 3	1 3 3
Sub divisional Officer Assistant Engineer Junior Engineer / Section Officer	Policy makers II	3 2 10	3 5 7	3 5 4
Work Inspector Grade I Work Inspector Grade II Technical Assistant Office Assistant	Workers I	22 21 4 9	17 17 6 9	2 10 1 0
Lascars / Mazdoors / Watchmen	Workers II	49	26	36
Total		**127**	**97**	**68**

Source: Personal survey, 2017.

occurred up to 1992 only at the workers level and not at the official level. Even in the case of workers the major reduction is at the lower level such as Lascars (local level PWD functionaries) and Watchmen. In 2017 there has been reduction at all levels including the official level whose number reduced to 19 and Work Inspectors to a level of 12 from the sanctioned strength of 43. More importantly, office assistant and technical assistant positions were almost nil, but it was 15 in 1992.

The head of the section office is usually the Junior Engineer (JE)/ Assistant Engineer (AE)/Section Officer. His main duties are to prepare the estimates for different works in his jurisdiction, attend the higher level meetings, supervise all the staff working in his section, inspect the length of channels in his charge once a fortnight and every tank once a month, and prepare estimates for repairs which may be necessary. As the Work Inspectors have generally been serving for many years in the same section, and know all the details about the section, they are an important staff category. Their main duty is to assist the JE/AE in all possible ways especially to prepare estimates, and supervise the maintenance works.

The Lascars are the "manual" workers who actually undertake all the duties as directed by the JE /Work Inspectors (WIs). Their main functions are to assist the WIs in all their duties. They have to maintain the tank bunds by clearing the bushes and unwanted growth; clean and oil the sluices and shutters to keep them in good working condition; perform duties (such as closing or opening the sluices to supply water to the *ayacut* during the tank supply season) assigned by the Junior Engineer; and maintain the records of daily tank water levels and the rainfall readings properly. Sometimes they need to supervise the construction works such as bund maintenance, bridge construction, and masonry works undertaken by the contractors.

It was reported by the PWD officials that the general functioning of the section offices is not satisfactory. This is for at least two reasons: (i) the fund allotted for repair works is far from adequate, and (ii) the available staff (in 2017) is not enough to oversee the system. In most of the section offices, the vacancies caused by retirement, death, transfer or promotion have not been filled. It may be seen from Table 4.3 that as of 1992 one out of the 12 posts of section officers, 9 out of 43 posts of work inspectors and 23 out of 49 post of watchmen/lascars were unfilled.

Likewise, in the year 2017, three out of the 12 posts of JEs / AEs, 12 out of 43 posts of work inspectors and 13 out of 49 post of watchmen/ lascars were unfilled. The most deficient staff category is of work inspectors and Luscars who play a critical role in managing the system. The number

Table 4.3: Staffing of PAS Section Offices, 2016-17

Section office	*Sanctioned staff as on 1992*						*As on April 2017 staff level*					
	1	*2*	*3*	*4*	*5*	*6*	*1*	*2*	*3*	*4*	*5*	*6*
Kalavai	1	1	2	4	-	1	1	-	1	3	-	-
Cheyyar 1	1	1	1	5	1	-	1	-	1	3	1	-
Cheyyar 2	1	2	2	3	-	1	Merged with Cheyyar 1					
Dusi	1	3	-	4	-	-	1	-	1	4	-	-
PA Section	1	2	2	9	1	1	1	-	2	6	-	-
Kaveripakkam	1	1	2	5	-	1	1	-	1	3	-	-
Panapakkam	1	1	2	3	-	-	1	-	1	3	-	-
Takkolam	1	2	2	2	-	-	1	-	-	4	-	-
Kancheepuram 1	1	4	2	5	-	1	1	2	1	3	-	-
Kancheepuram 2	1	3	3	5	-	1	Merged with Kancheepuram 1					
Sriperumpudur 1	1	1	1	2	1	2	1	-	2	7	-	-
Sriperumpudur 2	1	1	2	2	1	1	Merged with Sriperumbudur 1					
Total	**12**	**22**	**21**	**49**	**4**	**9**	**9**	**2**	**10**	**36**	**1**	**0**

Note: Staff categories are:
1 = Junior Engineer / Assistant Engineer / Section Officer.
2 = Work Inspector Grade I; 3 = Work Inspector Grade II.
4 = (Luscars) Irrigation Assistants (including Watchman / Mazdoors and helpers).
5 = Technical Assistant; 6 = Office Assistant.
Comparing 1992 section offices, in 2017 three sections (Cheyyar 2, Kancheepuram 2 and Sriperumbudur 2) have been merged with the offices of the same name presently functioning.
Source: Data collected from each section office, 2017.

of persons employed in these categories in the system as a whole is three-fourth of the sanctioned strength in 1992 and barely less than a half in 2017.

4.1.1 Tenure of Staff

Generally all the staff working in different categories are transferable. However the 'tenure of stay in a section before transfer' varies among different categories. For instance, the Junior Engineer/Assistant Engineer/ Section Officer is supposed to be transferred once in three years though in some specific cases such as request on personal and medical grounds transfer may be effected even before the tenure period. For Work Inspectors and Lascars, transfers are not compulsory. In some sections it was reported that the Work Inspectors have been working for many years in the same section office and have not been transferred except on personal request.

4.2 Tank Level Organisation

The system level management under the control of the PWD determines the amount of water to be released from the Anicut to the various channels, regulates supplies to the individual tanks and maintains the anicut, the main distribution network (including inlet channels delivery of water to the tanks), and the bunds and associated structures of tanks served by the system. The subsequent operations, including maintenance and water distribution, are handled by the tank and village level organisation. Most tanks are single village tanks and the village irrigation committee deals directly with the system. But in the case of KPT and DMT, being large multi-village tanks, the PWD is directly involved in the tank operation. To look after the tank level operations four *Neerkatties* (irrigation workers) for KPT and two for DMT are appointed and paid by the government. The title of "Neerkatties" has been changed in the late 1990s as "irrigation assistants." These *irrigation assistants* are responsible for maintenance of the tank bund and associated structures and release of water. The latter in consultation with tank level management Committee. As indicated, due to retirement of irrigation workers, only temporary persons are inducted as and when required to look after irrigation assistants' duty in both tanks.

4.2.1 The Organisation before Constitution of Tank Board

Available historical records give hardly any information on the existence of tank level organisation prior to the construction of Palar Anicut. However, the *Mamulnamas* (see Appendix 3.1) indicate the "informal involvement" of the beneficiaries of the villages served by each tank, for appropriation of water to the tank and its distribution to the *ayacut*. After the Anicut was constructed in 1858 the PWD took a direct responsibility for managing the system.

Before the Irrigation Board (IB) came into existence in 1917 the irrigation management was jointly done by the PWD, the Revenue Department and the local village authorities, each with well-defined functions and responsibilities. The village *ayacutdars* and the Revenue Department were responsible for *Kudimaramath* and for the smooth distribution of water. The regulation of tank water to the *ayacut* was governed by certain procedures. According to these procedures, during the irrigation season every month the *Maniagar* (village headman) in each village under the tank *ayacut* should send to the tank Superintendent two statements one on 15th of each month indicating the area actually under irrigation and another on 25th indicating the amount of water required during the first half of the next month. Copies of these statements should also be submitted to the Tahsildar. The *Maniagar* should report to the Superintendent and the Tahsildar whether any fluctuation of water supply in the areas under their control due to heavy rain in the *ayacut* which necessitate modification in quantity of water to be released by the Superintendent at any time. The discharges at the main tank sluices should be regulated by the Superintendent according to the needs of the *ayacutdars* through requests made to the *Maniagar* and reported to him to increase or reduce the quantum of supply released from the tank.

4.2.2 Constitution of Irrigation Boards

For reasons which are not clear - available archival information is also meagre - working of the aforesaid distribution system was changed in 1917 when the government decided to vest the responsibility for management of all major tanks with a Board comprising elected representatives of the *ayacutdars*. Under this new arrangement the ryots owning or cultivating lands under the tank or their nominees were to elect delegates from whom the *Panchayatdars* to serve the Board were also elected. The Revenue Divisional Officer was expected to convene a meeting of the delegates representing various villages under the *ayacut*, electing not less than five persons from among them to serve as panchayatdars. The election was to be held once in three years. The President and the Secretary of the IB were to be chosen by the *panchayatdars*. Copies of the minutes of the meetings were to be sent to the Collector and the Executive Engineer for information. The Collector was given the power to dissolve the Board without giving any reason.

For proper distribution of water within the *ayacut* the control of channels was given to the Board with certain conditions. The Board decided the time for commencing the supply of water, duration and the quantity to be supplied. The government sanctioned an annual grant for the up-keep of the inlet channels which the Board utilised under the supervision of the PWD officials. The PWD attended to the clearance of silt and ordinary repairs (Administration Report of the PWD, 1927-28).

Though the rules and regulations for the constitution of Irrigation Boards came into force in 1917, they were not evenly enforced in all the tanks. The formation of IBs was not compulsory, and the IBs were not uniformly formed throughout the State (GO 2459, BOR, 7-7-1933). The IB in KPT was constituted in 1923 and functioned continuously till 1969. After that no election has taken place. But the Board has not been disbanded. The President, Secretary and the delegates elected in the 1960s were continued to function till 1990s. After thatelected members of Water Users Association (WUA) under TNFMIS Act-2000 handled these matters (see Table 4.4).

Table 4.4: Composition of KPT Irrigation Board, 1969-1990

Sl. no.	Name of villag	Ayacut (Ha) (1983)	Delegates by Caste, 1969							Total	Delegates, 1990
			Brahmin	Nayakar	Mudaliar	Naidu	Reddi	Pillai	Scheduled Castes		
1	Alapakkam	85.63	-	-	-	-	1	-	-	1	1
2	Athipattu	113.72	1	-	1	-	-	-	-	2	1
3	Cheri	298.02	-	1	-	-	-	1	-	2	1
4	Eralacheri	180.16	-	-	1	-	-	-	1	2	-
5	Kadaperi	27.09	-	1	-	-	-	-	-	1	-
6	Kattalai	129.88	-	1	-	-	-	-	-	1	1
7	Kaveripakkam	590.65	1	2	1	1	-	-	-	5	3
8	Kilveeranam	37.25	-	-	1	-	-	-	-	1	-
9	Kondapuram	213.68	-	-	2	-	-	-	-	2	2
10	Maganipattu	204.01	-	-	-	1	-	-	-	1	-
11	Panniyur	29.23	-	-	-	-	1	-	-	1	-
12	Puthupattu	96.15	-	1	-	-	-	1	-	2	2
13	Sirukarumbur	173.56	-	-	1	-	-	-	-	1	1
14	T.P.Pakkam	190.61	-	-	2	-	-	-	-	2	2
	Total	2369.64	2	6	9	2	2	2	1	24	14
	Percentage		8.3	25.0	37.5	8.3	8.3	8.3	4.2	100	

Note: Dash indicates nil.

Source: Interview with the Secretary, Irrigation Board, Kaveripakkam, May 1990.

Out of the 24 members elected in 1969, only 14 were available in 1990; eight were not alive and two were not residing in their villages. The caste composition shows the predominance of Mudaliars (37.5%) followed by Nayakars (25%) and the negligible presence of representatives of Naidus, Reddis, and Pillais (8.3% each) and the Scheduled Castes (4.2%) in the IB. All the remaining board members who continued their position informally were replaced in 2004 by the elected members of WUAs.

Details of the origin of the IB in DMT could not be traced. However, there is evidence of the satisfactory working of the Board during 1930s (GO 2404 I, BOR, PWD, 17-5-1937) and 1940s (see Thomas and Ramakrishnan, 1940: 186). The last election to the Board was conducted by the Revenue Divisional Officer (RDO) during 1969. Later the Board became defunct.

One of the major reasons for defunct was frequent squabbles due to political involvement and party factions among the members. Another reason was the inefficient functioning of the Board under the last elected president. In 1969 Mr. Kannappa Mudaliar (an active Diravida Munnetra Kazhagam (DMK) functionary and well-to-do landlord in Menallur village) was elected president. After his election the Board's functioning was not normal; regular Board meetings were not held and the farmers' problems remained unattended. As a result, most of the *ayacutdars* were reported to be dissatisfied with the functioning of the Board. There was a feeling that the Board did not exert pressure on the officials/ politicians to implement a long pending proposal to construct an anicut across the Cheyyar to feed some of the PAS tanks including DMT, or to maintain and strengthen the Rajakkal and Dusi-Thennampattu inlet channels to get enough supply to DMT from the Palar.

4.2.3 Constitution of Informal Irrigation Organisation

Consequently, water distribution came to be handled wholly by the *ayacutdars* themselves. Since there was no active IB to take collective action to augment supply, especially from the Rajakkal channel, and to manage water distribution, the tank was fed inadequately. The *ayacutdars* themselves organised informally in 1975 and elected a few efficient and knowledgeable persons from different *ayacut* villages to apprise the officials of their problems. This informal "Dusi-Mamandur Hissa Tank Farmers' Association" was gradually strengthened and in September 1980, was registered under the Societies Registration Act. After the emergence of this Association the formal IB has become more or less defunct. Interestingly, a few delegates who were members of the formal IB (including the then president Mr. Kannappa Mudaliar) were also members of that Association.

Table 4.5 gives the then existing structure of the Farmers Association. As may be seen from it, out of 57 members two are auditors and one a life-time member. The number of representatives for each village is related to the area irrigated and number of beneficiaries. The representatives from different villages are also selected on a preferential basis, that is, if a person is actively engagedin public welfare activities and also interested in tank management he would be given preference to represent the village. This accounts for the difference in the number of delegates from different villages. The Association conducts election regularly once in three years.

Table 4.5: Composition of Hissa Tank Farmers' Association, DMT, 1990

Designation	*No. of persons*
President	1
Vice President	5
Secretary	1
Deputy secretaries	3
Treasurer	1
Auditors	2
Life time member	1
Specialist members	4
Representatives	39
Total	**57**

Source: Sivasubramaniyan, 1995.

The caste composition (below the Table 4.6) of the Association shows the predominance of Nayakar delegates (61%), followed by Mudaliars (25%). During the settlement period (1882) Brahmins dominated most of the villages and controlled more than 50% of the total DMT *ayacut*. Over the years their number has declined, and they account only six per cent of the total representation in the Farmers' Association.

4.2.4 Main Functions of the KPT and DMT Organisations

The main functions of the KPT Irrigation Board and the DMT Farmers Association are to: (i)enforce the rules and regulations to maintain and distribute water during normal and lean supply periods; (ii) solve major problems in the system collectively; (iii) hold meetings with all the representatives once a month to discuss the irrigation matters and other

Table 4.6: Composition of Hissa Tank Farmers' Association, DMT, 1990

Sl. no.	Name of village	Ayacut (acres) (1983)	Delegates by Caste						Delegates, 1990
			Brahmin	Nayakar	Mudaliar	Naidu	Pillai	Scheduled Castes	
1	Vagai	26.19	-	3	-	-	-	-	3
2	Dusi	180.65	2	5	-	-	-	-	7
3	Kurankanilmuttam	72.59	-	1	-	-	-	1	2
4	Pallavaram	163.44	-	2	2	-	-	-	4
5	Kanikiluppai	44.94	-	-	2	-	-	-	2
6	Kilnayakanpalayam	102.67	-	-	2	-	-	-	2
7	Menallur	66.80	-	-	3	-	-	-	3
8	Seniyanallur	41.54	-	1	-	1	-	-	2
9	Poonaithangal	59.15	-	-	2	-	-	-	2
10	Mamandur	401.09	1	5	-	-	-	-	6
11	Narasamangalam	44.70	-	5	-	-	-	-	5
12	Valavandal	95.51	-	1	1	-	-	-	2
13	Vadakalpakkam	93.24	-	1	1	-	-	-	2
14	Girijapuram	47.69	-	1	-	-	1	-	2
15	Sothiyampakkam	54.53	-	2	-	-	-	-	2
16	Bagavanthapuram	43.16	-	2	-	-	-	-	2
17	Elacheri	132.39	-	2	-	-	-	1	3
18	Alinjilpattu		-	-	-	-	-	-	-
	Total	1675.79	3	31	13	1	1	2	51
	Percentage		6	61	25	2	2	4	100

Note: Dash indicates nil.

Source: Interview with the Treasurer of the Association, Dusi, May 1990.

activities; and (iv) distribute the tank supply fairly among the *ayacutdars* located in different reaches of the villages. The DMT Farmers' Association also helps in augmenting water flow to the tank from all the sources, especially from the Rajakkal channel which is more dependable. It mobilises community labour to form *mud-kondam* (temporary earthen weir) across the Palar river to divert the floods effectively through the Rajakkal channel.

One of the important functions of tank level organisation in both the tanks is to release the required quantity of water from the tank to the *ayacut* villages. For this a request is made by the representative of the particular village to the Secretary of the IB/Farmers' Association by giving *yadas* (written slips by the representative) stating the amount of water (in terms of sluice screws to be raised or lowered) to be released in the channel. This time honored practice still continues in both the tanks, even after defunct of farmers associations in the 1990s and this job is handled by the elected representatives of WUAs in both the tanks.

In KPT the responsibility of releasing required quantity of water is vested with the secretary of the IB, whereas in DMT it is shared by a group of six members of the Farmers' Association. In DMT, in the monthly meeting during the supply period, the Association regularly decides the maintenance of the main outlet channels, the quantity of water to be released in each channel and other related matters. The type of crops to be grown by *ayacutdars* is decided by farmers themselves; neither the PWD nor the IB interferes in it. Sometimes the DMT Farmers' Association settles some inter-village disputes also.

When the Irrigation Boards were set up initially, the government appointed *Neerkatties* (four for KPT, two for DMT) to help the Board by performing following functions: To watch the inlet and main outlet channels, report immediately to the Board about channel breaches for taking necessary action through the PWD, protect the bunds from cattle during the supply season, report encroachment, if any, in the channels, prevent tampering with the sluices, and carefully watch the channels during the lean periods of supply to prevent illegal tapping. However, they are not going to irrigate the fields. Their salary is paid by the government. In 1992 they were paid a consolidated salary of Rs. 612 per month. Since the IB of KPT is not active and DMT Farmers' Association is an informal one, the *Neerkatties* appointed by the government do not properly discharge their duties. At present, their services are utilised mostly by officials of the Revenue Department. However, once retired no new appointment was made for this category of workers.

4.2.5 Problems Faced by the Organisations

The organisations functioned in both the tanks faced many problems. Though the informal Farmers' Association was effective in DMT, the issues raised by it had not received adequate attention from the government mostly because it lacked formal status. The amount of money collected from the *ayacutdars* for various purposes on an acreage basis was too meagre to meet the expenses necessary for lobbying and exerting pressure on the government.

Though the IB was inactive in KPT, the functions of water release and closure from the tank were still handled in accordance with certain procedures. The then Secretary (elected in 1969) of the IB, took active interest in the tank management. His post was honorary. Though he did not have any land in the tank *ayacut*, he continued to hold the post because of his knowledge and experience in irrigation matters. Though the farmers were generally in favour of reconstituting the Board, the election was to be held by the Revenue Divisional Officer and Tahsildar. When these two officials were contacted, they informed that they were unaware of any IBs and rules and procedures. This situation continued in both the tanks till 1990s. After that the govt. enacted TNFMIS ACT-2000 (see Appendix 1.1) to enable these tank associations to work with a formal status in the name of WUAs.

4.2.6 Present Functioning of Tank Level Organizations

To streamline the functioning of informal farmers' associations in all surface irrigation systems in Tamil Nadu, the government has enacted a Rule in 2000, called Tamil Nadu-Farmers Management of Irrigation Systems Act-2000 (Tamil Nadu Act-7/2001) to enable all farmers associations to function with a formal status under the umbrella of Water Users Associations (WUAs). Implementation of this Act based on the procedures laid down was started in 2001 but it took four to five years to complete the procedural formalities such as preparation of voters list based on landholding in each irrigation system. After completing the procedures, elections for all WUAs were held in 2006-07. Once the elections were completed all the elected representatives were appointed to carry forward their responsibilities to lead the irrigation systems. The same process was also followed in the tanks of KPT and DMT.

As per the TNFMIS Act-2000, the following representatives as per the Territorial Constituencies were elected in 2007 to manage the water distribution across the ayacut/villages in both tanks. Below the tables 4.7 and 4.8 provide the details of formation of WUAs in KPT and DMT respectively.

As per this new arrangement the tank ayacut is divided into small constituencies ranging from 15.14 ha (KPT - Alapakkam - Constituency no. 2) to 137.49 ha (Eralacheri - Constituency no. 6 of Lion Sluice in KPT) under each sluice. As per the Act, the minimum area for a WUA is 40 ha and above but it is not followed. Each constituency is managed by an elected representative called TerritorialConstituency Member (TCM). A group of 10 constituencies is headed by a President, with forming managing committee and 4 sub-committees.

Managing Committee of the WUA may constitute specific sub-committees to carry out specific functions as assigned by the managing committee.

(i) Finance and Resources Sub-Committee.

(ii) Works Sub-Committee.

Table 4.7: Formation of Water Users Associations in KPT Villages as per TNFMIS Act- 2000.

Sl. no.	WUA No.	Name of the WUA	Name of Sluice & No. of Villages Covered	Total No. of TCs	Area (Ha)
1	82	Kadaperi, KPT Gaffor sluice, Sumaithangi sluice 1 and 2 WUA	Gafoor Sluice - 1. Kadaperi	1	26.60
2	83	KPT Fox sluice WUA	Fox Sluice - 1. Kaveripakkam (2) . 2. Kondapuram (3). 3. Athipattu (2). 4. Sirukarumbur (2). 5. Eralacheri (1).	10	693.85
3	84	KPT Lion sluice WUA	Lion Sluice - 1. Kaveripakkam (1 to 5). 2. Eralacheri (1)	6	453.12
4	85	KPT Moolai sluice WUA	Moolai Sluice - 1. Cheri (2). 2. Kattalai (2). Kaveripakkam (2).	6	329.19
5	86	KPT Cheri Low Level and High Level Sluices WUAs	Cheri LL & HL sluices - 1. Cheri (4). 2. Maganipattu (3&4)#	4	147.80
6	87	KPT Maganipattu WUA	Maganipattu sluice - 1. Cheri (2). 2. Maganipattu (2). 3. Pudupattu (2)	6	226.62
7	88	KPT T.P.Pakkam WUA	Thuraiperumpakkam TPP) sluice - 1. Maganipattu (2). T.P.Pakkam (4)	6	381.37
8	89	KPT Paracheri and Panniyur sluices and Panniyur Tank WUA	Paracheri sluice - 1. Cheri (1). 2. Panniyur Sluice and 3. Panniyur Tank (3). 3. Kilveeranam (2).	6	82.96
9	97	KPT Alapakkam and Sathuvanatham Tank	Alapakkam Tank (2) and Sathuvanatham Tank (2)	4	80.50
		Grand Total	14 Villages	49	2422.01

Note: WUA No = Water Users Association Number. TC = Territorial Constituency

# indicates among the 4 TCs of Cheri ayacut Maganipattu also shares 2 TCs of 3 and 4.

Figures in brackets indicate number of Territorial Constituencies allotted in respective villages.

Source: GoTN, Registers of Constituency wise Electoral list of Ayacutdars, PWD, Kaveripakkam section (no date)..

Table 4.8: Formation of Water Users Associations in KPT Villages as per TNFMIS ACT-2000.

Sl. no.	*WUA No.*	*Name of the WUA*	*Name of Sluice & No. of Villages Covered*	*Total No. of TCs*	*Area (Ha)*
1.	TVM - 87	Dusi-Mamandur Tank WUA	DMT-Sluice No: 1 - 1. Vagai (1). 2. Dusi (3). 3. Kurankanilmuttam (2). 4. Pallavaram (2)	8	248.64
2.	TVM - 88	Dusi-Mamandur Tank WUA	DMT Sluice No: 2 - 1. Dusi (2). 2. Mamandur (4). 3. Pallavaram (3). 4. Kanikiluppai (1).	10	357.92
3.	TVM -89	Dusi-Mamandur Tank WUA	DMT Sluice No. 3 - 1. Mamandur (5). 2. Vadakalpakkam (1). 3. K.N. Palayam (1). 4. Menallur (1). 5. Senianallur (1). 6. Girijapuram + Punaithangal (1)	10	644.12
4.	TVM - 90	Dusi-Mamandur Tank WUA	DMT Sluice No: 4 - 1. Alingilpattu + Mamandur (1). 2. Narasamangalam (1). 3. Sothiyampakkam (1). 4. Bagavanthapuram (1). 5. Valavandal (2). 6. Elacheri (4).	10	405.99
		Grand Total	18 Villages	38	1656.67

Note: WUA No = Water Users Association Number. TC = Territorial Constituency

Figures in brackets indicate number of Territorial Constituencies allotted in respective villages.

Source: GoTN, Registers of Constituency wise Electoral list of Ayacutdars, PWD, Kaveripakkam section.

(iii) Water Management Sub-Committee.

(iv) Monitoring Evaluation & Training Sub-Committee.

Each Sub-Committee shall consist of a convener and four members.A Convener shall be a member of the Managing Committee.In each WUA, members not exceeding four to be drawn for each Sub-Committee from among the land holders within its jurisdiction.The Sub-Committee shall be constituted by the General Body of the respective Committee and no member shall represent more than one Sub-Committee. Sub-Committees should meet once a month to deal with its functions.

Although elaborate details are specified in the TNFMIS ACT – 2000 **(for more details see Appendix 1.1)** to carry out the functions of WUAs effectively, only a fewfarmers under KPT villages and about half of them in DMT villages reported that these WUAs works are considered satisfactory. Contrary to this the ayacutdars specifically telling, as a result of formation of the WUAs, the informally working farmers associations that were functioned hitherto abstained from carrying out the tank related functions, although they are more familiar with these aspects and these activities are new to the newly elected members. Due to these problemsfull satisfaction of water management work was not carried out by the TCMs of the WUAs. However, they are reported as functioning, even after their expiry of their term.

One more reason for their unsatisfactory work effort is the long gap in the availability of Tank water supply (last filling of water in both tanks was in 2005-06 and the latest was only in 2015-16). So, during their potential tenure, the TCMs have not experienced any tank water management due to no/little supply in both tanks. Another problem is that even after completion of the members' 5-year term no re-election was held to elect new members of WUAs. Consequently, the associations are not working satisfactorily and they are simply considered as TCMs in both tanks now.

4.3 Village Level Organisation

The village level organisations are responsible for various tasks of irrigation management below the tank sluices. These include organizing the *ayacutdars* for maintenance of the channels, and to divert water through the inlet channels and allocating water among the *ayacutdars* both during normal and lean supply periods.

Scrutiny of the *Mamulnamas* indicates that each village had a well-articulated organisation during the early 19th century for appropriation of tank water. The organisation consisted of a headman (*Nattanmaikar*), *Kavaimaniyam* (person responsible for maintenance of channels) specialised in tank management, an accountant and a *Munsiff* as first tier, and below it the *Thotti, Kammukutti, Talaiyari* and the *head fisherman* to perform irrigation tasks in the field. But very little is known about the details of tank operation.

According to villagers up to 1960s the IIs at the tank and village levels functioned actively and in an organised manner but their functioning has gradually weakened and at present the working is not so satisfactory. In the past land control was in the hands of a relatively small, and homogeneous upper caste groups whose economic power reinforced their superior status in the hierarchy. The leadership and centre of power in the villages were with *Nattanmaikars* who were invariably from the upper castes. The *Kavaimaniyams* - who specialised in irrigation management - were also from upper castes. They commanded the authority and the power to ensure that the rules regarding management of tank water were effectively enforced. Since then there have been far reaching changes in the land ownership pattern in both the tank *ayacuts*. These include increase in the number of *ayacutdars*, changes in land ownership pattern, diminishing storage capacity of tanks due to silting, encroachment in the channels, deficient supply from the source for a long time, growing value of water due to changes in agricultural technology, and increasing private ownership of wells in the *ayacut*.

4.3.1 Land Control and its Transformation

The population of the villages served by both the tanks has increased over the years. The population of KPT villages has more than doubled between 1881 and 1981, and that of DMT villages has increased by 150 per cent. Between 1981 and 2011 it has trebled in KPT and more than trebled in DMT (Table 4.9). The available land being more or less fixed, there has been a progressive subdivision of holdings, and a steady increase

Table 4.9 Population, Ayacut Subdivisions and Pattadars in KPT and DMT, 1881, 1983 and 2017

Name of the tank	*Population*			*Ayacut Subdivisions*			*Total Pattadars*		
	1881	*1981*	*2011*	*1882*	*1983*	*2017*	*1881*	*1983*	*2017*
KPT	16075	33042	47091	6085	12266	13287	2131	10286	11437
DMT	7534	18775	24264	8638	14378	15931	1573	6277	7813

Source: Census registers 1881, 1981 and 2011; Settlement (1882), Updated settlement (1983) and Adangal (2017) registers of KPT and DMT Villages.

in the number of *ayacutdars*. The total number of subdivisions in the KPT villages has more than doubled, and that in DMT has increased by 66% between 1882 and 1983. Between 1983 and 2017 further increase of ayacut subdivisions are noticed. This increase is 8.3% in KPT and 10.8% in DMT. The total number of *Pattadars* (including both wet and dry lands) has increased about five fold in KPT and nearlyquadrupled in DMT. Between 1983 and 2017 this Pattadars increase is 11.2% in KPT and 24.5%in DMT. With perpetual increasing number of plots and farmers to be served by tank water, the potential for options and the difficulties of mobilising people for maintenance, and enforcing allocation rules also increase, often more than proportionately.

As Tang has observed that it is easier to organize collective action in irrigation systems of smaller sizes and with fewer users:

> Even if individual irrigators are willing to contribute to collective endeavours, they have to expend resources to organize among themselves to assign responsibilities and undertake water allocation and maintenance jobs. Both the size of irrigation system and the number of users of the system may affect farmers' actions... all other things being equal, information gathering, communication, decision making and monitoring costs increase as the size of a resource increases. By the same token, various kinds of transaction costs increase as the number of irrigators increases (1992: 23).

The social components of land owning groups have changed dramatically: Thus in KPT in 1882 Mudaliars were the dominant land owners in most of the villages controlling about 64% of the total *ayacut*. Brahmins and Pillais dominated a few villages and owned about 13% of the total land holdings (below the Table 4.10). Nayakars, who were far more numerous, owned only about 13% and they were not dominant in any of the KPT villages.

In 1983 this pattern has changed drastically: the share of Mudaliars in the tank *ayacut* has been reduced by less than a third and that of Brahmins from 10% to a mere 2.5%. Mudaliars, however own about 30% of the *ayacut* and have retained their dominance in six villages. The land ownership by Brahmins is not significant in any village. The land control of other upper castes (namely Reddies and Pillais) improved from about six per cent in 1882 to 13% in 1983; but they are not dominant in most of the villages. Over the years Nayakars, who were earlier mostly tenants of the upper-castes, have acquired greater control over land and now own 31% of the *ayacut*. They are dominant owners in four of the head and middle reach

Table 4.10: Caste wise Distribution of Land Holdings by Pattadars in KPT Villages, 1882, 1983 and 2017

(Area in Hectares)

Village and reach		Ayacut			Castewise Distribution of Land Holdings, 1882							Castewise Distribution of Land Holdings, 1983							Castewise Distribution of Land Holdings, 2017						
		1882	1983	2017	MR	Nayakar	Brahmin	Reddy	Pillai	SCs	Others	MR	Nayakar	Brahmin	Reddy	Pillai	SCs	Others	MR	Nayakar	Brahmin	Reddy	Pillai	SCs	Others
Kadaperi	HR	26.9	27.1	26.6	22.6	-	0.7	-	0.5	0.3	2.8	7.2	14.5	-	-	1.6	1.1	2.7	6.2	19.3	-	-	-	1.1	-
					(83.9)	-	(2.5)	-	(1.9)	(1.1)	(10.5)	(26.6)	(53.7)		-	(5.8)	(4.4)	(9.9)	(23.3)	(72.6)	-	-	-	(4.1)	-
Kaveripakkam	HR	766.0	590.4	590.4	302.3	286.0	47.5	5.5	9.3	5.1	110.3	80.7	382.3	28.9	-	11.4	6.2	80.8	94.8	360.4	-	-	40.8	52.6	41.8
					(39.5)	(37.3)	(6.2)	(0.7)	(1.2)	(0.7)	(14.4)	(13.7)	(64.8)	(4.9)	-	(1.9)	(1.1)	(13.7)	(16.1)	(61.0)	-	-	(6.9)	(8.9)	(7.1)
Kondapuram	MR	215.8	213.6	211.4	179.4	6.6	7.8	4.5	0.3	3.5	13.7	45.2	80.8	2.1	6.5	11.3	13.4	54.3	90.4	63.4	4.2	-	2.1	42.2	9.1
					(83.2)	(3.1)	(3.6)	(2.1)	(0.2)	(1.6)	(6.3)	(21.2)	(37.8)	(1.0)	(3.0)	(5.3)	(6.3)	(25.4)	(42.7)	(30.0)	(2.0)	-	(1.0)	(20.0)	(4.3)
Sirukarumbur	TR	177.7	173.5	175.8	152.2	3.6	5.1	-	12.4	1.7	2.8	109.0	17.9	-	-	4.2	20.9	21.4	70.2	24.5	-	19.5	8.6	24.0	29.0
					(85.6)	(2.0)	(2.9)	-	(7.0)	(1.0)	(1.5)	(62.8)	(10.3)		-	(2.4)	(12.1)	(12.3)	(39.9)	(14.0)	-	(11.1)	(4.9)	(13.6)	(16.5)
Athipattu	TR	115.4	113.7	114.8	90.0	-	19.3	-	1.8	2.6	1.7	55.3	8.7	5.1	-	7.1	16.5	21.0	51.0	28.2	2.8	-	5.7	16.6	10.5
					(78.0)	-	(16.7)	-	(1.5)	(2.2)	(1.5)	(48.7)	(7.7)	(4.5)	-	(6.2)	(14.5)	(18.5)	(44.4)	(24.6)	(2.5)	-	(4.9)	(14.4)	(9.2)
Eralacheri	TR	185.1	180.1	180.1	158.6	5.1	3.5	-	11.2	3.9	3.0	36.6	23.5	-	5.5	14.4	87.7	12.4	32.3	12.4	3.2	8.3	64.6	47.9	11.4
					(85.6)	(2.7)	(1.9)	-	(6.0)	(2.1)	(1.6)	(20.3)	(13.1)		(3.0)	(8.0)	(48.7)	(6.9)	(17.9)	(6.9)	(1.8)	(4.6)	(35.9)	(26.6)	(6.3)
G.nellur.	TR	43.5	*	*	14.5	9.1	16.7	-	-	0.5	2.7	Merged with Puthupattu							Merged with Puthupattu						
					(33.3)	(20.9)	(38.4)	-	-	(1.2)	(6.2)														
Cheri	HR	296.7	297.9	296.4	236.7	7.2	12.3	17.0	10.1	4.0	9.3	121.6	76.1	5.4	19.7	10.4	17.8	47.0	84.7	160.9	8.5	-	3.2	39.1	-
					(79.8)	(2.4)	(4.1)	(5.7)	(3.4)	(1.4)	(3.1)	(40.8)	(25.5)	(1.8)	(6.6)	(3.5)	(6.0)	(15.8)	(28.6)	(54.3)	(2.9)	-	(1.1)	(13.2)	-
Kattalai	HR	130.3	129.8	128.3	108.8	1.4	5.1	4.6	1.3	2.2	6.9	19.1	88.4	0.8	3.6	-	10.6	7.4	22.0	91.8	-	-	-	14.5	-
					(83.5)	(1.0)	(3.9)	(3.5)	(1.0)	(1.7)	(5.3)	(14.7)	(68.1)	(0.6)	(2.7)	-	(8.2)	(5.7)	(17.1)	(71.6)	-	-	-	(11.3)	-
T.p.pakkam #	TR	200.0	190.5	238.5	165.0	3.6	4.0	11.7	1.1	3.1	11.4	120.8	17.0	-	32.7	0.2	16.0	16.0	70.1	76.1	-	-	-	91.2	1.2
					(82.5)	(1.8)	(2.0)	(5.8)	(0.5)	(1.6)	(5.7)	(57.0)	(8.9)		(17.2)	(0.1)	(8.4)	(8.4)	(29.4)	(31.9)	-	-	-	(38.2)	(0.5)
Maganipattu	MR	210.9	203.9	203.4	179.0	-	15.8	-	2.1	4.0	10.0	117.9	11.0	8.8	22.9	18.1	17.9	7.3	24.4	71.2	-	-	91.5	16.3	-
					(84.9)	-	(7.5)	-	(1.0)	(1.9)	(4.8)	(57.8)	(5.4)	(4.3)	(11.2)	(8.9)	(8.8)	(3.6)	(12.0)	(35.0)	-	-	(45.0)	(8.0)	-
Alapakkam	MR	103.4	85.6	80.5	12.0	-	82.8	3.7	0.7	1.4	2.9	3.7	5.5	4.8	13.4	28.6	23.1	6.4	3.7	18.5	4.8	12.4	19.6	15.1	6.4
					(11.6)	-	(80.0)	(3.6)	(0.6)	(1.3)	(2.8)	(4.4)	(6.5)	(5.6)	(15.7)	(33.4)	(27.0)	(7.5)	(4.6)	(23.0)	(6.0)	(15.4)	(24.3)	(18.8)	(8.0)
Puthupattu	TR	53.6	96.1	97.6	-	-	29.8	8.0	12.1	0.9	2.8	-	9.2	1.9	52.5	22.4	6.3	3.8	-	10.5	-	65.3	21.4	0.4	-
					-	-	(55.6)	(14.9)	(22.6)	(1.6)	(5.3)	-	(9.6)	(2.0)	(54.6)	(23.3)	(6.5)	(4.0)	-	(10.8)	-	(66.9)	(21.9)	(0.4)	-
Panniyur	HR	26.3	29.2	29.2	4.2	-	0.5	17.9	3.3	-	0.4	1.3	-	-	24.5	-	1.6	1.8	-	-	-	29.2	-	-	-
					(15.6)	-	(2.0)	(68.0)	(12.7)		(1.5)	(4.4)	-		(84.0)	-	(5.5)	(6.1)	-	-	-	(100)	-	-	-
Kilveeranam	TR	37.2	37.2	37.2	28.3	-	4.9	-	2.5	0.4	1.1	15.4	7.0	0.8	1.0	2.6	8.9	1.5	11.7	-	-	13.7	6.0	5.8	-
					(76.0)	-	(13.2)	-	(6.8)	(1.0)	(3.1)	(41.5)	(18.8)	(2.2)	(2.7)	(6.9)	(23.8)	(4.1)	(31.4)	-	-	(36.9)	(16.0)	(15.6)	-
Total #		2589	2369	2410	1653.6	322.5	255.9	72.9	68.6	33.4	181.8	721.8	741.9	58.6	182.2	132.4	247.8	283.8	561.3	937.3	23.6	148.4	263.5	366.8	109.4
Percentage		(100)	(100)	(100)	(63.9)	(12.5)	(9.9)	(2.8)	(2.7)	(1.3)	(7.0)	(30.5)	(31.3)	(2.5)	(7.7)	(5.6)	(10.5)	(12.0)	(23.4)	(38.9)	(1.0)	(6.2)	(10.9)	(15.2)	(4.5)

Note: MR = Mudaliar, SCs = Scheduled Castes; Dash indicates nil; HR = Head reach; MR = Middle reach; TR = Tail reach.

Others include Carpenter, Blacksmith, Glodsmith, Potter, Pandaram, Dhobi, Pandithar, Karnam, Fishermen, Panchangam, Talaiyari, Manyam and Temple lands, Udaiyar, Chettiar, Naidu & Muslim.

Figures in the brackets represent percentage to the total. * denotes 40 ha of Mundivakkam Kondam ayacut is merged with KPT ayacut.

*During the re-settlement Gengatharanellur was merged with Puthupattu village. # 48 hectares of local tank ayacut of Mundivakkam has been included in the KPT ayacut area between 1983 and 2017.

Source: Settlement (1882), Updated settlement (1983) and Adangal / Chitta registers (2017) of KPT villages.

villages. Significantly, the Scheduled Caste *Pattadars* have also acquired a sizeable proportion of land - their share in total *ayacut* rising from a mere 1.3% in 1882 to 12 % in 1983. They now rank third in the overall possession of land holdings among the different castes.

In 2017, further changes in the caste structure is noticed that also occurred in favour of Nayakars and Scheduled Castes. The upper caste dominance of Mudaliars and Brahmins have reduced to less than a fourth (Brahmins are practically disappeared in most villages), but Mudaliars are still dominant in three villages but not even one in the head reach of the tank. Nayakars have gained their status from 31% in 1913 to 39% in 2017 and they are dominant in four villages. Reddies and Pillais have improved marginally from 13% to 17% over the period and each one has dominant in three villages. The SCs have also improved their land holding status from 10% in 1983 to 15% in 2017 and they are now dominant in one tail end village.

It is also noteworthy to mention that, while the Nayakars who have bought lands closer to the tank which have abundant and assured water supply, the Scheduled Castes have acquired lands mainly in the tail reaches which have less assured tank water supply.

In DMT villages the shift in land ownership from upper castes to the lower castes is even more pronounced. In 1882 Brahmins and Mudaliars owned more than four-fifths of the total ayacut. Nayakars and Scheduled Castes owned very little land. Land owned by Scheduled Castes was mostly given in return for maintenance of the channels and distribution of water in the village.

The area owned by Brahmins, declined from 760hectares in 1882 to 112ha in 1983 and further to just 13 ha in 2017; and that under Mudaliar ownership from 645ha to 372 ha and further down to 187 ha in the same periods. In contrast, the area owned by Nayakars increased from a mere 54 ha in 1882 to 855 ha in 1983 and further increased to 1008 ha (see below the Table 4.11). As a result, Nayakars have emerged as the most significant land owning group controlling three-fifth of the total *ayacut*. They also constitute the single largest land owning caste in 11 villages. In recent years the Scheduled Castes have also acquired considerable portion of land, whose share has gone up from 1%in 1882 to 7.5% in 1983 and further gained to 18% in 2017 in the tank *ayacut*.

The changes in caste-class and land holding structure have affected the working of IIs. Though the upper caste land holders are present even now

Table 4.11: Caste wise Distribution of Land Holdings by Pattadars in DMT Villages, 1882, 1983 and 2017.

(Area in Hectares)

Village and reach		Ayacut 1882	Ayacut 1983	Ayacut 2017	1882 Brahmin	1882 MR	1882 Nayakar	1882 Pillai	1882 SCs	1882 Others	1983 Brahmin	1983 MR	1983 Nayakar	1983 Pillai	1983 SCs	1983 Others	2017 Brahmin	2017 MR	2017 Nayakar	2017 Pillai	2017 SCs	2017 Others
Dusi Sluices 1 and 2																						
Vagai	HR	26.2	26.2	26.1	7.0	13.4	2.9	-	0.8	2.2	0.3	9.1	15.4	-	-	1.3	-	6.4	16.3	0.8	-	2.7
					(26.9)	(51.0)	(10.9)	-	(2.9)	(8.3)	(1.1)	(34.8)	(59.0)	-	-	(5.1)	-	(24.3)	(62.5)	(2.9)	-	(10.3)
Dusi	HR	175.1	180.6	180.6	153.1	18.5	-	-	1.1	2.5	19.3	-	147.3	-	1.8	12.2	6.1	4.0	170.5	-	-	-
					(87.4)	(10.5)	-	-	(0.6)	(1.4)	(10.7)	-	(81.6)	-	(1.0)	(6.7)	(3.4)	(2.2)	(94.4)	-	-	-
K.muttam.	HR	57.4	72.5	72.5	53.9	-	0.9	0.5	0.7	1.4	18.5	2.1	10.0	-	36.4	5.6	4.8	-	25.6	-	42.1	-
					(93.9)	-	(1.5)	(0.8)	(1.2)	(2.5)	(25.4)	(2.9)	(13.8)	-	(50.1)	(7.7)	(6.7)	-	(35.3)	-	(58.0)	-
Pallavaram	TR	162.3	163.4	164.9	3.1	142.1	1.4	7.0	0.9	7.8	-	30.4	80.7	6.2	27.1	19.0	-	24.3	96.2	-	29.6	14.8
					(1.9)	(87.6)	(0.9)	(4.3)	(0.5)	(4.8)	-	(18.6)	(49.4)	(3.8)	(16.6)	(11.6)	-	(14.7)	(58.3)	-	(18.0)	(9.0)
Kanikiluppai	MR	45.7	44.9	44.9	0.5	40.6	0.2	-	1.0	3.5	-	34.9	-	-	4.3	5.7	-	24.6	-	-	16.2	4.1
					(1.1)	(88.7)	(0.5)	-	(2.1)	(7.6)	-	(77.7)	-	-	(9.6)	(12.7)	-	(54.8)	-	-	(36.1)	(9.1)
K.n.palayam.	MR	102.2	102.6	102.0	-	90.1	1.1	1.6	1.8	7.6	-	67.6	10.2	12.0	5.3	7.4	-	20.2	2.4	-	40.5	38.9
					-	(88.2)	(1.1)	(1.5)	(1.7)	(7.5)	-	(65.9)	(10.0)	(11.7)	(5.2)	(7.2)	-	(19.8)	(2.4)	-	(39.7)	(38.1)
Menallur	MR	66.3	66.8	66.6	-	61.8	-	-	0.5	3.9	-	43.1	7.0	10.4	1.8	4.4	-	28.8	18.9	7.9	5.7	5.3
					-	(93.2)	-	-	(0.8)	(6.0)	-	(64.6)	(10.5)	(15.6)	(2.7)	(6.6)	-	(43.2)	(28.4)	(19.9)	(8.6)	(8.0)
Senianallur	TR	41.6	41.5	41.5	IST SETTLEMENT WAS MADE IN 1964						-	16.3	19.0	4.2	-	2.0	-	-	17.7	11.4	-	12.4
											-	(39.2)	(45.7)	(10.2)	-	(4.9)	-	-	(42.7)	(27.5)	-	(29.9)
Punaithangal	TR	59.9	59.1	58.8	-	52.9	0.4	0.5	0.6	5.6	-	43.5	5.9	-	6.7	3.1	-	46.2	-	-	5.4	7.2
					-	(88.3)	(0.6)	(0.8)	(0.9)	(9.4)	-	(73.6)	(9.9)	-	(11.3)	(5.2)	-	(78.6)	-	-	(9.2)	(12.2)
Total		736.8	757.7	757.9	217.7	419.4	6.8	9.5	7.3	34.5	38.1	247.0	295.6	32.9	83.4	60.7	10.9	154.5	347.6	20.1	139.5	85.4
		(100)	(100)	(100)	(29.5)	(56.9)	(0.9)	(1.3)	(1.0)	(4.7)	(5.0)	(32.6)	(39.0)	(4.3)	(11.0)	(8.0)	(1.4)	(20.4)	(45.9)	(2.6)	(18.4)	(11.3)

(Contd...)

Village and reach	Ayacut			Castewise Distribution of Land Holdings 1882						Castewise Distribution of Land Holdings 1983						Castewise Distribution of Land Holdings, 2017					
	1882	1983	2017	Brahmin	MR	Nayakar	Pillai	SCs	Others	Brahmin	MR	Nayakar	Pillai	SCs	Others	Brahmin	MR	Nayakar	Pillai	SCs	Others
Mamandur Sluices 3 and 4																					
Mamandur HR	407.9	400.8	400.0	359.7	6.7	3.8	2.7	3.0	32.1	68.8	-	257.3	-	16.1	58.7	2.5	-	378.1	-	6.5	12.9
				(88.2)	(1.6)	(0.9)	(0.7)	(0.7)	(7.9)	(17.2)	-	(64.2)	-	(4.0)	(14.6)	(.6)	-	(94.5)	-	(1.6)	(3.2)
Narasamangalam HR	44.1	44.7	44.7	31.1	-	3.1	1.7	0.4	7.9	4.9	-	38.3	-	-	1.5	-	-	35.0	1.4	-	8.2
				(70.5)	-	(7.0)	(3.8)	(0.9)	(17.8)	(11.0)	-	(85.7)	-	-	(3.3)	-	-	(78.4)	(3.2)	-	(18.4)
Valavandal MR	96.5	95.5	95.5	73.2	10.1	2.1	1.5	0.6	9.0	-	64.9	24.3	-	3.0	3.3	-	2.7	39.6	-	53.2	-
				(75.9)	(10.5)	(2.2)	(1.5)	(0.6)	(9.3)	-	(68.0)	(25.5)	-	(3.1)	(3.5)	-	(2.8)	(41.5)	-	(55.7)	-
Vadakalpakkam MR	96.5	93.2	92.4	13.6	62.6	11.1	0.7	0.8	7.7	-	26.6	50.3	-	7.7	8.6	-	3.1	44.5	-	39.6	5.2
				(14.1)	(64.8)	(11.5)	(0.8)	(0.9)	(8.0)	-	(28.6)	(53.9)	-	(8.3)	(9.2)	-	(3.4)	(48.2)	-	(42.9)	(5.6)
Girijapuram MR	46.4	47.7	47.9	0.3	39.1	-	4.2	0.6	2.2	-	19.5	13.1	12.1	1.7	1.3	-	13.8	6.5	8.1	10.1	9.4
				(0.7)	(84.1)	-	(9.1)	(1.3)	(4.8)	-	(40.8)	(27.5)	(25.4)	(3.5)	(2.8)	-	(28.8)	(13.6)	(16.9)	(21.1)	(19.6)
Sothiyampakkam TR	57.4	54.5	55.9	37.3	7.3	1.3	7.0	0.5	4.0	-	-	40.8	8.1	2.3	3.2	-	-	32.2	7.2	15.1	1.4
				(65.0)	(12.7)	(2.2)	12.3)	(0.8)	(7.0)	-	-	(74.9)	(14.9)	(4.3)	(5.9)	-	-	(57.6)	(12.9)	(27.0)	(2.5)
Bagavanthapuram TR	42.7	43.2	43.9	11.8	-	23.7	3.7	0.6	2.9	-	-	42.2	-	-	1.0	-	-	39.5	-	-	4.4
				(27.6)	-	(55.4)	(8.8)	(1.3)	(6.8)	-	-	(97.7)	-	-	(2.3)	-	-	(90.0)	-	-	(10.0)
Elacheri TR	131.5	132.3	131.3	9.5	99.4	1.0	7.8	4.7	9.2	-	13.5	87.9	14.7	10.6	5.6	-	13.2	81.1	-	37.0	-
				(7.2)	(75.5)	(0.8)	(5.9)	(3.6)	(7.0)	-	(10.2)	(66.4)	(11.1)	(8.0)	(4.3)	-	10.1	(61.8)	-	(28.2)	-
Alingilpattu HR	6.7	5.5	6.0	5.7	-	0.7	-	-	0.3	-	-	5.0	-	0.4	0.1	-	-	4.0	-	1.0	1.0
				(85.0)	-	(11.1)	-	-	(3.9)	-	-	(91.0)	-	(6.4)	(2.6)	-	-	(67.1)	-	(16.8)	(16.1)
Total	929.7	917.3	917.5	542.1	225.1	46.7	29.4	11.2	75.2	73.7	124.5	559.2	34.9	41.7	83.3	2.5	32.8	660.5	16.7	162.5	42.5
Percentage	(100)	(100)	(100)	(58.3)	(24.2)	(5.0)	(3.2)	(1.2)	(8.1)	(8.0)	(13.6)	(61.0)	(3.8)	(4.5)	(9.1)	(0.3)	(3.6)	(72.0)	(1.8)	(17.7)	(4.6)
DMT as a Whole	1666.5	1675.0	1675.5	759.8	644.5	53.6	38.9	18.5	109.7	111.8	371.5	854.8	67.8	125.1	144.0	13.4	187.3	1008.1	36.8	302.0	127.9
Percentage	(100)	(100)	(100)	(45.6)	(38.7)	(3.2)	(2.3)	(1.1)	(6.6)	(6.7)	(22.2)	(51.0)	(4.0)	(7.5)	(8.6)	(0.8)	(11.2)	(60.2)	(2.2)	(18.0)	(7.6)

Note: MR = Mudaliar, SCs = Scheduled Castes; Dash indicates nil; HR = Head reach; MR = Middle reach; TR = Tail reach.

Others include Carpenter, Blacksmith, Glodsmith, Potter, Pandaram, Dhobi, Pandithar, Karnam, Fishermen, Panchangam, Talaiyari, Manyam and Temple lands, Udaiyar, Chettiar, Naidu, Muslim and Unidentified.

Figures in the brackets represent percentage to the total.

Source: Settlement (1882), Updated settlement (1983) and Adangal / Chitta registers (2017) of DMT villages.

in a few villages under both the tanks, their position as land owners as well as their authority in the community has declined. With the diffusion of land ownership, and the passing of land control to lower castes the ability of the village leaders to coordinate, and to enforce rules on maintenance and water distribution has weakened. One could expect that strong IIs in villages where a single caste dominates land control than those where ownership is diffused among several castes. When different villages under a tank have two or more dominant castes, the problem gets compounded. Actually, the latter is the case in the villages studied under both the tanks.

There are also clash of interest between Mudaliars and Nayakars within villages. In KPT, in the head reach (Kaveripakkam) village of Lion sluice and Fox sluice, Nayakars have acquired a lot of lands in recent decades and in Eralacheri the tail reach village of these sluices, the Scheduled Castes have acquired a lot of land holdings. Whenever the tank storage reaches less than full tank level it is alleged that the Nayakars of the upper reaches do not let water to flow into the tail reach village. This has resulted in several water disputes between these two castes. Due to this persistent problems, during 2017 supply year, it is reported that Eralacheri village was fully excluded from getting KPT supply. Another problem for this no tank supply to Eralacheri is reported that full encroachment of the main supply channel leading to Eralacheri village. Similar disputes were also reported from Punaithangal village in DMT. Thus, the changes in the pattern of land transfers from the upper castes to lower castes within an *ayacut* do make severe problems in the smooth functioning of IIs especially to enforce the regulations for maintenance of channels and management of tank supply.

4.3.2 The Growing Demand for Water

The growing demand for water in agriculture is mainly due to three factors: (i) increased use of HYV seeds and chemical fertilizers; (ii) changes in cropping pattern and (iii) introduction of multiple cropping. Before the 1960s farmers mainly raised crops by using local variety seeds which did not require much water. Water intensive crop of paddy was grown mainly in the rainy season (except in well irrigated areas). In other seasons dry crops were grown. Well irrigation was not developed much before the 1960s.

This situation has changed drastically in the last 4-5 decades, with the substitution of local varieties by HYVs; seasonal crops to annual crops; increased use of fertilizers, and introduction of multiple cropping. These changes have led to higher demand for water and made control over timing and volume of irrigation crucial. As available supply from the common sources such as tanks and canals has not increased - if anything the neglect of system maintenance mainly have reduced supplies - there is greater potential for conflicts over water allocation. Controlled irrigation is difficult, more wells have been dug and the existing ones have been deepened.

To some degree this has been taken care of by expansion of wells in both the tank *ayacuts* (as discussed in detail in chapter III). But this process is badly to have weakened the working of institutions for tank management. Those who own wells, depend less on tanks and if they get adequate supplies from their wells their interest in tank management weakens. If thewell farmers happen to be large in number and they possess, more pump sets in the villages, this further weaken the tank institutions.

To capture the gradual changes that happened in the village level irrigation institutions, following gist of the functioning of village institutions that existed in both KPT and DMT in the year 1990 is more useful to understand how these institutions have gradually and steadily declined over a period of time. In short, it should be mentioned that these village level organisation was functioning with 40-50% capacity in 1990 but nearly in the last three decades most of the village irrigation institutions have collapsed/disappeared and its present functioning is only 20 to 30% capacity, where it exists. Further,in many of the villages surveyed in 2017 in both the tanks these village institution is not at all existing.

Generally, the village organisation taken care of maintenance of channels and distribution of water. The lower level irrigation workers undertook distribution of water (see Table 4.12) at the field level and organised most of the works. They were the main irrigation functionaries. The head of the Supervisory category was the *Nattanmaikar* who represented

Table 4.12: Composition of Village Organisation in KPT and DMT, 1990

Personnel	Caste Composition of the Personnel												
	Kaveripakkam Tank						Dusi-Manamdur Tank						
	Naya-kar	Muda-liar	Reddy	Pillai	Sche-duled Caste	Total	Brah-min	Naya-kar	Muda-liar	Pillai	Naidu	Sche-duled Caste	Total
SUPERVISORY LEVEL													
Nattanmai	17	12	4	2	3	38	4	13	9	2	1	2	31
Kavaimaniyam	1	1	1	0	0	3	0	1	2	0	0	0	3
Delegates of IB	5	7	1	1	0	14	3	35	14	2	1	2	57
LOWER LEVEL													
Kammukutties	9	0	0	0	14	23	0	0	0	0	0	17	17
Thotties	0	0	0	0	20	20	0	0	0	0	0	16	16

Source: Sivasubramaniyan, 1995.

the village as a whole or the *Kavaimaniyam* (usually one person from a village specialized in canal management).

During 1990, in most villages the leading *Nattanmaikar* or the delegates to the Irrigation Board/Farmers' Association supervised the tank-related activities. Most of the IB delegates were drawn from the *Nattanmaikars*. A few villages had*Kavaimaniyams* specialised in tank management. Each three villages in KPT and DMT had active *Kavaimaniyams* at that time. All these villages are located at the tail end of the tanks and have separate branch distributaries, proper maintenance of which is more important for the free flow of water.

Nattanmaikars were responsible for not only the irrigation tasks but also for several other duties in the village community including organisation of festivals, and settlement of disputes. The *Kavaimaniyam's* main function was to upkeep and maintain the channels serving the village *ayacut*. Traditionally, the Nattanmaikars and Kavaimaniyams were drawn from rich upper caste farmers in the villages. Both are hereditary and honorary positions. However, at the time of 1990 it was reported thatin several villages (especially in Kaveripakkam) hereditary appointment had ceased and persons from lower castes like Nayakars had also been appointed by the villagers as *Nattanmaikars*.

The actual situation that existed in 1990 is reflected in Table 4.12 that indicate only 20 out of 55 *Nattanmaikars*/*Kavaimaniyams*/IB delegates in KPT are Mudaliars, whereas 23 are Nayakars. Three *Nattanmaikars* are also from Scheduled Castes. In DMT, nearly half the higher level village functionaries are Nayakars and Mudaliars. Brahmins account for only 12% of the total.

The *Nattanmaikar* or *Kavaimaniyam* usually called a meeting of *ayacutdars* in their village as soon as the tank started receiving supply, with the *Thotties* announcing the information through tom-toming. In that meeting it was decided when to start the maintenance work of main channels and the distributaries passing through their village, and the share of contribution of labour by each *ayacutdar* in the village. Usually three persons per hectare was required in the head reach villages. However, in most tail end villages, the allotment per hectare went up to 8 to 10 persons because of the long distance they had to cover during the maintenance work. If anyone of the *ayacutdars* failed to contribute his own labour, he was substituted it with hired labour. If a farmer failed to do both, he was fined. Some *Kavaimaniyams* maintained good records of the extent of labour involved in the maintenance work. The *Kavaimaniyams* and the

Nattanmaikars also had to allot the duration of supply of water to each *ayacutdar* in times of scarcity. The *Kavaimaniyams* supervised the observance of rotation for rationing water and arbitrated over disputes, if any, arose out of water sharing.

Under both the tanks the Nattanmai and *Kavaimaniyam* were assisted by a group of local level irrigation workers called *Kammukutties* and *Thotties*. Their number in a village varied from one to six (excluding *Neerkatties*, who were appointed by the government) depending upon the size of the *ayacut*. For instance, in large *ayacut* villages such as Kaveripakkam and Kondapuram the number that existed in 1990 was six each. In smaller *ayacut* villages such as Kadaperi, Kattalai, Panniyur and Kilveeranam it was one to two. In the DMT villages the maximum number available in a village was only four in Pallavaram in sluices 1 and 2, whereas there was none in three villages in sluices 3 and 4. In those villages, farmers themselves taken care of the irrigation activities. The number of *Kammukutties* and *Thotties* available in the villages served by the tanks was not proportionate to the *ayacut* of the villages. In some villages there were no *Kammukutties*. Where they were absent, their work was done by either the *Thotties* or a couple of persons temporarily appointed by the village organisation.

The Kammukutties and *Thotties* were expected to organize the *ayacutdars* for channel maintenance. In most DMT villages where *Oru Madai Paichal* (channel-wise turn distribution of water supply) is widely prevalent, the *Kammukutties* had to regulate water supply to various plots according to rules during the lean supply periods. In some long distance channels passing through their villages, they had to supervise the channels for illegal tapping and preventing misuse of water by the upstream farmers both during normal and lean supply periods.

Thotties were village level workers performed traditional tasks such as beating drums on important occasions, removing carcasses, and carried messages to relatives of dead persons. They also had to assist farmers to build *Kondam* (a temporary earthen embankment in the bed of the channel to divert water to the upland fields) across the channel during lean supply periods.

It may be seen from Table 4.12 that in both the tanks, the upper caste members predominated the Supervisory level, whereas mostly Scheduled Caste members were appointed at the lower level. However, in KPT the Nayakar caste members also hold the lower level workers position like

Kammukutties (their main duties are distributing tank water and to supervise the channels). By and large, Nayakars and Mudaliars predominate the supervisory category and the Scheduled Castes were only in the lower category in the villages under both the tanks.

4.3.3 Mode of Selection of Irrigation Functionaries

In most villages the *Nattanmaikars* or *Kavaimaniyams* have held their position for many years, and for life time in some villages. The duration generally depends upon how well they perform these functions. So long as the villagers are satisfied, they can continue indefinitely. The *Kammukutties* and *Thotties* were selected on hereditary basis mainly from among the Scheduled Caste households. Usually every year they were selected and assume charge in the Tamil month *Aadi* (July).

4.3.4 Remuneration to Irrigation Functionaries

While the working of higher level functionaries such as *Nattanmaikar*, *Kavaimaniyam* and IB delegates was honorary, the lower level workers were paid in kind by the farmers. Their remuneration varies across seasons and between KPT and DMT. The following points will help in having a better understanding of the remuneration to irrigation functionaries.

(1) In KPT no remuneration was paid when the tank does not supply water. In DMT even in the first season (July-October when the crop depends entirely on rainfall) *Kammukutties* and *Thotties* were paid.

(2) In KPT payments were made only when the *ayacutdars* raise paddy and other seasonal crops such as groundnut, ragi, cumbu and so on. In DMT where paddy cultivation predominates in all the three seasons, in each season the functionaries were paid. If the farmers cultivate annual crops other than paddy, the services of the *Kammukutties* were discontinued in KPT. This is especially so in the case of users of well water. In some cases where the tank water was supplemented with well water to raise banana or sugarcane, the *Kammukutties* had to be paid in cash at the rate of Rs. 40 to 50 per hectare. This was reported in KPT villages where a significant shift from paddy to sugarcane and banana crops has taken place.

(3) In KPT the functionaries were paid only during the tank supply years. In DMT whenever a crop is raised in the *ayacut* by using either the tank supply or the monsoon showers the payments were

compulsory. However, if no crop was raised or if the raised crop fails due to pest attack or monsoon failure the payments were not made even in DMT villages.

(4) The development of well irrigation under the tank *ayacut* has also made the work of *Kammukutties* less effective in recent decades. In both tanks if the crop, especially annual crops such as banana and sugarcane, was raised exclusively by using well water in the tank *ayacut* the functionaries were not usually paid because the farmers themselves irrigate the crop. It was pointed out by the functionaries that whatever little cash paid for banana and sugarcane crops was mostly on the generosity of the *ayacutdars*. As a consequence, their remuneration was very poor, and in some drought years, they actually get very little (For more details see Sivasubramaniyan 1995).

The preliminary survey conducted in all the KPT and DMT villages in 1990 provides the details of remuneration given to irrigation workers (IWs) in the supply years 1988 in KPT, and 1985 in DMT. The tanks were dry at the time of the survey. Hence the preceding supply year was chosen as the reference period for the survey. Since both the tanks did not get the supply evenly in a single year before 1990, the year 1988-89 for KPT and 1985-86 for DMT was chosen. The following account is based on the rapid survey of all the villages based on interviews with village leaders and *Kammukutties/ Thotties*.

KPT: Four out of the 14 villages (one in head reach and the remaining three at the tail end) in the KPT *ayacut* do not seem to have any *Kammukutties* or *Thotties* for irrigation management. Of these four villages Sirukarumbur and Kilveeranam have not made use of tank water since 1981. In Athipattu the irrigation workers have not functioned since 1979 due to disintegration of the village level organisation. In Kattalai there is no *Kammukutti* and management of irrigation is done by the *ayacutdars* themselves. If the tank receives supply the *ayacut* of Kattalai which is close to it, gets copious supply due to seepage from the sluices and continuous flow of water passing to lower level villages. For any help during the supply season the *ayacutdars* temporarily appoint a person and pay him only for that season. Hence the *Kammukutties* presence is not needed. In the remaining ten villages the available number of irrigation workers (IWs) ranged from one in Kadaperi to six in Kaveripakkam and Kondapuram.

The sluice wise and reach wise analysis shows that in all the elevated sluices the number of IWs per acre of *ayacut* is relatively high - twice or more of the tank average. This may be because water supply is lighter and more uncertain in those sluices, and therefore, demands intensive vigil over water allocation to ensure fair distribution (see below the Table 4.13). This supposition is reinforced by the fact that villages served by the deeper sluices or those located at the head reach of a sluice (eg. Kaveripakkam and Cheri) have fewer IWs per acre. Under these conditions water supply is more abundant and assured, and the need for maintenance and water allocation is correspondingly less. In several cases the *ayacutdars* themselves irrigate their landsand the IWs only supervise the channels to ensure proper supply. Though the two tail end villages of Eralacheri and Thuraiperumpakkam are affected by inadequacy of water supply the number of IWs per hectare is also relatively low. The extent and frequency of such shortage is so severe that the problem is not one of augmenting surplus as of rationing available supplies. Having more irrigation workers does not solve the problem. In general the middle reach villages, where supply is better but regulating supply to ensure fair distribution is impossible, have a higher number of IWs per hectare of *ayacut*.

DMT: In this tank in all the head reach villages (except Vagai and Kuranganilmuttam) there are no permanent irrigation workers; given the location and relative assurance of supplies, they do not feel the need for these workers. The number of IW per hectare of *ayacut* is more in DMT than in KPT. The number of IW per hectare of *ayacut* is more in villages served by sluices 3 and 4 than sluices 1 and 2. Within each of the two sluice groups, one of the head reach villages under sluices 1 and 2 and all the head reach villages in sluices 3 and 4 do not have any irrigation workers. On an average there are fewer IWs per hectare in the head reach villages than in the middle and tail reaches (see below the Table 4.14).

4.4 Detailed First Round Survey, 1991-92

A farm level survey was conducted in selected villages under both the tanks during the full tank supply year 1991-92 to ascertain the details of actual payments made to irrigation workers. The villages chosen were five in KPT and four in DMT. Also, the same sample villages are taken for the present survey in 2016-17. Information on payment by the sample farmers - by season-wise work and amount - was obtained for 1991-92, which happened to be a good year.

It may be seen from Table 4.13 that the importance of irrigation workers in the villages declined gradually in recent years. For instance, it was reported that some 10 years ago Athipattu village had two *Kammukutties* whereas it

Table 4.13: Remuneration of Irrigation Workers in KPT Villages, 1988-89

Sluice	Village	Reach	Ayacut hectares 1983	No. of irri-gation workers	Wage rate Kg/ hectare	Total wage (kg)	Rate per kg (Rs)	Total amount (Rs)	Amount/ person/ crop	Amount/ person/ month	Amount per hectare (Rs)	Ayacut/ irriga-tion worker
1	2	3	4	5	6	7	8	9	10	11	12	13
Moolai	Kattalai	HR	*									
	Cheri	MR	73.08	1	15	1083	2	2166	2166	542	30	73
	Kaveripakkam	TR	114.05	1	10	1127	2	2254	2254	563	20	114
Cheri LL	Cheri	HR	88.30	1	15	1309	2	2617	2617	654	30	88
Lion	Kaveripakkam	HR	318.10	3	10	3143	2	6286	2095	524	20	106
	Eralacheri	TR	130.93	1	10	1294	2	2587	2587	647	20	131
Maganipattu	Cheri	HR	*1									
and	Maganipattu	MR	204.01	4	17	3527	2	7055	1764	441	35	51
Thurai-	Alapakkam	MR	85.63	4	15	1269	2	2538	635	159	30	21
perum-	Puthupattu	TR	96.15	4	15	1425	2	2850	713	178	30	24
pakkam	Thuraiperumpakkam	TR	190.61	2	10	1883	2	3766	1883	471	20	95
Cheri HL	Cheri	HR	92.51	1	15	1371	2	2742	2742	686	30	93
Fox	Kaveripakkam	HR	129.68	2	10	1281	2	2562	1281	320	20	65
	Kondapuram	MR	213.68	6	15	3167	2	6334	1056	264	30	36
	Eralacheri	TR	49.23	1	10	486	2	973	973	243	20	49
	Sirukarumbur	TR	*2									
	Athipattu	TR	*3									
Paracheri	Cheri	ES	16.48	1	15	244	2	488	488	122	30	16

(Contd...)

Panniyur	Panniyur	ES	29.23	2	10	289	2	578	289	72	20	15
	Kilveeranam	ES	*2									
Gaffor	Kadaperi	ES	27.09	1	15	401	2	803	803	201	30	27
Total			1858.74	35		23299	2	46598	1331	333	25	53

Note: In Kaveripakkam village 28.83 hectares of Surplus Kondom and Catchment ayacut has been excluded.

1988-89 represents the latest supply year in KPT.

HR=Head reach, MR=Middle reach, TR=Tail reach, ES=Elevated sluice.

Column 5 : Kammukutties and Thotties (excluding Neerkatties, who are paid by the government).

Column 8: Paddy rate per kilo gram is the average wholesale price in North Arcot district during 1988-89.

Column 11: This is calculated by dividing by four of col. 10. Assuming that the crop period is four months duration and the total wage relates to one crop season (November-February) only.

* This village ayacut is located very close to the tank, hence there is no permanent irrigation worker.

*1 In Cheri village under this sluice 27.65 hectares have been irrigated without appointing the irrigation workers.

*2 Water has not been taken for the past nine years; hence it is excluded from this analysis.

*3 After 1970s irrigation workers have not functioned.

Source: Survey of Irrigation workers (Kammukatties and Thotties) in KPT villages, 1990.

Table 4.14: Remuneration of Irrigation Workers in DMT Villages, 1985-86

Sl. no.	Name of village	Reach	Ayacut Ha 1983	Irrigation workers (No.)	Wage rate kg / Ha	Total wage (kg)	Rate per Kg (Rs)	Total amount (Rs)	Amount/ person/ crop (Rs)	Amount/ person/ month (Rs)	Amount per Ha (Rs)	Ayacut/ irrigation worker
1	2	3	4	5	6	7	8	9	10	11	12	13
Elevated and Deep Sluices 1 and 2												
1	Vagai	HR	26.19	1	20	520	1.70	884	884	221	33.7	26
2	Dusi	HR	-	-	-	-	-	-	-	-	-	-
3	K.muttam.	HR	72.58	2	15	1074	1.70	1826	913	228	25.2	36
4	Kanikiluppai	MR	44.92	2	10	444	1.70	755	377	94	16.8	22
5	K.n.palayam.	MR	102.68	3	15	1524	1.70	2591	864	216	25.2	34
6	Menallur	MR	66.81	2	10	660	1.70	1122	561	140	16.8	33
7	Pallavaram	TR	163.44	4	20	3232	1.70	5494	1374	343	33.6	41
8	Senianallur	TR	41.55	2	10	412	1.70	700	350	88	16.9	21
9	Punaithangal	TR	59.16	2	15	876	1.70	1489	745	186	25.2	30
	Total		577.33	18		8742	1.70	14861	826	206	25.7	32
Deep and Elevated Sluices 3 and 4												
10	Mamandur	HR	-	-	-	-	-	-	-	-	-	-
11	Narasamangalam	HR	-	-	-	-	-	-	-	-	-	-

12 Alingilpattu	HR	-	-	-	-	-	-	-	-	-	-
13 Valavandal	MR	95.50	3	20	1888	1.70	3210	1070	267	33.6	32
14 Vadakalpakkam	MR	93.22	3	20	1840	1.70	3128	1043	261	33.6	31
15 Girijapuram	MR	47.70	2	15	708	1.70	1204	602	150	25.2	24
16 Sodiyampakkam	TR	54.52	2	15	810	1.70	1377	689	172	25.3	27
17 Bagavanthapuram	TR	43.17	2	20	856	1.70	1455	728	182	33.7	22
18 Elacheri	TR	-	-	-	-	-	-	-	-	-	-
Total		334.109	12		6102	1.70	10373	864	216	31.0	28
DMT as a Whole		911.441	30		14844	1.70	25235	841	210	27.7	30

Note: Dash indicates that no irrigation workers were appointed in the villages; and hence the ayacut of these villages is not included for analysis.

1985-86 represents the latest supply year under DMT before 1990; HR=Head reach, MR=Middle reach, TR=Tail reach.

Column 5: Kammukutties and Thotties (Excluding Neerkatties, who are paid by the government).

Column 8: Paddy rate per kilogram is the average wholesale price during 1985-86 in Chengalpattu district.

Column 11: This is calculated by dividing by four of column 10. Assuming the crop period is four months duration and the total wage relates to one crop season (November to February) only.

DMT is situated at the border of North Arcot and Chengalpattu districts, and it is situated about 7 kms from the Kancheepuram Town. Most of the transaction in DMT villages have been undertaken in this town itself,hence the wholesale price of Chengalpattu district is taken into account for the analytical purpose.

Paddy Price Conversion = Rice Price - 1/3 of Rice Price (for Paddy Husk).

Source: Survey of Irrigation workers (Kammukatties and Thotties) in DMT villages, May 1990.

has none at present. So, for our analysis, this village was not taken into account.

The farm level survey (1992) indicates that in KPT, in Panniyur village apart from getting payments in kind, IWs were provided daily food (morning and night) in all the three seasons. This was not seen in any other villages. It may be seen from Table 4.13 that the IWs were paid only during the tank supply seasons inKPT villages whereas they were paid in all three seasons during normal year in the DMT villages.

Payments by the *ayacutdars* depend upon the supply of water available from the tank and also the productivity of land. Further, if the crop does not give yield (due to deficit water supply or severe pest attack) they do not get anything. For instance, in KPT where in the third season, 1991-92, almost all villages suffered severe pest attack, a majority of farmers could not get any yields, and therefore, did not pay anything to the functionaries.

Table 4.15 and 4.16 represent the money value of all the kind payments given by the respondents in the supply year 1991-92. The Table is based on weighted averages to the total extent owned by the sample farmers, i.e., the actual amount paid by the respondents per hectare of the sample holdings multiplied by the total extent of land owned by each type of (well, head and tail) farmers. From this imputed value the annual average amount per hectare in the village as a whole is derived. In the KPT villages only in the second season the amount paid was significant, whereas in the other two seasons the *ayacutdars* did not pay at all (except in Kattalai served by the deep sluice). In DMT, the estimated contribution per hectare in 1991-92 supply year was several fold more than that of KPT sample villages irrespective of reaches from the tank. This means that good tank water supply leads to good remuneration which in turn makes smooth functioning of irrigation institutions in the villages under DMT.

Across villages there was a slight difference in per hectare amount in both the tanks. In KPT the total imputed value of contribution per ha in Alapakkam was a little more than in Kadaperi. But between the elevated and deep sluice villages the variation is not much. In DMT, between the head reach villages Vagai and Narasamangalam the value of contribution per ha in Vagai is about two fold more than in Narasamangalam. In the tail end village of Punaithangal, the total amount paid per ha was a little more than that of Sothiyampakkam.

In KPT within the village non-well farmers contributed more amount than the well farmers. In Alapakkam, the KPT supply was easily accessible

Table 4.15: Payments of Irrigation Functionaries in KPT 1991-92

Name of sample villages	Sample hhs by type	Sample extent, hectares	Total extent, hectares	Actual payment latest year						Imputed value of contribution (Rs)	
				Amount paid by sample farmers			Imputed value for all farmers			for all farmers	per ha
				S-1	S-2	S-3	S-1	S-2	S-3		
							5/3X4	6/3X4	7/3X4	8+9+10	(11/4)
1	2	3	4	5	6	7	8	9	10	11	12
Kadaperi	Well 8	7.64	9.73	0	98	0	0	125	0	125	12.8
E (H)	Head 8	4.68	9.28	0	116	6	0	230	12	242	26.1
	Tail 8	3.62	8.57	0	106	0	0	251	0	251	29.3
	Total	15.93	27.58							618	22.4
Panniyur	Well 8	4.88	6.94	0	111	0	0	158	0	158	22.8
E (H)	Head 8	2.17	14.33	0	41	0	0	271	0	271	18.9
	Tail 8	1.54	8.55	0	120	0	0	346	0	346	40.4
	Total	8.59	29.83							775	26.0
Kattalai	Well 8	10.76	43.51	0	138	29	0	558	117	675	15.5
D (H)	Head 8	3.17	52.52	0	89	13	0	1474	215	1690	32.2
	Tail 8	3.45	36.90	0	98	19	0	1047	203	1250	33.9
	Total	17.38	132.93							3615	27.2
Alapakkam	Well 8	15.60	22.12	0	339	0	0	481	0	481	21.7
M (T)	Head 8	2.35	28.09	0	138	0	0	1651	0	1651	58.8
	Tail 8	7.16	36.48	0	168	0	0	856	0	856	23.5
	Total	25.11	86.70							2987	34.5
Athipattu	Well 8	8.99	59.55	0	0	0	0	0	0	0	0.0
D (T)	Head 8	1.89	23.86	0	0	0	0	0	0	0	0.0
	Tail 8	2.56	30.91	0	0	0	0	0	0	0	0.0
	Total	13.45	114.33							0	0.0

Note: Payments relate to the irrigation functionaries of Kammukutties and Thotties. Table 4.13 contd ...

E (H) = Elevated sluice head reach village; M (T) = Middle sluice tail reach village.

D (H) = Deep sluice head reach village; D (T) = Deep sluice tail reach village.

In Athipattu no irrigation worker was appointed and no payment was given by the ayacutdars.

In Kattalai village, one person was temporarily employed by the ayacutdars to help them during irrigation season. Actually there is no permanent IW in this village.

S-1 = Season 1; S-2 = Season 2; S-3 = Season 3.

Source: Sivasubramaniyan, 1995

to the head reach farmers than to the tail reachers. This situation helped the IWs get more wages and the IWs rendered more work to the head reachers. Actually in Alapakkam, in 1991-92 the supply available was more than 3 months to the head reachers and only two months to the tail reachers. Most of the latter let the crop wither away due to non-availability of water. But the head reachers reaped a good harvest, and paid well to the IWs. In Panniyur the tail reachers were in a disadvantageous position regarding availability of tank supply. As a result, they were assisted more by the IWs and paid them more than the head reachers. Hence, within the villages also wide variation in the availability of supply was reported.

In DMT between the non-well farmers head reachers contributed more amount in all the villages. This is to some extent due to the satisfactory work done by the irrigation workers in the head reach of the village *ayacut*.

Table 4.16: Payments of Irrigation Functionaries in DMT 1991-92

Name of sample villages	Sample hhs by type	Sample extent, hectares	Total extent, hectares	Actual payment latest year: Amount paid by sample farmers			Imputed value for all farmers			Imputed value of contribution (Rs)	
				S-1	S-2	S-3	S-1 5/3X4	S-2 6/3X4	S-3 7/3X4	for all farmers 8+9+10	per ha (11/4)
1	2	3	4	5	6	7	8	9	10	11	12
Vagai	Well 2	4.79	4.79	143	250	133	143	250	133	526	109.8
E (H)	Head 8	1.88	9.84	67	89	41	350	465	214	1030	104.6
	Tail 8	5.28	11.31	136	224	176	291	480	377	1148	101.5
	Total	11.95	25.94							2704	104.2
Punaithangal	Well 8	8.27	28.71	225	493	243	781	1711	843	3336	116.2
E (T)	Head 8	2.94	22.18	44	216	174	332	1628	1311	3270	147.5
	Tail 8	3.33	7.44	32	51	54	72	114	121	306	41.2
	Total	14.54	58.33							6912	118.5
Narasamangalam	Well 8	6.27	13.01	66	254	62	137	527	129	792	60.9
D (H)	Head 8	4.45	16.84	40	182	43	151	689	163	1003	59.6
	Tail 8	2.29	14.72	7	90	29	45	578	186	810	55.0
	Total	13.01	44.57							2605	58.4
Sothiyampakkam	Well 8	10.34	29.05	205	620	165	576	1741	463	2780	95.7
D (T)	Head 8	3.55	15.30	26	268	38	112	1155	164	1431	93.5
	Tail 8	5.29	9.50	73	361	57	131	648	102	882	92.8
	Total	19.19	53.85							5092	94.6

Note: Payments relate to the irrigation functionaries of Kammukutties and Thotties.

E (H) = Elevated sluice head reach village; E (T) = Elevated sluice tail reach village.

D (H) = Deep sluice head reach village; D (T) = Deep sluice tail reach village.

S-1 = Season 1; S-2 = Season 2; S-3 = Season 3.

Source: Sivasubramaniyan, 1995

Mostly within the village in the head reach of the *ayacut*, to some extent the availability of water supply is secured and the IWs easily manage the supply in that portion. This is not possible in the tail reach of the village, especially during deficit supply periods. This induces the farmers to pay more to IWs. Within the village, in the non-well farmers group, only in Punaithangal a wide variation in the amount paid was noticed. This is due to the fluctuations in the availability of tank supply. That is, in this village, the tail reachers did not get proper supply for the past 10 years but the head reachers received regular supply.

The analysis shows that the amount paid for the IWs is far better in the villages where the irrigation institution is more active than in those where it is less active. In Panniyur (active) in order to maintain the tradition, each household was providing some food twice a day (morning and night) to the IWs. This shows that the *ayacutdars* do induce the IWs to work better by providing extra remuneration. In Contrast, in Athipattu (less active) as the functionaries were not at all paid even during the tank supply season they refused to work for the past two decades. It should, however, be noted that in Panniyur tank water is the main source for cultivation and the available groundwater is very limited. But in Athipattu groundwater is abundant and those farmers who have wells are not bothered about the tank supply.

4.4.1. Detailed Second Round Survey, 2016-17

It is important to note that the present survey conducted in 2016-17 in the same sample villages could not provide such an elaborate details of the functioning of irrigation institutions at the village level in both the tanks. The simple fact is that in no village irrigation functionaries are working now and the top level management functionaries such as Nattanmaikar and Kavaimaniyam are also ceased to function. This is simply because of three reasons: 1. Unusual long gap in getting the tank supply in both the tanks (after 2005-06 only after 10 years in 2015-16 both tanks got full storage); 2. Between the tank supply years, the TNFMIS Act-2000 based WUAs elections were held in 2006-07 in all system tanks in Tamil Nadu including KPT & DMT that made the appointment of new tank representatives called Territorial Constituency Members (TCMs) for all the sluices/ villages served by both tanks. The induction of new TCMs have made the work of the traditional informal irrigation functionaries futile in all the village. As a result,the traditional members abstained to do any tank related activities in recent years; and 3. Due to changes in the land holding structure, increasing factions in the villages, development of wells in the ayacut, intermittent and/or improper remuneration to village level functionaries; easy accessibility of jobs in towns and cities to the so called traditional irrigation workers with good remuneration all these get automatically stopped to do these irrigation related activities in both the tanks. Consequently, the aforesaid details were not simply applicable in both the tanks. In short, the water management and maintenance activities at present are solely undertaken by ayacutdars themselves and wherever assistance is required temporary people's services within villages or from outside are utilised in all the villages in both tanks. Also, machinery use for initial channel maintenance was utilised in both tanks where the main channels are passing for a long distance for this contributions are jointly made in cash by all users of the tank supply.

4.5 Functioning of Irrigation Institutions

4.5.1 Criteria of Effectiveness

The effectiveness of irrigation institutions (IIs) at various levels can be measured by two important activities: proper maintenance of all the channels supplying water to the tank and to the *ayacut* and allocation/ distribution of water from the tank. These two activities are very important in deciding the availability of tank water to individual farmers located in various reaches of the system. Proper maintenance of inlet/main/branch/ field channels enhances the storage available to the tank and facilitates smooth flow of water to all parts of the ayacut. Fair allocation of available tank supplies among different villages especially during scarcity periods ensures a fair share of water to the villages located in the tail end of the system. Fair distribution of available supply within the village during scarcity period reduces conflicts among the farmers', ensures more equitable sharing of available supply and may enhance the productivity of water.

The effectiveness of functioning of IIs may be judged in two ways: one is to ascertain the opinion of the *ayacutdars* about the working of irrigation institutions at different levels. The other and more objective way is on the basis of the number and character of irrigation functionaries, and the extent to which specific maintenance and allocation functions are in fact performed. The survey tried to assess the former with the help of an opinion survey of a sample of farmers in selected villages under both the tanks. In doing so, since it is expected the assessment to differ between head and tail reach and well, and non-well farmers it is divided the farmers in the *ayacut* of each village on this basis and chose eight farmer households from each stratum randomly.

As already indicated effective functioning of irrigation system requires multi-level cooperative efforts (cooperation at village, tank and bureaucracy levels). In order to understand the actual functioning of irrigation institutions (IIs) it is elicited the opinion of the respondents on the level of functioning as they experienced it. The responses are categorized in four types as shown below the Table 4.17.

Between the tanks, the functioning of village level institutions and the elected body of WUA is seen to be better in DMT than in KPT. In DMT the bureaucratic institution (PWD) is not functioning well. A few respondents did not know about the functioning of either the WUA or the PWD to take care of tank irrigation matters. This lack of knowledge was relatively more in KPT. This is mainly because the WUA was not functioning for the past few years since there was no tank water available. However, the village elders and the former irrigation management functionaries are informally helping the farmers to handle the release of tank water for the requirements of the *ayacutdars* in accordance with conventional procedures which was followed in the past. Only the TCMs of WUA have direct contact with the PWD for the release or closure of tank water supply. As the *ayacutdars* did

Table 4.17: Functioning of Irrigation Institutions in KPT and DMT sample Villages, 2017

Name of tank and sample villages	Sample hhs by type	Opinion on the Functioning of										
		Village Organisation			Water Users Associations				PWD			
		GF	NS	NF	GF	NS	NF	NC	GF	NS	NF	NC
Kaveripakkam tank												
Kadaperi	Well 8		3	3		4	4		4	4		
E (H)	Head 8	1	3	4		5	1	2	6	1		1
	Tail 8		2	6		4	4		6	1		1
Panniyur	Well 8	4	3	1	2	4	2		8			
E (H)	Head 8	3	3	2	3	2	1	2	6	1		1
	Tail 8	3	3	2	5	1	1	1	8			
Kattalai	Well 8		5	3		3	5				8	
D (H)	Head 8		4	3		4	4				8	
	Tail 8		4	4		4	4			1	4	3
Alapakkam	Well 8		5	3			4	4		7	1	
M (T)	Head 8		4	4		2	5	1		6		2
	Tail 8		4	4			6	2		2	1	5
Athipattu	Well 8		5	3			7	1			7	1
D (T)	Head 8			8			8				8	
	Tail 8			8			8				8	
Dusi-Mamandur tank												
Vagai	Well 2	1	1		2					2		
E (H)	Head 8	7	1		7	1				6	2	
	Tail 8	4	4		5	3				7	1	
Punaithangal	Well 8	1	4	3	4	2	2			5	2	1
E (T)	Head 8	6	2		7	1				3	5	
	Tail 8		3	5	1	4	3			5	3	
Narasamangalam	Well 8	1	6	1	2	6				5	3	
D (H)	Head 8	3	1	4		7		1		5	2	1
	Tail 8		5	3	1	5	1	1		1	4	3
Sothiyampakkam	Well 8	6	2		5	1		2	1	6	1	
D (T)	Head 8	5	2	1	8					5	2	1
	Tail 8	4	4		8					5	1	2

Note: GF=Good functioning; NS=Not Satisfactory; NF=Not Functioning; NC=Not consulted or No idea.

E (H) = Elevated sluice head reach village. E (T) = Elevated sluice tail reach village.

D (H) = Deep sluice head reach village. D (T) = Deep sluice tail reach village.

M (T) = Middle sluice tail reach village.

Source: Field survey data, 2017.

not have direct contact with the PWD, the frequency of these reporting "not functioning" or "not consulted" is more in KPT. In DMT, whenever water is available from the Palar, the local Farmers' Association and also the elected WUAs are keen to divert it into the tank through the Rajakkal sluice. Most of the respondents reported good functioning of WUAs.

Across villages, regarding village level organisations and WUAs, "good functioning" was reported only in Panniyur village. Even in Kattalai village the opinion is on the negative side in all three institutions. This is happened because, although KPT has got full storage but the deep sluice of Moolai was fully damaged due to broken down of sluice shutter. Consequently, the entire sluice was closed for irrigation. Even after the water level receded, either the PWD or the delegates of WUA or the village functionaries took efforts to restore the sluice. So, the villagers are deprived of KPT supply throughout the year. In Kadaperi and Alapakkam the response "not satisfactory" and "not functioning" are high. The tail end of Athipattu is an exception where the respondents opined that all the three levels of institutions are "not functioning" (see Table 4.14). Regarding performance of PWD "good functioning" was reported in the two villages under the elevated sluices, but "not satisfactory" and "not functioning" in Kattalai and Alapakkam villages. This is because in 2015-16 in KPT the FTL supply was maintained for more than a month and during this period supply was also released for cultivation. As a result the Kadaperi farmers reaped one full crop without facing any shortage of water and Panniyur farmers received full supply for more than six months.

Alapakkam villagers reported specific problems which were not resolved by the PWD. For example, the main supply channel which passes to Alapakkam was blocked while constructing a bridge (to lay a road to Maganipattu village) across this supply channel by the Block Development Officials of Kaveripakkam. Repeated requests to both the PWD and the BDO to remove the obstruction and to make way to supply water as usual did not yield any result. Consequently, even in the first tank supply season, the villagers did not get adequate supply. As a result, some of the villagers were not satisfied with the PWD's functioning.

In DMT in all the villages (except Narasamangalam) most of the respondents reported "good functioning" of village level institutions and WUAs. Very effective functioning of village level institution was reported in Sothiyampakkam. However, in all the villages, irrespective of reaches (whether head or tail) the responses regarding the functioning of PWD was "not satisfactory". This collective response about the PWD is mainly due to its inefficient functioning both for maintenance of supply channels and allocation of water through the sluices.

In DMT maintenance of supply channels is supposed to be done by the PWD. But for a long time it did not do this. In 2015-16, a large quantity of floods received by the Palar could not be adequately diverted by the *ayacutdars* as usual to DMT. This was not because of any labour shortage for the diversion of river flow, but owing to the PWD's negligence to supervise the inlet channel. For instance, at the intake point of the supply channel, immediately below the Rajakkal head sluice, a lot of debris was accumulated that blocked the flow of supply heavily. This blockagewas persisted for more than a week. Only by the ayacutdars efforts, this block was removed to let water to DMT. A knowledgeable person (one of the contractors of this section office) who has close contact with the PWD, attested that the diversion of water from the Palar river to DMT was fully undertaken only by the *ayacutdars* under the supervision of the prominent land holders as well as the TCMs of WUAs. However, after the completion of the diversion work, the PWD prepared an estimate for the cost of diversion work, submitted it to the government and the PWD officials themselves allegedly misappropriated the sanctioned amount.

Another important reason reported by the respondents for the "not satisfactory" functioning of PWD is that in the 2015-16 supply year, DMT got FTL storage. This is adequate to feed the *ayacut* for about 12 to 14 months. To utilize this water proper maintenance of main channels are a must. But in almost all the main channels thick bushes are grown and in many places the breadth of the channels were encroached heavily all these problems led to inadequate supply to the tail end villages and elevated lands. These situations forced the farmers to think of the poor working condition of the PWD in tank management.

In Punaithangal the respondents, especially at the head reach, reported good functioning of the PWD. Punaithangal is a tail end village. Normally it gets supply from the tank surplus and the drainage water available from the upper reach *ayacut*, and only a limited quantity of supply through the tank sluice reaches them. In the supply year 2015-16, due to heavy encroachment of channels only limited supply was available that water finally reached the *Anjuran Kondam* located seven kms away from DMT. From this *Kondam* only Punaithangal and a few other tail end villages get water supply to their *ayacut*. Punaithangal villagers (only the head reachers) received some tank supply (for two crop seasons with facing some water shortage) in 2015-16. This fact probably induced the respondents to report the functioning of PWD is not satisfactory.

Wherever there is either deficit or excess tank supply available, farmers attribute it mainly to inefficient functioning of the WUA or the PWD; whereas those villagers who received adequate supply generally replied that it was due to good functioning of the institutions. The main objective of the institutions is proper maintenance of irrigation facilities and equitable distribution of available water for crop production across different reaches of the *ayacut*. When this primary objective is not met satisfactorily, negative answers are possible from the *ayacutdars*, especially those, whose location is disadvantageous. On the whole, the village level institutions and the WUAs are to some extent functioning better in DMT than in KPT, while the functioning of PWD is seen to be better by KPT *ayacutdars* than by the DMT *ayacutdars*.

The effectiveness of IIs is also assessed on the basis of whether maintenance of inlet/main/field channels was done by the villagers in the supply year 2015-16 when water was available: whether *murai* system is in vogue; and whether irrigation functionaries such as *Nattanmaikar, Kavaimaniyam, Kammukutti* and *Thotti* are active. The relevant information was obtained from the village leaders / knowledgeable persons and also from the TCMs. As may be seen from the Tables 4.18 and 4.19 IIs are active in three KPT villages (one each in head and middle reaches and in head reach of elevated sluice); less active in six villages (three in head reach; two in middle reach, and one in tail reach) and in the remaining five

Table 4.18: Functioning of Irrigation Institutions in KPT, 2015-16

Name of village and reach	*Functioning of I Is*					*Number of Functionaries*				*Activeness of IIs*
	Maintenance of Inlet / Main / Field ---- Channels -----				*Preva-lence of murai*	*Kav-aima-niyam*	*TCMs of WUA**	*Kam-mukut-ties*	*Thtti-ies*	
Kadaperi	HR	N	Y	Y	N	0	1	0	0	A
Kaveripakkam	HR	N	N	Y	N	0	9	0	0	LA
Kondapuram	MR	N	N	Y	N	0	3	0	0	LA
Sirukarumbur	TR	N	N	N	N	0	2	0	0	D
Athipattu	TR	N	N	N	N	0	2	0	0	D
Eralacheri	TR	N	N	N	N	0	1	0	0	D
Cheri	HR	N	Y	Y	N	0	9	0	0	A
Kattalai	HR	N	N	N	N	0	2	0	0	LA
T.P. Pakkam	TR	N	Y	N	Y	0	4	0	0	LA
Maganipattu	MR	N	Y	Y	Y	0	4	0	0	A
Alapakkam	MR	N	Y	Y	Y	0	2	0	0	LA
Puthupattu	TR	N	Y	N	N	0	2	0	0	D
Panniyur	HR	N	Y	Y	N	0	3	0	0	A
Kilveeranam	TR	N	N	N	N	0	2	0	0	D

Note: N indicates that no maintenance was work done by the villagers and no murai system was followed during2016; A = Active(3/4); LA = Less active (2/4); D = Defunct (0/4). Y indicates maintenance work was done in the supply year 2015-16 and the functioning of murai system was effective in 2015-16 wherever tank supply was available. HR = Head reach; MR = Middle reach; TR = Tail reach.

* indicates Territorial Constituency Members of the WUAs elected in 2006-07.

Source: Interview with Knowledgeable farmers in the KPT villages, December 2016.

Table 4.19: Functioning of Irrigation Institutions in DMT2015-16.

Name of village and reach		*Functioning of I Is*				*Number of Functionaries*				*Active-ness of IIs*
		Maintenance of Inlet / Main / Field ----- Channels -----			*Preva-lence of murai*	*Kav-aima-niyam*	*TCMs of WUA **	*Kam-muk-utties*	*Thot-tiies*	
DUSI SLUICES 1 and 2										
Vagai	HR	Y	Y	N	Y	0	1	0	0	A
Dusi	HR	Y	Y	Y	Y	0	3	0	0	A
Kurankanilmuttam	HR	Y	Y	Y	Y	0	2	0	0	A
Pallavaram	TR	N	Y	Y	Y	0	2	0	0	LA
Kanikiluppai	MR	Y	Y	Y	Y	0	1	0	0	A
K.N. Palayam.	MR	Y	Y	Y	Y	0	1	0	0	A
Menallur	MR	Y	Y	Y	Y	0	1	0	0	A
Senianallur	TR	Y	Y	Y	Y	0	1	0	0	A
Punaithangal	TR	N	Y	N	Y	0	2	0	0	LA
MAMANDUR SLUICES 3 and 4										
Mamandur	HR	Y	N	Y	N	0	6	0	0	LA
N. Mangalam.	HR	Y	N	Y	N	0	1	0	0	LA
Valavandal	MR	Y	Y	Y	N	0	2	0	0	A
Vadakalpakkam	MR	Y	Y	Y	N	0	1	0	0	A
Girijapuram	MR	Y	Y	Y	N	0	1	0	0	A
Sothiyampakkam	TR	N	Y	Y	N	0	1	0	0	A
Bagavanthapuram.	TR	N	Y	Y	N	0	1	0	0	A
Elacheri	TR	N	N	N	N	0	4	0	0	D
Alingilpattu	HR	N	N	N	N	0	1	0	0	D

Note: N indicates that no maintenance work was done by the villagers and no murai system was followed during 1988; A = Active (4/4); LA = Less active (2/4); D = Defunct (0/4). Y indicates maintenance work was done in the last supply year (1985) and the functioning of murai system was effective in 1985. HR = Head reach. MR = Middle reach. TR = Tail reach.

* indicates Territorial Constituency Members of the WUAs elected in 2006-07.

Source: Interview with irrigation workers in the DMT villages, December 2016.

tail end villages it is defunct. They are active in 11 DMT villages (three each in head and tail reaches, five in middle reach); less active in fourvillages – two each in head and tail reaches and defunct in three villages one each in head, middle and tail end villages.

The preliminary survey indicates that the functioning of village level institutions varies between the two tanks. In KPT, after the construction of Palar Anicut, the maintenance of inlet channel was undertaken by the PWD. Hence the farmers have not been involved in that work. However, in DMT the ayacutdars still deploy large labour force to divert the flood flows from the Palar river to the tank through the Rajakkal channel.

Although in theory the PWD is wholly responsible to divert the flows, its efforts have been inadequate to fill the tank in any supply year. Therefore, the farmers themselves gather under the leadership of village headman/

TCMs of WUA and the members of WUA to perform this important task during every supply year. In KPT no village has taken part in this task since there was no necessity. In DMT most of the villages (except the tail reach farmers in Mamandur - deep – sluices and two tail end farmers in Dusi sluices) have usually actively participated in this task. Since the ayacut in Alingilpattu is very less (less than 5 ha) they were not shown interest in participation. However, in 2015-16 supply year two of the head reach villages, viz., Mamandur and Narasamangalam had also participated in the diversion of the Palar water to DMT.

Non-participation by farmers in one head and tail villages under sluices 3 and 4 is mainly due to location factor. It was reported by the tail end farmers that in the past even after they had performed their share of work, they did not get adequate supply which led to crop failure. Hence they are not interested to participate. On the other hand, the head reach farmers in these sluices are in a favourable position regarding tank water supply. Even if the tank gets only half of FTL supply the lands in these head reach villages receive ample supplies. Hence they do not bother about maintenance or diversion work. The actual problem of these head reachers is drainage during the supply period. However, since there was a long gap in the availability of supply to the tank these two villagers (Mamandur & Narasamangalam) also participated in 2015-16 supply year. In Alingilpattu village the farmers do not have the right to irrigate lands directly from the channel, but can bale (lift) water to irrigate some 4.52 hectares of land. Hence they do not participate in the maintenance either.

All villages under KPT, except three tail end villages Sirukarumbur, Athipattu and Kilveeranam reported participating in maintenance of main and branch channels during every supply year. However, the head reach of Kattalai village is not done the maintenance. This is due to the specific problem pertaining to the village of sluice closure due to break down of the shutter. Hence no tank was possible and hence farmers did not do the work. This is only a temporary phenomenon. In DMT all the villages reported involvement in maintenance during every supply season. However, if the supply is inadequate, the tail-enders do not participate as they are not likely to get enough supply for the entire crop season.

4.6 Murai System and Effectiveness of IIs Under the Tanks

Table 4.18 provides that in KPT none of the head reach villages have any regulation to ensure equitable distribution of water during the scarcity period. The only exception is Panniyur, irrigated by an elevated sluice, where a turn system (called *Murai*) is in vogue. In all the middle reach and in one of the tail reach villages, the *murai* system is practicedwhen there is shortage of water available to the villages. The practice has been discontinued in the other five tail end villages, Athipattu, Sirukarumbur, Eralacheri, Puthupattu and Kilveeranam.

In DMT under sluices 1 and 2, even the head reach villages follow a system called *Oru Madai Paichal*. Although these villages are located at head reach, the supply of water available is poor compared to the head reach villages in sluices 3 and 4. Further, available groundwater is also limited compared to the needs of *ayacutdars*. Hence farmers take extra efforts to distribute the available water fairly among the *ayacutdars*. In contrast, none of the head reach villages under sluices 3 and 4 follow the *murai* or *Oru Madai Paichal*; two of the head reach villages being located very close to the tank get plenty of supply; the third village does not have the right of irrigating lands directly (see Table 4.19).

All the middle and tail reach villages (except Elacheri) actively follow the *murai* system. It is less active in Elacheri because of very poor availability of tank water during the crop season. Even in normal supply years, there was practically no tank supply towards the end of the crop period which led to crop failure. It should be mentioned that in both the tanks wherever the *Kavaimaniyam* exists the *murai* institution was also functioning well except in Athipattu, and Sirukarumbur in KPT. There access to tank water is poor due to the tail end location of the village in the Fox sluice. However, as it is situated adjacent to the Palar river the available groundwater is tapped to raise crops. As a result, the farmers do not bother too much about the supply of tank water. This situation also weakens the interest in rationing tank water.

4.7 Physical Factors Affecting Irrigation Institutions

Apart from the institutional factors, some of the physical factors come in the way of effective functioning of IIs. Important among these factors are the location and sources of supply; siltation and growth of weed in the tank and in supply channels; encroachments in the supply channels, foreshore area, tank bed, and in the catchment area.

4.7.1 Location and sources of supply

One of the factors affecting the functioning of IIs is the situation of tank from its supply source. The availability of supply is more dependable from tanks (like KPT) located close to the river or its main or branch channel than from tanks (like DMT) situated at the tail reach of the system. This is true of KPT and DMT. For instance, KPT is situated at the head reach of its single source supply channel, the Kaveripakkam main channel, taking off from the Palar Anicut system, whereas DMT is fed by three sources. However, DMT is located at the tail end of its three supply sources. The supply available from these channels is also inadequate to fill the tank. Although the PWD is responsible to divert the supply to both the tanks, it does not discharge this responsibility completely and DMT has not been filled frequently. Hence the *ayacutdars* themselves divert the flood flows into the tank. As this necessity does not arise for KPT *ayacutdars,* the II in it is less active.

The crucial point is that if the *ayacutdars* (in DMT) properly contribute their labour to divert the flood flows during the flood season, they will definitely get enough water to the tank. If this is not done in time (within the flood period) the tank will not get enough supply. Moreover, in DMT while the ayacutdars divert the flood flows from the Palar through the Rajakkal channel, they do not do so from the other two sources, namely, the Palar anicut (Dusi-Thennampattu) source and the Thandarai source as these sources supply very limited water to DMT after filling all the upstream tanks. This clearly indicates that the IIs are active and work well if the benefits arising out of their efforts are substantial.

4.7.2 Siltation and Weed Growth in the Tank and in the Supply Channels

Accumulation of silt and the growth of weeds in the supply channels are major problems inhibiting the smooth flow of water to the tanks. If silt is not removed at least once a year the volume of water flowing into the channel would be reduced considerably.

To cite an illustration, a water course (one foot width) passing a length of 270 meters from the supply source (5 Horse Power Electric motor fitted well) would take about 15 minutes to reach the land when the channel is free from silt and weed. If the channel is not desilted and deweeded for two tothree months (assuming that every day a few hours supply goes through the channel), the time required for water to reach the land gradually increases during the three months period, andit takes about 45 to 50 minutes to cover the same distance at the end of three months. Further important point is that in the former case, if the motor is stopped immediately after the water reaches the land, the passing water in the channel would irrigate six to eight cents of land. But in the latter case, if the same is done, it would irrigate only four to six cents. This clearly shows the enormity of water absorbed by the channel perpetually because of the non-removal of silt and weed growth.

In the light of the above information, it may be stated that water passing through the wide and long supply channels like Rajakkal and Kaveripakkam (about 24 to 27metres width) for a distance of about 21 kms and 8 kms respectively from the supply sources, when not desilted and deweeded for many years could cause enormous wastage of water. Maintenance of these channels is vested with the PWD. But it has not performed its function properly for several years. The reported reason is that available funds are grossly inadequate. In the case of DMT though the *ayacutdars* are active to divert the flood flows into the Rajakkal channel, they do not attend to any maintenance work such as desilting and deweeding in the Rajakkal supply channel. A visit along this channel in April 2017 showed that most parts of the distributary are silted and huge bushes cover the channel reflecting the prolonged neglect of maintenance. The KPT supply channel was not as bad as the DMT supply channel. The siltation problem is not only in the supply channels but also in the tank beds (as discussed in chapter 3). KPT is silted up by more than 5 metres at the deepest sluice. In DMT there has not been any systematic silt survey so far. But enquiries with the TCMs of WUA and visit to the tank bed confirm that this tank is also silted to a depth of about 2.5 metre around the deepest sluice.

Although the IIs are changing gradually but working favorably at the village level in both the tanks, they did not make much effort to maintain the tanks and its supply channels against siltation and weeding. Only at the village level, the distributaries and channels are periodically desilted whenever the tanks received enough water. In a few tail end KPT villages (e.g. Kilveeranam,Athipattu, Sirukarumbur, Puthupattu and Thuraiperumpakkam), and also in DMT (Elacheri village) the distributaries are fully silted up, because between the period 2006-2015 for 10 years the water available in the tanks was so much below average. Thus the tail end villages were unlikely to get any water and therefore did not do any maintenance. In all the other villages under both the tanks maintenance is done regularly to get the tank water for irrigation and to raise the water table in the wells; and also, to some extent, to solve the drinking water problem in the villages.

4.7.3 Encroachment in the Tank and Channels

The encroachment problem was not so severe some five or six decades ago. The interviews with the then *Kavaimaniyam, Nattanmaikars*, the IB delegates and the TCMs of the WUAs confirmed that only after the 1960s, the encroachment problem increased in the *ayacuts* under both the tanks. It was pointed out earlier that the increasing demographic pressure led people to encroach upon the adjacent distributaries. This reduced the width of channels over time. It was reported by all the villagers in both the tanks that the width of present distributaries/channels is only half of their original width. In some tail end villages the entire field channels were included as land by the adjacent land owners. As a result of this encroachment, the tail end villagers face more problem as the channels are not able to carry even half the quantity of supply which they did earlier. This has reportedly aggravated conflicts between the head and tail reach farmers under both the tanks. Due to removal of even field channels by adjacent farmers the actual ayacutdars did not irrigate their land even though the tank has its Full Storage in 2015-16.

These encroachments are not confined to the channels but also spread over in the tank bed and foreshore area. This is especially so under Kaveripakkam tank bed, where about half the extent of tank bed is cultivated with banana routinely (see photos in p. 423). Neither the PWD nor the local village administration is powerful enough to prevent this problem. The head reach villagers mostly do the tank bed and foreshore area cultivation.

4.7.4 Effectiveness of IIS with Sluicewise and Villagewise Density of Wells under KPT

Below the Table 4.20 the villages have been arranged into head, middle and tail reaches and the sluices are arranged into deep, middle and elevated, according to their elevation in the tank. As may be seen from the Table the two tail reach villages, Athipattu and Sirukarumbur, are located close to the Palar river which helps recharge the wells effectively. This reliable well irrigation induced the farmers not to depend much on tank water. However, other two tail end villages, Eralacheri and Puthupattu do not have other source facility like the river supply. Even if they work hard to get the tank supply the head and middle reach farmers prevented to get the required supply.So, the II is defunct in these two villages.However, they depend only on their well supply for cultivation. But the available groundwater is inadequate from most of the villages. Hence the well density is moderate.

The three elevated sluice villages, Kadaperi, Panniyur and Kilveeranam, always get inadequate tank supply. Even if the water reaches its FTL they

Table 4.20: Sluicewise, Reachwise Ayacut and Wells Particulars and the Activeness of Irrigation Institutions in Kaveripakkam Tank, 2017

Name of sluice	Reach	Village	Head Reach			Reach	Village	Middle Reach			Reach	Village	Tail Reach			Total ayacut Ha	Total wells	Wells/ 100 ha
			ayacut Ha	Total wells	Wells/ 100 ha			ayacut Ha	Total wells	Wells/ 100 ha			ayacut Ha	Total wells	Wells/ 100 ha			
Moolai	HR	Kattalai (LA)	129.88	74	57	MR	Cheri (A)	73.08	54	74	TR	K.v.p. (LA)	114.05	34	30	317.00	162	
Cheri low level	HR	Cheri (A)	88.30	42	48		-	-	-			-	-	-		88.30	42	
Lion	HR	K.v.p. (LA)	318.10	102	32	MR	-	-	-		TR	Eralacheri (D)	130.93	41	31	449.03	143	
Total			536.28	218	41			73.08	54	74			244.98	75	31	854.33	347	41
M.pattu & T.p.p.	HR	Cheri (A)	27.65	21	76	MR	Maganipattu (A)	204.01	129	63	TR	Puthupattu (D)	96.15	103	107	327.81	253	
Cheri high level	HR	Cheri (A)	92.51	58	63	MR	Alapakkam (LA)	85.63	76	89	TR	T.p.p. (L.A)	190.61	260	136	368.74	394	
Fox	HR	K.v.p. (LA)	129.68	49	38	MR	Kondapuram (LA)	213.68	162	76	TR TR	Eralacheri (D)	49.23	27	55	392.59	238	
											TR	S.k.bur. (D)	173.56	238	137	173.56	238	
												Athipattu (D)	113.72	148	130	113.72	148	
Total			249.84	128	51			503.32	367	73			287.29	386	134	1376.4	881	64
Paracheri	HR	Cheri (A)	16.48	14	85		-	-	-			-	-	-		16.48	14	
Panniyur	HR	Panniyur (A)	29.23	56	192		-	-	-		TR	K.veranam (D)	37.25	65	175	66.48	121	
Gaffor	HR	Kadaperi (A)	27.09	37	137		-	-	-			-	-	-		27.09	37	
Total			72.79	107	147		-	-	-				37.25	65	175	110.04	172	156
Surplus kondam			15.51	2	13		-									15.51	2	13
Catchment			13.32	-												13.32	-	-
Grand total			887.73	507	57		-	576.40	421	73			569.51	526	92	2369.6	1795	76

Note: LA=Less Active; A=Active and D=Defunct functioning of irrigation institution.

HR=Head reach; MR=Middle reach; TR=Tail reach.

T.P.P = Maganipattu and Thuraiperumpakkam; K.v.p. = Kaveripakkam; E.cheri = Eralacheri. S.k.bur = Sirukarumbur; K.veranam = Kilveranam.

Source: Adangal Registers (2017) of the KPT villages.

get supply only for one crop season. So, there is a need to depend on wells. Hence the well density is more. In these villages higher density combined with poor tank water supply makes the village level tank management institution less active in Kadaperi, active in Panniyur and defunct in Kilveeranam.

Some other villages located at the tail and middle reaches, however, have very low well density. For instance, Alapakkam, Maganipattu, Puthupattu and Thuraiperumpakkam are situated mostly in the tail region and the well density ranges from 63 to 136 per 100 ha of ayacut. The moderate to high growth of wells in these villages is due to very limited availability of sub-surface water. This situation has necessitated the villagers to work actively to get the available water from the tank. So the IIs in these villages are active to less active compared to other villages.

Among the head reach villages, Kaveripakkam has a low well density (32 per 100 ha of ayacut) but is very favourably located to get water supply from the three deepest sluices. Hence the dependence on well water is less. Since the available supply from the tank is assured the IIs are less active.

The foregoing analysis shows that if a village is located far from the tank or irrigated by an elevated sluice, the well per acre is more. However, there are exceptions: the density is low in a few tail end villages and high in a few head reach villages. Generally, the functioning of IIs varies considerably according to the availability of tank supply.

4.7.5 Effectiveness of IIs under DMT

The reach wise analysis (below the Table 4.21) shows that the well density gradually increases from head to middle to tail reaches in both the Dusi and Mamandur sluices, except in the head reach of Mamandur sluices. However, the density of wells in the Dusi sluices (46 per 100 hectares) is lower than in Mamandur sluices (65 per 100 hectares). Further Dusi sluices are located three feet above the sill of the Mamandur sluices. As a result comparatively less duration of tank supply is available to the villages under the former sluices. These two (low well density and the elevated sluices) are perhaps main reasons for the active cooperation among *ayacutdars* in all the villages under Dusi sluices to get enough tank supply.

In all the head reach villages served by the Mamandur sluices the tank management institution is less active and it is defunct in Alingilpattu, where the well density is the highest (272/wells 100 ha) compared to all the villages

Table 4.21: Sluicewise, Reachwise Ayacut and Wells Particulars and the Activeness of Irrigation Institutions in the Villages under Dusi-Mamandur Tank, 2017

Sl. no. Name of village and reach	Reach	Ayacut in 2017 <--- Hectares --->		Wells 2017	Total wells	Wells/ 100 Ha	Avg. no. of wells	Act- ive ness of IIs
Elevated and Deep Sluices 1 and 2								
1 Vagai	HR	26.19 }		5 }		19 }		A
2 Dusi	HR	180.66 }		43 }		24 }		A
3 Kurankanilmuttam	HR	72.58 }	279.433	37 }	85	51 }	31	A
4 Kanikiluppai	MR	44.92 }		14 }		31 }		A
5 Kilnayakanpalayam	MR	102.68 }		34 }		33 }		A
6 Menallur	MR	66.81 }	214.413	75 }	123	112 }	59	A
7 Pallavaram	TR	163.44 }		71 }		43 }		LA
8 Senianallur	TR	41.55 }		31 }		75 }		A
9 Punaithangal	TR	59.16 }	264.13	37 }	139	63 }	60	LA
Total			757.98	347	347	46		
Deep and Elevated Sluices 3 and 4								
10 Mamandur	HR	400.98 }		133 }		33 }		LA
11 Narasamangalam	HR	44.68 }		43 }		96 }		LA
12 Alingilpattu	HR	5.52 }	451.2	15 }	191	272 }	134	D
13 Valavandal	MR	95.50 }		70 }		73 }		A
14 Vadakalpakkam	MR	93.22 }		39 }		42 }		A
15 Girijapuram	MR	47.70 }	236.4	39 }	148	82 }	66	A
16 Sothiyampakkam	TR	54.52 }		76 }		139 }		A
17 Bagavanthapuram	TR	43.17 }		61 }		141 }		A
18 Elacheri	TR	132.40 }	230.1	123 }	260	93 }	125	D
Total			917.69	599	599	65		
Grand total			1384.21		946	68		

Note: A=Active; LA=Less active; D= Defunct. HR=Head reach; MR=Middle reach; TR=Tail reach.

Source: Updated settlement (1983) and Adangal (2017) registers of the DMT villages.

in Mamandur sluices. This is mainly because all these villages are located very close to the tank and get favourable tank supply. For instance, as the Mamandur village is located very close to the tank and access to tank supply is favourable to the *ayacutdars* hence they do not follow any *murai* system. In two of the tail end villages the institution is active. In the remaining one it is defunct. The villages Narasamangalam and Alingilpattu being situated at elevated levels depend more on wells for water supply. However, if the tank gets adequate storage they also get enough supply first. As a result, channel maintenance is less important, as is the need for any *murai* system. So, the working of IIs for managing tank water is less active.

The tail end villages being located far away from the tank, they get only a season's tank supply if the tank storage reaches between 7.5 and 9.0 metres. Two of these villages, Sothiyampakkam and Bagavanthapuram, though locationally disadvantageous, are fed by a separate supply channel from DMT. They can get enough supply at least for a crop whenever DMT reaches FTL storage. Local involvement in maintenance and water allocation is therefore more than in Elacheri village which gets a season's supply only when the tank surpluses.

In all the middle reach villages the local arrangements for maintenance and for water allocation are active. The middle reach villages are situated below Mamandur and form a chain. Supply of tank water to these villages is significantly affected by the maintenance of channels. Although well density in a few villages in Mamandur sluices is higher than in Dusi sluices, it does not affect the functioning of institutions. The main reason is that these wells get recharged only with tank supply. So, the farmers do their best to maintain the channels. Most of these villages follow either the *murai* or the *Oru Madai Paichal* system during scarcity period, when the tank supply gets decreased below normal level.

Summary

In this chapter we have discussed the structure and functioning of IIs in KPT and DMT. The discussion has shown that the system level organisation (PWD) does not discharge its responsibility for proper maintenance of main channel of the system and the inlet channel of particular tanks to ensure adequate supply. In both the tanks, the informal organisation at the tank level has been replaced by the formal WUAs which are represented by the TCMs. When there is enough supply in both the tanks, majority of the village level organisations function actively.

There have been considerable changes in the village level organisation and of both the tanks and their functioning. The transfer of land from the upper castes to the lower castes has meant the disappearance of the traditional congruence between social and land holding hierarchy at village level, and to some extent, tank level organisations.

Subdivision and fragmentation of land holdings, and encroachment of water courses, tank beds and catchment areas - all of which reflect growing demographic pressure - have affected the functioning of the tank irrigation system. The prolonged neglect of maintenance of supply channels together with encroachment have considerably reduced the storage capacity and supply of water to the *ayacut*.

As a result of transfer of a major portion of land from the landlords to the tenants, the number of owner cultivators has increased substantially. This has had its effects on the organisational set-up in the villages. Upper caste dominance in land control has significantly declined in both the tanks; with the Nayakars emerging as the single largest caste, controlling of (two-fifth in KPT and three-fifth in DMT) total *Pattas*. The Scheduled Castes acquired a sizeable proportion of land ownership in a few tail end villages under both the tanks. Changes in land control, and the large increase in number of *ayacutdars* must have made enforcement of rules, previously vested in the dominant and powerful upper castes more difficult.

In the decade 2005-2015 supply of water in the tanks was much below average. Therefore, the level of activity in collective maintenance and regulation of water allocation in this period were much weaker than normal. The reduction of flow to the tank has also resulted in some changes in cropping pattern, with the farmers growing only paddy whenever the tank has got full storage. With the expansion of well irrigation there has also been a shift in cultivation from paddy to sugarcane and banana.

This chapter also presents the findings of a detailed survey of a sample of farmers in selected villages: The survey shows that participation in maintenance is widespread in both the tanks; but more in DMT than in KPT; and farmers contributed substantial amount to secure water supply both for maintenance and for payment to irrigation functionaries. However, the concept of remuneration to be given to IWs fully disappeared in both tanks and only ayacutdars themselves maintaining the channels based on their contributions. In both tanks both types of irrigation functionaries – supervisory and lower level - ceased to function, mainly because of long gap in the availability of tank supply, changes in crop pattern and irregular payments to IWs. Also, easy availability of job works in towns and cities altogether abolished these IWs activities.

Some of the physical factors such as location of tanks from the supply sources, siltation and encroachment in the supply channels, foreshore area and in the tank bed which affect the functioning of IIs have been discussed in detail. It is found that the tail end location of DMT from all its supply channels resulting to very poor supply to the tank. But in KPT the favourable location benefits to get adequate supply always. The only advantage of tail end location of DMT is that the siltation in the tank bed is comparatively less than that KPT. Since KPT is situated at the head reach of the chain system it has far more siltation problem. The tank bed at present is silted up to 5 metres at the deepest sluice which considerably reduced the storage capacity.

The development of wells in tank *ayacut* has been a major factor in the working of IIs. Both the KPT and DMT confirm the conventional expectation of increase in well density as one moves along the head reach to the tail end. Further, if a village is situated too close or too far from the tank, the institution is less active according to the availability of supply. This is true in most KPT and DMT villages. One of the main reasons for the maintenance of channels by the *ayacutdars* even when the density of wells is high is to get at least a meagre supply from the tank, which helps recharge the wells.

Notes

1. For instance, the Kaveripakkam tank should not be filled above two feet below Full Tank Level until after 15th December. But there was seldom any appreciable supply after this date. So, the tank was being kept at less than its full capacity. For several years after the escapes (Surplus Weirs) were improved and fitted with screw gearing shutters, these principles as existed before were not changed and only in September 1899 filling of the tank up to one foot below FTL before 15th December was permitted. The *Mamulnamas*, i.e., the rules and regulations stipulated by the *ayacutdars* for proper distribution of tank water below the tank sluices as well as for the augmentation of water to the tank from the sources also show these principles. The *Mamulnamas* actually came into existence during the 18th century and these rules and regulations are not suited to the present conditions. Nevertheless, the government has not altered these rules to suit the changed circumstances (GO 270 I, PWD, 24-08-1910, p. 7).

Chapter - 5

Maintenance of the Systems

Management of any irrigation system depends upon proper maintenance of the physical facilities including channels carrying water from the supply source to the individual plots, drainage networks, and regulatory structures. This task requires careful attention and cooperation of both the users and the administrators of the system. In this connection the following observations by Vaidyanathan are relevant:

> Inefficient maintenance could adversely affect the water deliveries by reducing the volume of water carried by the canals; slowing down the speed of water flows; increased waste due to leakage and spills; and in the extreme case a partial or total breakdown. All of these reduce the volume of water made available to the field, and hence the feasible level of production compared to the potential of the system (given the design). The quality of maintenance affects the interests of both the organisation as a whole (which is presumably interested in getting the maximum production with the available water) and the users (whose output and incomes are directly affected by it) (1985: 56).

Historically, the user communities themselves maintained tanks and other small water works in condition. In Tamil Nadu, under *Kudimaramath*, users were expected to contribute labour for upkeep and repair of irrigation facilities from time to time. Information available on the arrangements for repair and maintenance of tanks and their effectiveness before the advent of the British rule is scanty. It is widely believed that *Kudimaramath* worked well till the advent of the British rule, progressively deteriorated thence, and became derelict in recent decades. Recent studies by Vaidyanathan

(1984b, 1985, 1991b), Vaidyanathan and Janakarajan (1989), Janakarajan (1993b), Sengupta (1991, 1993) Sivasubramaniyan (1995, 2007, 2013, 2017), Rajendran (2018) and Venkatachalam and Kulbhushan Balooni (2018) suggest that though the traditional irrigation institutions have undergone major transformations, they are still active in a number of places. This is corroborated by this study also.

5.1 A General Historical Perspective

One of the important factors accounting for the regular up-keep of channels by the ryots under the ancient Kingdoms was the cultivators' obligation to pay the land rent or tax imposed by the rulers (the ryots were, in fact, the tenants of the State), and subsequently by intermediaries (especially the *Zamindars*) to whom the rulers assigned the lands. This placed the ryots in a position where they had to undertake a good deal of maintenance work of channels to get proper supply of water for irrigation of their lands. It continued till the abolition of the *Zamindari* Settlements. The *Zamindars* were believed to have enforced maintenance and restoration of tanks and channels:

> Of course the security of the revenue demanded that irrigation works should be attended to, and this attention took the form of enforcing the corvee, and if the official summons did not prove effective, the Amildar's "slippers, rattans, and jairbunds" were employed to bring a refractory ryot to a proper realization of his responsibilities (Raju 1941: 124).

The *Zamindars* initially excavated the irrigation tanks in many villages and kept them in repair at their own cost. This cost was later recovered from the ryots. The recovery was at the rate of 1 *toom* (a local measure of grains equal to 48 Seers) per *candy* of gross produce (45 Kg: 1 Seer = 0.933 Kg). When repairs were executed, half the cost was contributed by the villagers and paid to the *Zamindar*; the other half was borne by the *Zamindar* himself. This payment was called *Maramat Nesuff*. The British government also followed the same system when the *Zamindari* estates came under its management. It introduced money rents, and undertook the major repairs at its own cost. The villagers were required to undertake the customary upkeep.

The state of *Kudimaramath* that existed under the Company rule was described by Raju as follows:

> Tahsildars being expected to enforce it. The village peon, toti, *taliari* and others had to watch the tanks and watercourses and as soon as any

> damage was discovered, the Tahsildars called out the ryots of the village or villages concerned... The labour, however, was not exacted strictly and whatever the power of the Tahsildars, it was apparently not exercised everywhere. The system even in pre-British times was based on voluntary action, - communal if not individual - and was managed by the village authorities who constituted the most natural agency. But with the decay of village corporate life and the weakening of the power of village officers, *Kudi-maramut* fell into desuetude (1941: 124-5).

After the introduction of ryotwari settlements by the middle of the 19th century, the effectiveness of the traditional system deteriorated progressively. The state of *Kudimaramath* which prevailed during the dawn of the 20th century, the Irrigation Commission of 1901-03 reported that it was repeatedly informed by the government that the maintenance of tanks was not satisfactory, *Kudimaramath* so valuable an institution was practically dead[1] and past restoration (see Krishnaswami 1947: 449).

The Company government's several attempts (including legislation) for the revival of the *Kudimaramath* system did not succeed. The element of obligation which was present in the traditional system was weakened by changes in land tenure. Earlier, bulk of the land used to be controlled by the Zamindars or by a few *Mirasidars* who were powerful in the village community. The British administration also could not enforce the *Kudimaramath*. After taking over the Zamindari estates and introduction of the Ryotwari tenure the traditional conditions in force gradually weakened. This naturally inhibited the capacity to enforce *Kudimaramath* obligations. The government, therefore, assumed more and more direct responsibility for maintenance and repair, only to find that it could not provide adequate resources for the purpose. Whatever resources it diverted were only for the maintenance of inlet channels, tanks and associated structures, and not for facilities within the *ayacut* of each tank. Despite this there was gradual deterioration in the inlet channels, sluices, regulatory structures and so on which indirectly affected irrigation from tanks.

As the government did not take much interest in the maintenance of irrigation facilities, especially in the local tank *ayacut*, the *ayacutdars* themselves took much care to maintain the irrigation networks. There is evidence to show that elements of *Kudimaramath* are still active in the tanks. On this the observations in a survey conducted in Dusi village:

> In Dusi, the *Kudimaramath* or the desilting of the main supply channels is undertaken every supply year and is normally completed in two or

three days. The norm is for the landowners to ensure the availability of one or two workers, depending on the quantum of silt clearing involved, per acre of land that they own. Some of them arrange for it by compelling their tenants to contribute free labour or labour at a token wage (Guhan and Bharathan 1984: 35-6).

Against the background of the foregoing discussion, the following account gives the maintenance of distribution networks in the KPT and DMT *ayacuts*, where the *ayacutdars* themselves contribute labour for cleaning the channels.

5.1.1 Maintenance in KPT and DMT

Traditionally, the *ayacutdars* of both the tanks were expected to participate in the construction of the *Korambus* to divert the river-flow and ensure that the water so diverted was brought to the tanks. Upkeep of the inlet channel, bund, sluices and other distribution networks was also the responsibility of the *ayacutdars*. According to the *Mamulnamas*, all the villages benefiting by the tank must jointly take part in the maintenance. The maintenance of main channels taking off from the tank sluices to the final distribution points in several KPT and DMT villages was to be jointly done by the *ayacutdars* proportionate to their limit specification, i.e., villagers were jointly responsible to maintain the channels passing through their village limit. In some cases, channels fully irrigating a single village had to be maintained only by the *ayacutdars* of that village.

The practice of maintenance changed after the PWD assumed the responsibility. While the *ayacutdars* continued to be responsible for the distribution networks, the government took over the maintenance/ management of inlet channels, tank bed, tank bund, surplus weirs, sluices and associated structures. In KPT and DMT, the latter work was far from regular or adequate. Even essential works were done piecemeal and inadequately, sometimes once in a decade or so. It was reported that the maintenance works on the inlet channels of Dusi-Thennampattu and Rajakkal were done some two to three decades ago. Occasionally the PWD took up major repairs such as lining of the head sections of the main channels of the Palar Anicut, and reconstruction/repair of sluices. For the most part maintenance mainly consists of removing small trees and other vegetation growth inside the channel. The entire length of the inlet channels of both the tanks still has sand accumulation of two to three feet depth.

Nothing has been done about encroachment or siltation, which must have resulted in reduction in the carrying capacity of the channels.

Ayacutdars reported that they were increasingly taking part in the construction of the *Korambus* to divert the river-flows into tank through the Rajakkal channel. The condition of the channel is bad but the *ayacutdars* have not done much about it. The reasons for this reluctance could not be ascertained. One possible explanation is the magnitude of the work and the fact that it involves numerous tasks. There is no single authority to assign the tasks equitably among the beneficiaries and to attend to any breach of obligation. The *ayacutdars* have no power either to evict the encroachers along the channels.

5.2 Maintenance Activity in KPT and DMT, 2015-16

5.2.1 Survey of Village Functionaries

In order to assess the manner and extent of *ayacutdars'* participation in maintenance, a survey was conducted during the latest supply years 2015-16 in a few selected villages under both tanks. This survey covered five villages in KPT (two irrigated by elevated sluices, two served by the middle level sluices and one head reach village served by the deepest sluice) and four villages in DMT (two in the head reach and two in the tail reach). In the first instance, the survey elicited information from the irrigation functionaries such as Territorial Constituency Members and other prominent land holders and also the then Kavaimaniyam / Nattanmaikar of each village about the frequency with which maintenance was done in the last ten years and details of the manner of organising maintenance (for survey related photographs see p. 426-27).

5.2.2 Maintenance in the last 10 years (2005-06 to 2015-16)

As may be seen from below the Table 5.1 except Kadaperi and Punaithangal, all the villages reported that maintenance was undertaken less than 10 times in a decade, ranging from 8 times in Sothiyampakkam in DMT to 0 time in Athipattu in KPT. Generally, during the first-time maintenance in a supply year both kinds of operations (desilting and deweeding) were jointly done. From the second time onwards in a particular year, only deweeding was undertaken. Whenever villagers undertook seasonal or monthly maintenance, the frequency of maintenance was more. Where maintenance was done annually, the frequency of maintenance was less.

5.2.3 Maintenance of Distribution Networks in the Ayacut

There are three types of channels under the tanks, viz., main, branch and field. The maintenance of main and branch channels was done jointly

Table 5.1: Extent of Encroachment and Frequency of Maintenance in the Channels Served by the Sample Villages, 2017

Name of tank and sample villages	Extent of encroachment				Effect of encroachment on the main channel	Frequency of mainte-nance. Main/Bra-nch chls.	Mainte-nance done in last 10 years	No. of years of tank supply in the last 10 years
	Main channel		Branch channel	Field channel				
	Outside village	Within village						
1	2	3	4	5	6	7	8	9
KPT								
Kadaperi E (H)	-	2' to 3'. 1.5 km from the sluice.	3' to 4' in many places	Nil	Breach immediately below the tank sluice affects the supply to tail enders	Once in 15 days	18-20 times in 2 supply years	2
Panniyur E (H)	Earlier 15'; now 7-8'.	Upto 3' in the bunds.	2' to 3' in many places	Nil	Only half the level of original supply is available now.	Seasonal	8 Times in 2 supply years	2
Kattalai D (H)	-	Earlier 25'-30'; now 10'-15'	2' to 3' in many places	No field channel	No problems reported. No KPT supply received in 2015-16.	Seasonal	3-4 Times during 2005-06	1
Alapakkam M (T)	Earlier 20'; now 5-10'.	Upto 4' in the bunds.	Limited Encroa-chment only.	Nil	Main channel supply is reduced cosiderably, water hardly reaches the tail end	Seasonal	6 Times in 2 supply years	2
Athipattu D (T)	Earlier 25'; now no channel	No Tank supply; no channel	No Tank supply no channel	No problem	Encroachment is a old story and no tank supply is possible now.	No tank supply in the last 2 decades	No tank supply received	0
DMT								
Vagai E (H)	6' Earlier; now 3-4'.	1' to 2' Encroa-chment.	No branch channel	No field channel	No problems reported	Yearly	6 Times	6
Punai-thangal E (T)	Bunds only encroached Upto 3-4'.	2' to 3' Encroa-chment.	No problems reported	No field channel	No problem in Palla-kalvai but no supply in the Mettukkalvai	Monthly	10 Times	3
Narasa-mangalam D (H)	-	Earlier 15'; now 5' to 7'.			Only tail end villagers are affected	Yearly	2 Times in 2 years	2
Sothiyam-pakkam D (T)	No problems reported (Separate main channel)	No problems reported	No problems reported	In some places	No problems reported since separate channel is existing.	Monthly	8 Times in 2 years	2

Note: Nil refers to no encroachment in the field channel; Dash indicates no main channel passes outside the village.
E (H) = Elevated sluice head reach village; E (T) = Elevated sluice tail reach village.
D (H) = Deep sluice head reach village; D (T) = Deep sluice tail reach village; M (T) = Middle sluice tail reach village.
Source: Village level survey, 2016-17.

by all the *ayacutdars* in a village, whereas the field channels were maintained individually. The responsibility for the maintenance of any main or branch channel passing through several villages vested with all the concerned villagers. For instance, consider a main channel passing through the villages A, B and C as shown below the figure 5.1 and finally reaching village D. During the maintenance period 'D' villagers have to maintain the main

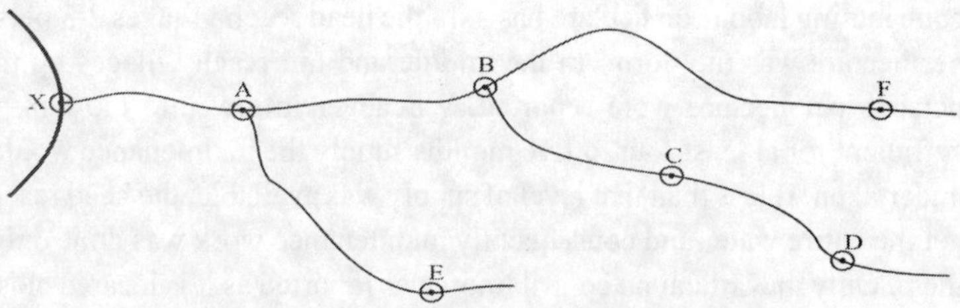

Note: X = Tank sluice; A, B, C and D are the location of villages in the main channel; E and F are villages under branch channels.

Fig. 5.1 Course of Main/Branch Channels and Location of Villages

channel separately up to the 'C' village limit and then jointly with the C, B and A villagers up to the tank sluice. In effect the 'D' villagers have to cover a longer distance and contribute more labour than all the other villagers. By contrast villages located very close to the tank cover the least distance, i.e., from the village limit to the tank sluice. One would expect this difference to be more marked in tanks like KPT and DMT where the main channels are long and serve many villages. Branch channels serve mostly one or two villages. Here again the responsibility of maintenance is shared by the *ayacutdars*. Clearly the *ayacutdars* of each village have to participate in maintenance work of both main and branch channels.

In all the sample villages, maintenance of channels was done only when the tank got supply of about 4 to 5 months storage. Since this was not available regularly, no maintenance work was undertaken in many years. The gap in maintenance was much wider in the tail end villages than in the head and middle reach villages. The reason for this was that some of the tail end villages did not take part in the maintenance unless the tank got more than half-full tank level storage. Whenever the channels were maintained regularly, the available tank supply was distributed fairly in the villages. If it was not done regularly, a lot of silt accumulated due to animal/people movements and natural calamities. Further, there was a lot of vegetative growth and in some places small trees were also found even in the middle of the channel. As a result, the maintenance of channels became very difficult at a later stage when the tank supply was available after a long break like the present situation.

In all the villages (except the head reach village of Kattalai in KPT) the maintenance of main/branch channels was done by the *ayacutdars* contributing labour on hectare basis. In the head reach villages 2-3 persons per hectare was the norm. In the middle and tail reach villages up to 10 persons per hectare were reported. Whenever inflow into the tank was sufficient for at least four to five months supply the maintenance would be undertaken. If less than that level of supply was available, the head reachers get the entire water and consequently maintenance work was done only by them. Only in Kattalai no contribution was reported as it is located close to the tank and gets abundant supply. However, unfortunately, during 2015-16 supply year Kattalai village did not get the tank supply fully due to closure of the deep sluice since its shutter was broken even when the KPT started getting its initial storage. Consequently, the number of maintenance done pertains only to the previous supply year (2005-06).

A system called the *Kolmurai Kanakku* (measuring by rod) is generally used for the maintenance of main and branch channels by the *ayacutdars* in each village. A five feet long stick is used for this purpose. The length of the channel for which the *ayacutdars* in each village have to work depends upon the breadth of the channel and the number of persons allotted to maintain the specific section of the channel. Previously the *Kammukutties/Thotties* measure the length of main/branch channels under the supervision of *Nattanmai/ Kavaimaniyam*. Now this practice is handled in some villages by the TCMs of the WUAs. For instance, in a village if 60 persons are available for maintenance, they have to divide themselves into three groups of 20 each. Then the TCMs or his representative allot the distance of three or four "stick" lengths (15 to 20 feet) to each group for maintenance.

The work is done in two ways. One is to divide the total number of effective workers (persons who actually participate) into several sub-groups and start the work to finish it in about a week before release of tank water. The other is to divide the total length of channel among the *ayacutdars*. For instance, a channel passing to a longer distance is divided into four parts. The total participants are also divided into four groups and each group is made separately responsible for each part. Which method is to be chosen would depend upon the urgency of water need faced by the *ayacutdars*?

In the case of main channels passing long distance from the tank, farmers located farther from head reach face more problems in receiving the supply of water. The head reachers (whenever they get good supply from that channel) do not always cooperate in maintenance work, because locational advantage enables them to get enough supply even when the tank is partially filled. As a result, the tail reachers are required to contribute more labour for the maintenance of channels. This situation was found in the Lion sluice and Fox sluice channels (between the head reach of Kaveripakkam and the tail end of Eralacheri villages) in KPT. However, this was the case during the 2005-06 supply year. After that Eralacheri was not getting any tank supply. In order to solve this more number of labour requirement for maintenance, in the 2015-16 supply year many villagers contributed money to do the initial maintenance by machine power called the Poclain to undertake the maintenance. Villagers informed that this new technology is the best option to complete the maintenance quickly and ayacutdars' also contribute money for this purpose without fail. In both the tanks, Poclain use was reported for maintenance purpose. In DMT, Sothiyampakkam ayacutdars used the Poclain initially during this supply year since their channel leads for a long distance of 4 to 5 kms.

The above practices are reported to be in operation only when the available water in the tank is sufficient to feed the entire command of a given sluice. If less than full supply (adequate for four to five months) is available, the tail end farmers tend to abstain from the maintenance work since they surmise that water would not reach their lands even for one month out of the four or five months of crop season. This "less-than-normal" supply is mostly utilised by the head and to some extent by middle reachers. When the prospects of getting supplies are not clear (because the tank level is inadequate to serve all sluices or all segments of particular sluices), there is little interest in collective maintenance. This affects normally the tail enders more than the head reachers.

When the supply from the tank is inadequate to feed the tail end of *ayacut* continuously for few years - as happened during the 1980s and also between 2005 and 2014 - the tail end farmers appear to lose interest in participating in maintenance. But it is not to be assumed that there is a permanent loss of interest. In such situations the head reach farmers contribute labour for maintenance, even if tail enders do not, since they only getting the benefits of meagre tank supply. Thus, it is found that farmers of Kaveripakkam and Kondapuram in the head reach of Fox sluice participated in maintenance while those of Sirukarumbur and Athipattu at the tail reach did not. It should be noted that although Athipattu got the tank supply in 2005-06 but it is not in 2015-16. The same problem was also reported in the Panniyur sluice, where the head reach of Panniyur villagers maintain channels routinely during supply years, whereas the Kilveeranam villagers abstain from taking water in the last 4 to 5 decades due to inadequate availability of tank supply even for one full crop.

In DMT also the tail end villagers of Elacheri, Bagavanthapuram and Sothiyampakkam did not get involved in maintenance whenever storage in DMT was less than 25 feet. Pallavaram also did not take part in maintenance work unless the tank received more than 18 feet of water. Moreover, mostly due to locational advantage, all the head reach villagers who followed the direct irrigation method did not participate in maintenance work.

More importantly, as the latest trend as noticed in the 2016-17 survey year, especially in Kaveripakkam tank six of the tail end villages (Sirukarumbur, Athipattu and Eralacheri irrigated by Fox sluice; Puthupattu and Thuraiperumpakkam served by Maganipattu & Thuraiperumpakkam sluices; and Kilveeranam served by Panniyur sluice) did not take tank water due to long (10 years) gap in the availability of tanks supply during this supply year and also all these villages located at the tail end of their respective sluices.

There were also instances of farmers in a village who did not use the tank water abstaining from the maintenance work. This was reported in Sirukarumbur, Athipattu and Kilveeranam in KPT and Elacheri in DMT where for the past 10 years between 2005 and 2014 the farmers did not avail tank water. The situation has been continuing even now in these villages even the tank got its full storage in 2015-16. Available tank supply being grossly inadequate, the *ayacutdars* in Sirukarumbur exclusively depended upon well irrigation to raise crops. Since the main Fox sluice supply channel was fully closed at Kaveripakkam Village Athipattu farmers also did not get the tank supply. In Kilveeranam more than half the total *ayacut* is owned by an individual farmer who was not interested to contribute his share of labour for maintenance but was ready to take water if others do the maintenance. As a result, none of the *ayacutdars* participate in the channel maintenance or take water.

Except Kadaperi and Athipattu, all the remaining villages in KPT undertook the channel maintenance once in a season (3 to 4 months duration). In Kadaperi it was done once in 15 days. In this village, as already mentioned, the farmers could be classified into two types: (i) direct KPT irrigation farmers (the head reachers) and (ii) the river channel farmers (tail enders). The latter get the KPT supply whenever the river channel fails to supply water. Only during that time, the river channel farmers join the maintenance work of the KPT main channel to get the supply.

However, the survey indicates that most of the lands of river channel farmers are sold out and made as residential plots in recent years. Consequently, only the head reach farmers strive hard to do the maintenance. In order to prevent the breach, which occurred immediately below the tank sluice, they were doing the maintenance properly once in 15 days. However, previous, the head reach farmers of Gaffor sluice usually did not take part in the channel maintenance because they got adequate supply. In DMT the channel maintenance is done once in a supply year in the head reach of Vagai and Narasamangalam villages, and once in a month in the tail reach villages which face water scarcity. The proximity of the head reach villages in DMT makes it unnecessary for the farmers to undertake the seasonal maintenance.

The situation at the time of our survey, reflected the non-availability of supply for several years continuously. Continued neglect has led to silting the channels. Whenever dry crops were raised in the *ayacut*, the adjacent *ayacutdars* tend to encroach the channels. This results in a distribution problem when the supply is reduced. Once the channel is encroached, it is difficult to evict the encroachers. Instances of this were found in almost all the KPT and DMT villages.

Field Channels: In the case of field channels, broadly speaking (a) if a channel exclusively irrigates one individual's land, the full responsibility for maintenance entirely depends upon that land holder. This type of channels was found in the *ayacut* in a few villages some decades ago, but is rare at present (because of the subdivision of holdings and plots); and (b) if two or more persons get supply from a channel then they jointly undertake the maintenance from the take off point to their land. Most channels carrying water from the branch channels to the fields in the *ayacut* of each village belong to this category. The maintenance of these channels is reported to be difficult now-a-days mostly due to non-cooperation among the *ayacutdars*.

In most of the KPT villages well demarcated field channels exist, whereas in a majority of the DMT villages (especially in the head reach) they do not exist. Generally, the pattern of irrigation in DMT villages is field to field, which according to villagers accentuates the problem of the tail enders in securing adequate water supply. This type of unscientific irrigation has resulted in poor yield creating an imbalance in the command. The efficiency expected to accrue from fertilizer application is not achieved and there is considerable wastage of water.

In almost all the villages, compared to the branch channels, the main channel (both within and outside the village) was considerably encroached by the adjacent farmers (Table 5.1). The severity of the problem in the main channel was specifically reported by Alapakkam and Athipattu villagers at the tail end of KPT. They reported that the head reach villagers had encroached the main channels to a large extent. Prominent land holders of Athipattu village reported that along the bunds of the main channel near Kaveripakkam village houses have been built over the last decade and the main channel is used as a latrine and for related purposes. As the channel is unclean, most of the farmers are unwilling to undertake maintenance in that portion (about 0.5 km) of the main channel. However, as cleaning is a necessity to get proper tank supply, previously the *Nattanmaikar* forced the *Thotties* to clean it, who did it under the influence of liquor. The PWD and the village administrators of Kaveripakkam have not attended to this problem even after several petitions were made by the *ayacutdars*. Although this was the case of 2005-06 supply year, the National Highways was developed around 2010 in the Chennai-Vellore Highways that passes across Kaveripakkam village, in which the KPT irrigation channels have been passing. Without bothering, the highways people fully closed that main channel during laying of extended highways that prevented the full supply to Athipattu. No authority, either from the PWD or from the Revenue bothered about this situation. The same problem also happened to Eralacheri village where a portion of main channel was encroached for this highway purpose. Consequently, many of these villages are removed from getting KPT supply in 2015-16 supply year.

In Sothiyampakkam, a tail end village, no problem of encroachment was reported, since this village is having a separate supply channel starting from the tank sluice. In all the other villages in both the tanks at least some extent of encroachment was reported. In the case of branch channels, the problem of encroachment was not severe. In DMT none of the sample villages reported any such problem in the branch channels. In the case of field channels, a limited number of farmers only reported the problem of encroachment in a few villages (more details are given in chapter 7).

Due to encroachment in the main/branch channels, the supply available to the tail end villages was reduced considerably. The head reach villages did not report such a problem (See Table 5.1). In Kadaperi due to encroachment in the bunds of main channel immediately starting from the sluice tended to breach whenever the supply was increased through the sluice. The closing of breach required a lot of labour, which was mobilised from the tail end of the *ayacut* by the *Kavaimaniyam* till 2005-06. Now, the work is fully done by the head reach ayacutdars themselves, since the tail end lands are sold out for housing purpose. This breach closing requires immediate remedial measures, but mobilising the ayacutdars was very difficult, but it is managed. This affected the supply considerably to the ayacutdars. In Panniyur and Alapakkam only less than half the quantity of supply is available now compared to the original designed level.

5.2.4 Ayacutdars Participation, 2015-16

Village level information was collected from the previous irrigation functionaries, prominent land holders, knowledgeable farmers and the TCMs of WUAs in the selected villages based on labour/money contribution, number of days/man-days work done, penalty levied, collected and utilised and the expectation regarding PWD's work on the main/branch channels. This information is summarised below the Table 5.2.

Except Kattalai and Athipattu in KPT all the remaining villages in both the tanks reported per hectare basis of labour contribution for maintenance work. The head reach villages contributed 3 persons per ha in both tanks (except in Vagai where it is 3 per ha). While the tail end village of Alapakkam in KPT contributed 5 persons per ha, the same in DMT being 6 persons in Punaithangal and 4 persons in Sothiyampakkam. The less number of contribution is compensated by more money contribution per ha (Rs. 450/ha) in Sothiyampakkam. Although in both the tanks the distance between the tank sluices and the location of tail end villages is about equal (5 to 6 Kms.) the basis of per hectare labour contribution varied between the tanks and was 2 to six persons per hectare. One probable reason for this variation is that in KPT the village level organisation is less active to persuade the *ayacutdars* to do the maintenance, due to better assured supply and more number of wells, whereas in DMT it is more active and levies a spot fine if the *ayacutdars* do not fulfill the norms and requirements. If the defaulters did not pay the fine, they would not be allowed to take water. It is also important to note that the number of persons required to maintain the channels depended on the breadth of the channels, number of hours work to be done in each day and the quality of work.

In the previous decade, the *Nattanmaikar/Kavaimaniyam* was an important person to enforce the rules and regulations for channel maintenance but now this activity is shifted to the TCMs of the WUAs in both tanks. In Kadaperi and Panniyur villages in KPT and Punaithangal and Sothiyampakkam villages in DMT the then *Nattanmaikars* were very

Table 5.2: Labour and Money Contribution for the MAintenance of Main/Branch Channels, 2015-16.

Name of tank and sample villages	Basis of contribution		Season II		Season III		wage rate/ man-day Rs.	Pena-lty per day Rs.	Amount collected (Rs) Season		Total Amount (5+7)X8+ (10+11)	Ave-rage Amount per ha. Rs.	Purpose of fine utilisation	Expectation of PWD's work
	per-sons per ha	Amt. per ha-Rs.	Times	Man-days	Times	Man-days			II	III				
1	2	3	4	5	6	7	8	9	10	11	12	13	14	15
KPT														
Kadaperi E (H)	3	NIL	9	75	-	-	300	100	400	-	22900	845	1 Chl.maintenance, 2 Temple festival.	1 Lining the main chl. 2 1st Time maintenance before the release of tank supply.
Panniyur E (H) E (H)	3	NIL	4	88	1	15	300	50	750	-	31650	1084	1 Chl.maintenance.	<---------- do ---------->
Kattalai D (H)	NIL	NIL	*	*	*	*	NIL	-	-	-	*	*	-	Sluice should be in good order.
Alapakkam M (T)	5	NIL	3	180	-	-	300	-	-	-	54000	631	-	1 To take steps to provide a separate main channel. 2 Lining the main chl.
Athipattu D (T)	NIL	NIL	*	*	*	*	NIL	-	-	-	*	*	-	1 Lining the main chl. Upto a certain distance and in some vulnerable places.
DMT														
Vagai E (H)	3	NIL	1	75	NIL	NIL	300	-	-	-	22500	859	-	Drainage facility must be created.
Punaithangal E (T)	6	300	3	130	2	50	300	100	1300	200	63000	1482	1 Temple festival, 2 Village expenses	Kondam and diversion weirs should be strengthened.
Narasamangalam D (H)	2	NIL	1	90	1	40	300	-	-	-	39000	872	-	Drainage facility must be created.
Sothiyampakkam D (T)	4	450	4	210	-	-	300	100	800	-	85100	1564	Equal distri-bution among participants.	Lining the main chl. Is a must.

Note: Dash indicates no supply in the third season and no penalty levied or collected and utilised.

* indicates no tank supply received in 2015-16 supply year.

Money contribution in Punaithangal (Rs. 7500) and Sothiyampakkam (Rs. 22100) is clubbed into the total amount.

Most of the well farmers and the very small land holders located at the tail end were not participated.

E (H) = Elevated sluice head reach village; E (T) = Elevated sluice tail reach village.

D (H) = Deep sluice head reach village; D(T) = Deep sluice tail reach village; M (T) = Middle sluice tail reach village.

Source: Village level survey, 2016-17.

effective to enforce the rules and levy the fines from the defaulters. The appreciable aspect is that still some farmers able to recall their work efforts and these persons also presently cooperate with the TCMs to carry out the maintenance adopted in the earlier pattern. Moreover, wherever the previous *Nattanmaikars* still alive their suggestions are also taken note of in maintenance activities. In Kadaperi, although the organisation was effective among *Athukal Paichal* farmers, who located at the tail end of the ayacut, most farmers sold out their land for housing. Hence only head reach farmers undertake this activity. The highest performance is seen in Sothiyampakkam village, where if a person does not take part in the maintenance in a day, he is required to pay the fine immediately before the next day's work starts. Otherwise he is not allowed to take part in the maintenance work later or to take water. This practice was also reported in Punaithangal village.

By contrast, the in Alapakkam village the TCM was less active. Though he is a prominent land holder in that village, elected in the 2006-07 election, he does not know much about water management and has very poor administrative/organisational skill. As a result, he was not able to organize the *ayacutdars* for effective channel maintenance. Further factionalism in the village between Reddis (the head and middle reach ayacutdars) and Scheduled Castes land holders prevented to do any joint activity to effectively maintain the main / branch channels to get the tank supply. Hence the villagers did not get adequate tank supply to the whole of the ayacut –but received only the head / middle reach farmers in a single season in the supply year 2015-16. However, when storage was at FTL, Alapakkam village could have got a supply for two seasons.

The two head reach villages, Vagai and Narasamangalam in DMT, did not deploy much labour for channel maintenance. This was due to locational advantage. Vagai did the maintenance only once at the beginning of the supply year, while Narasamangalam did so only a month before depletion of tank supply, that too with a limited number of persons to utilise it. Although this pattern prevailed even now, but in KPT the Kattalai village did not get the tank supply wholly due to collapse of Mulai sluice, hence no maintenance was done by the ayacutdars in 2016-17 supply year.

Compared to the two elevated sluice head reach villages, Kadaperi and Panniyur, the contribution from Panniyur village is more. This is mainly because assurance in tank water supply for full crop if the maintenance is carried out. This is because, the other tail end village of Kilveeranam village of the same Panniyur sluice did not take water for many decades. Hence if the maintenance is done the full benefit is only for Panniyur farmers. So their contribution for maintenance is more.

In KPT, two of the selected villages did not get the tank supply wholly in 2015-16 supply year. In three villages, Panniyur in KPT and Punaithangal and Narasamangalam in DMT, the *ayacutdars* were involved in the maintenance in both the supply seasons (viz season II and season III). All these 3 villages contributed labour for third season also. In season II, the contribution was the maximum in all the labour participating villages. The maximum labour contribution is in Sothiyampakkam (210 mandays) in DMT followed by Alapakkam in KPT. In the remaining villages in both tanks the contribution was between 75 and 130 mandays. Apart from labour contribution, in Punaithangal (Rs. 7500) and Sothiyampakkam (Rs. 22100) money contribution was also collected. In the third season, mostly one-time maintenance was reported and the maximum labour used was in Punaithangal village (50 man-days) followed by Narasamangalam (40 man-days) in DMT. The least contribution of 15 mandays was reported in Panniyur (15 man-days in KPT). The maximum number of (nine) times work done was reported in Kadaperi village in KPT. Minimum labour input was used in Vagai village of about 75 man-days in only one time and no penalty is levied for non-participants for maintenance (see Table 5.2).

Levy of penalty for the non-performance of maintenance was reported only in four villages, two each in KPT and DMT. However, only in the DMT villages penalty was strictly enforced and collected. The penalty amount was utilised in three ways: (1) Channel maintenance; (2) Temple and other village level festivals; and (3) Equal distribution of the balance amount among all the participants. The nature of utilisation of the penalty was, however, decided by the village level meeting.

Overall, valuing labour contributions at the going wage rate (2015-16) in the tanks the *ayacutdars* contributed between Rs. 631 and Rs. 1084 per hectare in KPT; whereas the same being Rs. 859 and Rs. 1564 in DMT. In KPT the tail end village contributed more than in elevated sluices. Much the same pattern of head-tail difference is noticed in DMT villages.

5.2.5 Expectation Regarding PWD's Work

According to government decision the PWD is responsible to maintain the system up to the tank sluice. However, enquiries with the prominent landholders / then irrigation functionaries / TCMs of WUA reveal that its work goes beyond the tank sluice, i.e. to construct different types of diversion structures in the main outlet channels passing through several villages.

Most of the tail end villagers expect that the PWD must do the lining of the main/branch channels up to a certain limit or if the lining is not possible, to undertake the maintenance of main/branch channels at least during the first time before the release of tank water during every supply year. To avoid water logging the head reach villagers insisted on drainage facilities to be provided in the entire village *ayacut*. More or less the same opinion was given by the sample farmers in both tanks during the 2016-17 survey.

5.2.6 Details for Two Villages

The data presented in Table 5.2 are based on information provided by the prominent landholders / then irrigation functionaries / TCMs. For the most part, it is based on recollection from their memory rather than records. None of the head reach villages (three in KPT and two in DMT) including those served by elevated sluices keep any records of maintenance work. Of the two tail end villages in KPT records for Alapakkam were inadequate and Athipattu had no records at all. However, in two villages (Punaithangal and Sothiyampakkam) of DMT detailed records of labour contribution, number of days work done and the total number of man-days of work are available. The following account is based on these records.

The total *ayacut* in Punaithangal village is 59.1 hectares, which gets DMT supply in two ways. The *Pallakkalvai* starting from the Dusi high level sluice irrigates the low-level lands of about 32 hectares. The remaining 27 hectares generally called *Mettukkalvai Paichal* lands are irrigated by the Mamandur high level sluice. Because of factionalism between the Mudaliars and Scheduled Castes, no maintenance was done in the *Mettukkalvai* and it did not get any water supply since 1981. This situation is also continuing even during the 2016-17 survey. The lands under this channel are divided equally between the Mudaliars and the Scheduled Castes; but as the Mudaliars have own wells they do not bother about the maintenance of the channel. Only the *Pallakkal* lands are getting regular supply. Hence the available maintenance details pertain to the *Pallakkal paichal* lands only.

As Punaithangal is the tail end village the *ayacutdars* need to contribute more labour for maintenance. During the maintenance period the villagers are expected to contribute 16 persons per day for maintenance. In the initial stage many labourers were required (see below the Table 5.3) since the channels were heavily silted and in many places, there was intensive vegetation. The maintenance of the main channel passing outside the village required two days work and it was repeated once a month. The main/branch channels passing within the village required the same labour input but took

Table 5.3: Labour Contribution for Maintenance of Channels in Punaithangal, 2015-16.

Name of village and month	*Main channel outside the village*			*Main/Branch channel within the village*			*Total man-*
	Days	*Per-sons*	*Man-days*	*Days*	*Per-sons*	*Man-days*	*days*
II Season: November-March							
1. November 2015							
Menallur }							
Senianallur }	1	28	28	-	-	-	28
Punaithangal }		16	16				16
Punaithangal	4	16	64	2	16	12	76
2. December 2015.							
Punaithangal	2	16	32	1	16	16	48
3. January 2016.							
Punaithangal	2	16	32	1	16	16	48
4. March 2016.							
Punaithangal	2	16	32	1	16	16	48
Total	**11**	**80**	**176**	**5**	**64**	**60**	**236**
III Season: April-June							
1. April 2016.							
Punaithangal	2	16	32	1	16	16	48
2. June 2016.							
Punaithangal	2	16	32	1	16	16	48
Total	**4**	**32**	**64**	**2**	**32**	**32**	**96**
Overall Contribution (II+III)	**15**	**112**	**240**	**7**	**96**	**92**	**332**

Note: Dash indicates nil; 1 man-day's work is of 6 hours, from 7 a.m. to 1 p.m. Season II total excludes the share of Menallur and Senianallur villages.
Source: From interviews with the then irrigation functionaries, prominent farmers and TCM of WUA in Punaithangal Village, 2017.

only one day. Overall, except the initial work at the beginning of the season, it took around 48 man-days to complete one round. However, it was reported by the TCM of WUA that by convention the allotted number of man-days (around 48) of work must be done during every maintenance. Further, he explained that while the number of persons required to work is fixed, the number of hours work to be done in each day is reduced after the first round. Generally, two to three hours work is sufficient in the second and subsequent rounds. Whether a person works effectively or not, he must be present during each time. If he is not, fine may be imposed. This peculiar system of labour contribution was reported only in this village which somehow seems to strengthen the spirit of organisational set up.

Data presented in Table 5.3 show that the labour requirement at the initial time was nearly twice (92 man-days) that of the subsequent maintenance (48 man-days). In the second season maintenance was undertaken four times. In the third season it was done only twice. By and large there has not been any deviation in the labour contribution for the maintenance of main/branch channels.

In Sothiyampakkam, even now the village level organisation was very strong and the then *Nattanmaikar* was also active and cooperate with the TCM of the WUA in enforcing the maintenance rules. The records were in good order, and a high degree of labour participation was observed in this village. Even at FTL storage in DMT this village can get only a single season's supply. This also motivates the villagers to work actively to get the available supply at least for that entire season. Further, although this village is located at the tail end of the tank, it has a separate supply channel called Sothiyampakkam channel (supplying water to Sothiyampakkam and Bagavanthapuram). As this is exclusively available for them and is not shared by the upstream villagers, it assures them that if they work actively they will get adequate tank supply.

During the single supply season in 2016-17 the villagers did the maintenance three times. The initial maintenance was done jointly with Bagavanthapuram villagers and the remaining two were done independently. The labour requirement at the initial stage was more than that of the subsequent stages (Table 5.4). Since the *ayacut* served by the main channel is more in Sothiyampakkam (56 hectares) than in Bagavanthapuram

Table 5.4 Farmers Participation in Maintenance of Main/Branch Channels in Sothiyampakkam, 2015-16

Name of village and month	*Main channel outside the village*			*Main/Branch channel within the village*			*Total man-days*
	Days	*Per-sons*	*Man-days*	*Days*	*Per-sons*	*Man-days*	
II. Season: November-March							
1. November 2015							
Bagavanthapuram	2	20	40	3	25	75	115
Sothiyampakkam	2	25	50	3	15	45	95
2. January 2016.							
Sothiyampakkam	1	40	40	2	22	44	84
3. March 2016.							
Sothiyampakkam	1	35	35	1	22	22	57
Total	4	100	125	6	59	111	236

Note: DMT supply to this village was available for only one season.
1 Man day is 7 hours of work from 9 a.m. to 4 p.m. with a meal break.
Total excludes the share of Bagavanthapuram village.
Source: From interviews with the then Kavaimaniyam, prominent farmers and TCM of WUA in Sothiyampakkam Village, 2017.

(44 hectares) the labour input is also high in the former. Generally, labour contribution norm of 4 persons per hectare was followed. For the second and third time maintenance the *ayacutdars* contributed 44 mandays and 22 mandays of work respectively. Overall, for the maintenance of main/ branch channels both within and outside the village, a total of 236 man-days (Sothiyampakkam only) of work was contributed to get the proper supply.

5.3 Farm Level Survey

Information on the maintenance of main/branch channels obtained from irrigation functionaries in selected villages was supplemented by a survey of selected farmers in each of these villages. This gives a more accurate idea of the extent of farmers' participation in the latest (2015-16) and the preceding (2005-06) supply years.

A summary account of the labour participation in maintenance obtained from the sample farmers is presented in Tables 5.5 to 5.8. These Tables show striking differences in labour participation in maintenance between KPT and DMT. As already noted, in KPT no farmer had involved in maintenance of inlet channel, whereas in DMT although the PWD is responsible for maintenance of all the inlet channels, the diversion of the Palar river floods into DMT through the Rajakkal sluice is routinely done by the *ayacutdars* themselves by constructing *Korambus* during every supply year. However, the maintenance of Rajakkal head sluice and desilting of the Rajakkal channel are occasionally done by the PWD.

5.3.1 Kaveripakkam Tank

In KPT in 2015-16, the proportion of sample farmers in three out of five villages, and most categories, participating in maintenance of main/ branch channels was very high (75%) and generally higher than the participation in maintenance of field channels. The difference is particularly marked in head and tail reach villages. In Panniyur and Alapakkam villages most participants reported contributing labour only once. In Kadaperi a relatively high proportion of participants reported contributing twice or thrice a year. In all sample villages, irrespective of reaches, contribution of own labour was more than hired labour. Only in Alapakkam, contribution of own labour was 100% in all category of farmers. Cash contribution was reported only Kadaperi and Panniyur villages, in which the former village contributed more than the latter. In Kadaperi instead of cash contribution by non-well head reach farmers they contributed more of labour compared to tail end category.

Except in Kadaperi, the proportion of sample well farmers participating in main/branch channels maintenance is not very different from non-well farmers, but the number of times they participate tend to be lower than non-well farmers. The former also tend to contribute in cash, when necessity arises. Contrary to expectation that the tail end farmers contribute more labour than the head/well farmers, the situation is against this view in 2015-16. The highest contribution of labour (in cash and cash equivalent of labour contribution at prevailed wage rate) for main/branch channels maintenance was reported in Kadaperi village (avg. Rs. 1278/ ha) followed by Alapakkam village (average Rs. 763 per hectare); here the head reach non-well farmers contributed more than well owners (except in Kadaperi). The lowest contribution of labour was in Alapakkam village (Rs. 311 / hectare); here again the non-well farmers participated more than the well farmers in the maintenance activity (see Table 5.5).

The estimates derived from the farm survey are broadly in line with those obtained from village functionaries. Elevated sluice villages contributed more to main/branch channel maintenance than the tail reach village. The absolute value of contribution in three of the five villages is also of the same magnitude.

Participation in maintenance of field channels varies considerably between tanks and across villages within the tank. All the sample farmers in Kattalai and Athipattu did not participate in this activity in 2015-16 since no tank supply was received. The extent of non-participation was also high in Kadaperi and Alapakkam. The high degree of non-participation in these villages was said to be the non-existence of field channels in Kadaperi and heavy encroachment (Alapakkam) of field channels in Alapakkam, especially in the tail reach of the village.

The survey shows that participation was widespread and maintenance work was done two or three times during the year in most of cases. The field channel maintenance work was exclusively done by the farmers themselves; no hired labour was reported in any of the sample villages. The number of man-days spent for field channel maintenance ranged from 3 mandays (tail reach) in Alapakkam to 10 mandays (well farmers) being the highest in the elevated sluice village of Panniyur. Invariably, in all the villages well farmers contributed more labour than the non-well farmers,

Table 5.5: Labour and Cash Contribution for Maintenance of Channels by Sample Households in KPT, 2015-16

(Cash in Rupees; Labour in Mandays)

Sample villages	Sample HHs by type	Extent of holdings		Main/branch channels								Field channels					Total Contribution			% of labour contribution (19/20) *100	Imputed value		Contribution for Main/ Br. Chls only	Imputed value	
		Sample	Total	No. of farmers participated				Labour		Total	Cash Only	No. of farmers participated				Own Labour	Labour only (11+ 17)	Converted into cash	By Sample farmers (19+ 12)		for all farmers in the ayacut (20/3)X4	Per Ha (22/4)		for all farmers in the ayacut (24/3)X4	Per Ha (25/4)
		<- Hectares ->		0	1	2	3	Own	Hired			0	1	2	3										
1	2	3	4	5	6	7	8	9	10	11	12	13	14	15	16	17	18	19	20	21	22	23	24	25	26
Kadaperi	Well 8	6.19	9.73	2	0	5	1	13	4	17	600	6	0	1	1	5	22	6600	7200	91.7	11308	1162	5100	8010	823
E (H)	Head 8	1.89	8.31	1	2	1	3	13	2	15	0	6	2	0	0	2	17	5100	5100	100.0	22458	2703	4500	19815	2385
	Tail 8	2.91	8.57	4	2	1	1	7	0	7	270	4	2	1	1	7	14	4200	4470	94.0	13143	1533	2100	6175	720
	Total	11.00	26.61																		46909	5399		34000	1278
Panniyur	Well 8	8.21	10.99	2	4	1	1	9	1	10	300		6	2	0	10	20	6000	6300	95.2	8431	767	3000	4015	365
E (H)	Head 8	3.42	9.64	2	4	1	1	9	2	11	120	1	6	1	0	8	19	5700	5820	97.9	16412	1703	3300	9306	966
	Tail 8	2.59	8.55	1	5	2	0	9	0	9	0		7	1	0	9	18	5400	5400	100.0	17856	2087	2700	8928	1044
	Total	14.21	29.18																		42699	4558		22249	763
Kattalai	Well 8	9.19	43.51	8	-	-	-	-	-	-		8	-	-	-	-	-	-			-	-	-	-	-
D (H)	Head 8	2.63	47.90	8	-	-	-	-	-	-		8	-	-	-	-	-	-			-	-	-	-	-
	Tail 8	3.32	36.90	8	-	-	-	-	-	-		8	-	-	-	-	-	-			-	-	-	-	-
	Total	15.13	128.31																						
Alapakkam	Well 8	16.34	22.12	3	4	1	0	6	0	6	0	2	4	2	0	6	14	4200	4200	100.0	5687	257	1800	2437	110
M (T)	Head 8	3.30	25.95	3	3	2	0	7	0	7	0	3	4	1	0	6	13	3900	3900	100.0	30669	1182	2100	16514	636
	Tail 8	4.79	32.43	5	3	0	0	3	0	3	0	5	3	0	0	3	6	1800	1800	100.0	12179	376	900	6089	188
	Total	24.43	80.50																		48535	1815		25041	311

(Contd...)

Athipattu	Well 8	22.26	59.55	8	-	-	-	-	-	-		8	-	-	-	-	-	-			-	-	-	-	-
D (T)	Head 8	4.59	24.35	8	-	-	-	-	-	-		8	-	-	-	-	-	-			-	-	-	-	-
	Tail 8	6.20	30.91	8	-	-	-	-	-	-		8	-	-	-	-	-	-			-	-	-	-	-
	Total	33.05	114.81																						

Note: In KPT the inlet channel maintenance has been fully undertaken by the PWD since 1858. Before the construction of Palar anicut (1858), the diversion of water from the Palar and the maintenance of the inlet channel was mainly undertaken by the ayacutdars themselves.

For field channel maintenance no cash contribution is reported. The Table is derived on the basis of weighted averages in each type of farmer groups.

0, 1, 2, and 3 as indicated in the Heading refers to the number of times the respondents participated in the maintenance activity. Dash indicates nil.

E (H) = Elevated sluice head reach village; D (H) = Deep sluice head reach village.

M (T) = Middle sluice tail reach village; D (T) = Deep sluice tail reach village.

Prevailed wage rate per manday in 2015-16 was: Rs. 300 for male members, who were participated in the maintenance activity.

Source: Field survey data, 2017.

Table 5.6: Labour and Cash Contribution for Maintenance of Channels by Sample Households in KPT, 2005-06

(Cash in Rupees; Labour in Mandays)

Sample villages	Sample HHs by type	Extent of holdings		Main/branch channels								Field channels					Total Contribution			% of labour contribution (19/20) *100	Imputed value		Contribution for Main/ Br. Chls only	Imputed value	
		Sample	Total	No. of farmers participated				Labour		Total	Cash Only	No. of farmers participated				Own Labour	Labour only (11+ 17)	Converted into cash	By Sample farmers (19+ 12)		for all farmers in the ayacut (20/3)X4	Per Ha (22/4)		for all farmers in the ayacut (24/3)X4	Per Ha (25/4)
		<- Hectares ->		0	1	2	3	Own	Hired			0	1	2	3										
1	2	3	4	5	6	7	8	9	10	11	12	13	14	15	16	17	18	19	20	21	22	23	24	25	26
Kadaperi	Well 8	6.19	9.73	0	6	1	1	11	6	17	500	0	4	1	3	15	32	4800	5300	90.6	8324	856	2550	4005	412
E (H)	Head 8	1.89	8.31	4	1	0	2	9	4	13	0	2	2	0	4	14	27	4050	4050	100.0	17834	2147	1950	8587	1034
	Tail 8	2.91	8.57	2	2	2	2	12	3	15	150	1	3	1	3	14	29	4350	4500	96.7	13231	1544	2250	6616	772
	Total	11.00	26.61																		39389	4546		19207	722

(Contd...)

Panniyur	Well 8	8.21	10.99	1	5	0	4	16	0	16	350	2	3	2	0	14	30	4500	4850	92.8	6491	591	2400	3212	292
E (H)	Head 8	3.42	9.64	0	4	2	2	14	0	14	120	0	6	2	0	10	24	3600	3720	96.8	10490	1089	2100	5922	615
	Tail 8	2.59	8.55	1	2	3	2	14	0	14	0	1	2	5	0	12	26	3900	3900	100.0	12896	1508	2100	6944	812
	Total	14.21	29.18																		29877	3187		16078	551
Kattalai	Well 8	9.19	43.51	2	1	3	2	13	2	15	0	1	1	3	3	16	31	4650	4650	100.0	22013	506	2250	10651	245
D (H)	Head 8	2.63	47.90	0	3	2	3	16	0	16	0	2	2	2	2	12	28	4200	4200	100.0	76571	1598	2400	43755	913
	Tail 8	3.32	36.90	2	2	1	3	13	0	13	0	1	2	3	2	14	27	4050	4050	100.0	45074	1221	1950	21702	588
	Total	15.13	128.31																		143658	3326		76108	593
Alapakkam	Well 8	16.34	22.12	1	2	3	2	14	0	14	250	0	1	2	5	20	34	5100	5350	95.3	7245	327	2100	2844	129
M (T)	Head 8	3.30	25.95	0	4	4	0	12	6	18	300	0	3	3	2	15	33	4950	5250	94.3	41285	1591	2700	21232	818
	Tail 8	4.79	32.43	2	5	1	1	10	8	18	0	3	2	3	0	8	26	3900	3900	100.0	26388	814	2700	18268	563
	Total	24.43	80.50																		74917	2732		42344	526
Athipattu	Well 8	22.26	59.55	4	3	1	0	5	17	22	400	3	2	3	1	11	33	4950	5350	92.5	14311	240	3300	8828	148
D (T)	Head 8	4.59	24.35	1	3	1	3	14	5	19	0	2	1	3	1	10	29	4350	4350	100.0	23090	948	2850	15128	621
	Tail 8	6.20	30.91	3	2	1	2	10	3	13	0	2	3	1	2	11	24	3600	3600	100.0	17941	580	1950	9718	314
	Total	33.05	114.81																		55343	1769		33674	293

Note: In KPT the inlet channel maintenance has been fully undertaken by the PWD since 1858. Before the construction of Palar anicut (1858), the diversion of water from the Palar and the maintenance of the inlet channel was mainly undertaken by the ayacutdars themselves.

For field channel maintenance no cash contribution is reported. The Table is derived on the basis of weighted averages in each type of farmer groups.

0, 1, 2, and 3 as indicated in the Heading refers to the number of times the respondents participated in the maintenance activity. Dash indicates nil.

E (H) = Elevated sluice head reach village; D (H) = Deep sluice head reach village.

M (T) = Middle sluice tail reach village; D (T) = Deep sluice tail reach village.

Prevailed wage rate per manday in 2005-06 was: Rs. 150 for male members, who were participated in the maintenance activity.

Source: Field survey data, 2017.

and between head and tail segment non-well farmers, the former group contributed more than the latter.

Overall, it is seen from Table 5.5 that the farmers contributed mostly by own labour and cash contribution is used for the initial maintenance especially for the use of machine power. Contribution of hired labour was very less (for main/branch channel maintenance) which is reported only in Kadaperi and Panniyur. The imputed cash value of the total contribution per hectare of *ayacut* is the highest in Kadaperi (Rs. 1278) followed by Panniyur (Rs. 763) served by the elevated sluices. The lowest is in the tail end villages of Alapakkam (Rs. 311). Within each village apparently well farmers contributed the lowest amount per hectare (except Kadaperi) than the non-well farmers. Among non-well farmers the head reachers contributed more per hectare of *ayacut* than the tail reach farmers (except Panniyur).

KPT: In 2005-06, the next preceding year in which supply was available, the farmers contribution to maintenance of main/branch and field channels was less than in 2015-16, (except in Alapakkam village). During this supply year all the sample villages got the tank supply and labour contribution was also moderate to good. The lower level of labour contribution was mainly due to lesser extent of maintenance requirement in the previous 10 years (1995 to 2005) where both tanks got filled 3 times compared to only once in the later 10-year (2005-2015).

Cash contribution for main/branch channels maintenance is negligible, but it is relatively more in the elevated sluice villages compared to tail end villages. The deep sluice village of Kattalai in 2005-06 did not contribute any cash. Total labour contribution by non-well farmers in all the villages (except Kadaperi and Athipattu) is more than that of well farmers. One time participation is more in all sample villages and two times participation is seen more in Kattalai and Alapakkam. Irrespective of farmer category, in Panniyur all three types contributing labour thrice in main/branch channel maintenance. A significant number of hired labour contribution is reported only in the tail end villages and also well farmers in Kadaperi (see the previous page Table 5.6).

For field channel maintenance well-farmers contributed more than the non-well farmers in all the villages. Between non-well farmers tail enders contributed more labour than the head reachers (except in Alapakkam). One of the striking features is that all the villagers and in all categories of farmers in the 2005-06 supply year contributed more labour (excluding hired labour) for field channel maintenance than in main/branch channels.

This was to ensure that the available supply was utilised properly through field channels to reduce water loss by flooding.

The imputed value of per hectare contribution (cash plus labour) in Kadaperi (Rs. 722) is more than twice the value of contribution in Athipattu. The reason for this high variation between the elevated and tail end sluices is that due to channel blockage in Kaveripakkam village adequate water was not reaching to Athipattu/ Consequently many farmers who expected that tank supply would not help to get full crop abstained from maintenance even from the beginning although the storage in KPT was at full level in 2005-06. But for Kadaperi a full crop season supply was possible. Hence, most of the farmers in the latter village were interested to contribute labour. Within each village the value of contribution per hectare in the head reach of non-well holdings is higher in the deep/tail end villages and lower in the elevated sluice villages. Interestingly, in all the sample villages (except in the elevated villages and in Alapakkam) only labour (but not much money) contribution was reported for channel maintenance. By and large, the estimated cash contribution for all the farmers is the highest (Rs. 1,43,658) in Kattalai and lowest (Rs. 29,877) in Panniyur. The full tank supply for the third crop season induced the Kattalai *ayacutdars* to contribute more labour for maintenance.

5.3.2 Dusi-Mamandur Tank

In DMT in the supply year 2015-16, except Sothiyampakkam, all the remaining sample villagers participated in the construction of *Korambus* to divert the Palar water to DMT. For this purpose, the contribution of labour was relatively more in the villages served by Dusi sluices than the villages under Mamandur sluice. Dusi deep sluice being three feet above the sill of the Mamandur deep sluice has a strong interest in ensuring sufficient diversion via Rajakkal. The same situation was also reported by respondents in the survey year 2015-16.

Unless DMT gets enough storage the villages under Dusi sluice will not get adequate water supply even in a normal year like 2015-16. Therefore, both head and tail reach villagers of the sluice contribute labour for diversion work. As evident from below the Table 5.7, under Dusi sluice more than one third of the sample farmers in Vagai and Punaithangal villages contributed labour. Farmers in the Narasamangalam, at the head of Mamandur sluice, also participated, though to a lesser degree than in Dusi sluice villages, in the *korambu* work. But in Sothiyampakkam which gets supply only when the tank receives more than 25 feet storage, none of the sample farmers contribute labour for the diversion work.

Table 5.7: Labour and Cash Contribution for Maintenance of Channels by Sample Households in KPT, 2015-16

(Cash in Rupees; Labour in Mandays)

Sample villages	Sample HHs by type	Extent of holdings: Sample (<- Hectares ->)	Extent of holdings: Total	Water diversion: No. of farmers partici-pated 0	Water diversion: No. of farmers partici-pated 1	Water diversion: Labour Own	Water diversion: Labour Hired	Water diversion: Labour Total	Main/branch channels: No. of farmers participated 0	1	2	3	Main/branch channels: Labour Own	Main/branch channels: Labour Hired	Main/branch channels: Total	Main/branch channels: Cash only Rs	Field channels: No. of farmers participated 0	1	2	3	Field channels: Own Labour	Total contribution: Labour only (9+16+22)	Total contribution: Converted into cash	Total contribution: by sample farmers (24+17)	% of labour contribution (24/25) X100	Imputed value: for all farmers in the ayacut (25/3)X4	Imputed value: Per ha (27/4)	Contribution for WD/ Main/ Br. Chls only	Imputed value: for all farmers in the ayacut (29/3)X4	Imputed value: Per ha (30/4)
1	2	3	4	5	6	7	8	9	10	11	12	13	14	15	16	17	18	19	20	21	22	23	24	25	26	27	28	29	30	31
Vagai	Well 5	9.06	9.06	3	2	2	4	6	0	0	2	3	13	14	27	650	5	0	0	0	0	33	9900	10550	94	4660	514	5600	2473	273
E (H)	Head 8	2.38	7.74	5	3	3	2	5	0	3	1	4	17	7	24	400	8	0	0	0	0	29	8700	9100	96	15290	1975	4750	7981	1031
	Tail 8	5.83	9.30	6	2	2	0	2	0	4	2	2	14	8	22	175	4	1	2	1	8	32	9600	9775	98	6707	721	3775	2590	278
	Total	17.27	26.10																							26657	1021		13045	500
Punai-	Well 8	10.45	28.71	5	3	6	0	6	0	5	2	1	12	14	26	4700	0	2	4	2	16	48	14400	19100	75	7314	255	9500	3638	127
thangal	Head 8	5.12	22.18	7	1	2	0	2	1	4	1	2	9	9	18	3350	0	2	2	4	18	38	11400	14750	77	11520	519	6350	4960	224
E (T)	Tail 8	3.49	7.89	8	0	0	0	0	8	0	0	0	0	0	0	0	8	0	0	0	0	0	0	0	0	0	0	0	0	0
	Total	19.05	58.77																							18834	320		8598	146
Narasa-	Well 8	7.19	13.01	5	3	3	1	4	4	1	2	1	8	3	11	700	4	2	1	1	7	22	6600	7300	90	4061	312	2950	1641	126
mangalam	Head 8	2.97	16.84	6	2	3	0	3	5	1	1	1	6	2	8	350	5	3	0	0	3	14	4200	4550	92	6125	364	2000	2692	160
D (H)	Tail 8	3.15	14.81	4	4	2	2	4	4	2	1	1	7	4	11	200	6	0	2	0	4	19	5700	5900	97	7502	507	2450	3115	210
	Total	13.31	44.66																							17688	396		7449	167

(Contd...)

1	2	3	4	5	6	7	8	9	10	11	12	13	14	15	16	17	18	19	20	21	22	23	24	25	26	27	28	29	30	31
Sothiyam-	Well 8	8.70	29.05	8	0	0	0	0	0	6	1	1	11	14	25	5800	3	2	1	2	10	35	10500	16300	64	7494	258	9550	4391	151
pakkam	Head 8	7.94	15.70	8	0	0	0	0	0	4	2	2	14	9	23	3350	4	2	0	2	8	31	9300	12650	74	6370	406	6800	3424	218
D (T)	Tail 8	3.93	11.16	8	0	0	0	0	0	5	1	2	12	11	23	3350	0	3	4	1	14	37	11100	14450	77	14718	1319	6800	6926	621
	Total	20.57	55.91																							28582	511		14741	264

Note: In Dusi-Mamandur tank the ayacutdars diverting the Palar river floods during north-east monsoon period through the (main) inlet channel called "Rajakkal" in every supply year. This work has been routinely followed by the ayacutdars, but the PWD has not effectively done this job till now.

For field channel maintenance no cash contribution is reported in any of the villages. The Table is derived on the basis of weighted averages in each type of farmer groups.

0, 1, 2, 3 and 4 as indicated in the Heading of the Table refers to the number of times the respondents participated in the maintenance activity.

E (H) = Elevated sluice head reach village; E (T) = Elevated sluice tail reach village; D (H) = Deep sluice head reach village; D (T) = Deep sluice tail reach village

Prevailed wage rate per manday in 2005-06 was: Rs. 300 for male members, who were participated in the maintenance activity.

Source: Field survey data, 2017

Table 5.8: Labour and Money Contribution for Maintenance of Channels by Sample Households in KPT, 2005-06

(Cash in Rupees; Labour in Mandays)

Sample villages	Sample HHs by type	Extent of holdings: Sample	Extent of holdings: Total	Water diversion: No. of farmers partici-pated 0	Water diversion: No. of farmers partici-pated 1	Water diversion: Labour Own	Water diversion: Labour Hired	Water diversion: Labour Total	Main/branch channels: No. of farmers participated 0	1	2	3	Main/branch channels: Labour Own	Main/branch channels: Labour Hired	Main/branch channels: Total	Main/branch channels: Cash only Rs	Field channels: No. of farmers participated 0	1	2	3	Field channels: Own Lab-our	Total contribution: Lab-our only (9+16+22)	Total contribution: Con-ver-ted into cash	Total contribution: by sample far-mers (24+17)	% of labour contri-bution (24/25) X100	Imputed value for all farmers in the ayacut (25/3)X4	Imputed value: Per ha (27/4)	Contri-bution for WD/ Main/ Br. Chls only	Imputed value for all farmers in the ayacut (29/3)X4	Imputed value: Per ha (30/4)
		<- Hectares ->																												
1	2	3	4	5	6	7	8	9	10	11	12	13	14	15	16	17	18	19	20	21	22	23	24	25	26	27	28	29	30	31
Vagai	Well 5	9.06	9.06	0	5	4	18	22	0	0	5	0	5	18	23	4050	5	0	0	0	0	45	6750	10800	63	4770	527	7650	3379	373
E (H)	Head 8	2.38	7.74	3	5	3	4	7	1	7	0	0	7	0	7	0	7	1	0	0	1	15	2250	2250	100	3781	488	1120	1882	243
	Tail 8	5.83	9.30	6	2	2	2	4	2	5	1	0	7	6	13	2700	4	2	2	0	6	23	3450	6150	56	4220	454	4060	2786	299
	Total	17.27	26.10																							12770	489		8046	308

(Contd...)

Punai-	Well 8	10.45	28.71	6	2	4	0	4	3	0	5	0	6	16	22	2250	3	1	4	0	9	35	5250	7500	70	2872	100	4330	1658	58
thangal	Head 8	5.12	22.18	7	1	2	0	2	4	3	1	0	5	3	8	450	7	0	1	0	2	12	1800	2250	80	1757	79	1250	976	44
E (T)	Tail 8	3.49	7.89	8	0	0	0	0	8	0	0	0	0	0	0	0	8	0	0	0	0	0	0	0	0	0	0	0	0	0
	Total	19.05	58.77																							4629	79		2634	45
Narasa-	Well 8	7.19	13.01	5	3	2	1	3	3	4	1	0	6	9	15	1200	2	6	0	0	6	24	3600	4800	75	2670	205	2640	1469	113
mangalam	Head 8	2.97	16.84	4	4	2	2	4	4	2	2	0	6	4	10	950	2	3	3	0	9	23	3450	4400	78	5923	352	2070	2786	165
D (H)	Tail 8	3.15	14.81	6	2	1	1	2	5	1	2	0	5	6	11	1050	2	5	1	0	7	20	3000	4050	74	5150	348	2090	2658	179
	Total	13.31	44.66																							13743	308		6913	155
Sothiyam-	Well 8	8.70	29.05	8	0	0	0	0	4	2	1	1	9	16	25	2600	2	5	1	0	7	32	4800	7400	65	3402	117	4600	2115	73
pakkam	Head 8	7.94	15.70	8	0	0	0	0	2	2	4	0	6	8	14	1200	2	5	1	0	7	21	3150	4350	72	2191	139	2320	1168	74
D (T)	Tail 8	3.93	11.16	8	0	0	0	0	4	2	2	0	4	3	7	450	4	2	2	0	6	13	1950	2400	81	2445	219	1010	1029	92
	Total	20.57	55.91																							8037	144		4312	77

Note: In Dusi-Mamandur tank the ayacutdars diverting the Palar river floods during north-east monsoon period through the (main) inlet channel called "Rajakkal" in every supply year. This work has been routinely followed by the ayacutdars, but the PWD has not effectively done this job till now.

For field channel maintenance no cash contribution is reported in any of the villages. The Table is derived on the basis of weighted averages in each type of farmer groups.

0, 1, 2, 3 and 4 as indicated in the Heading of the Table refers to the number of times the respondents participated in the maintenance activity.

E (H) = Elevated sluice head reach village; E (T) = Elevated sluice tail reach village; D (H) = Deep sluice head reach village; D (T) = Deep sluice tail reach village

Prevailed wage rate per manday in 2005-06 was: Rs. 150 for male members, who were participated in the maintenance activity.

Source: Field survey data, 2017

In Vagai, hired labour contribution was more than that of own labour. This is mainly because the land holdings by Mudaliars, who control most of the land, send their tenants for maintenance work. No cash contribution for this work was reported in any of the villages. Since the floods in the Palar last for about a week or two, the diversion work needs to be done as quickly as possible, and most of the *ayacutdars* participated in this activity. Between well and non-well farmers, participation by the former is more than the latter. This could be because recharge of DMT wells depends critically on the tank supply.

For main/branch channels maintenance, labour contribution was reported by all sample farmers in Vagai and Sothiyampakkam villages (see Table 5.7). In Punaithangal a part of the *ayacut* served by the Mamandur sluice did not get supply for more than two decades. As a result, the farmers did not participate fully. In Narasamangalam, situated adjacent to the tank, most of the farmers did not participate. In Punaithangal all the well farmers sent only hired labourers.

In Sothiyampakkam, where the village level organisation is strong, the *ayacutdars* contributed labour properly. If they do not contribute labour they are fined. In Vagai, though situated at the head, farmers actively participated in channel maintenance not so much to get proper tank supply as to prevent the water logging/drainage problem. Whenever water is released from the tank it seeps through the channels and makes the *ayacut* water logged. Hence all the farmers are alert in keeping the main/branch channels clean for free flow of water.

Labour participation for field channel maintenance was very limited in DMT. This is mainly due to non-existence of field channels, especially in Vagai village. In the remaining three villages, Sothiyampakkam contributed more labour than Narasamangalam. Within each village well farmers' share was the highest. Cash contribution for channel maintenance was reported in all the sample villages.

The imputed value per hectare of labour contribution indicates that ayacutdars in Sothiyampakkam contributed more amount than those in all the remaining villages. Between the head reach villages of Vagai and Narasamangalam, the former village contributes more because they have to tackle drainage problem for that more labour contribution required. Within each village, well farmers contributed less than that of non-well farmers. The overall value of labour contribution in Sothiyampakkam was Rs. 28,582 followed by 26,657 in Vagai. In the other two sample villages Punaithangal and Narasamangalam the average amount of contribution is about Rs. 18000 in both villages. However, there is no clear pattern emerging from the labour contributions.

DMT: In 2005-06, the tank has got full storage like the latest supply year. Consequently, all the sample villages received tank supply as per the regular existing pattern that more or less resembles to that of the present supply year (2015-16). For water diversion work from Palar river to DMT through Rajakkal Supply channel, except Sothiyampakkam farmers, all others participated and contributed labour. Maximum number of persons participated was from Vagai village followed by Narasamangalam. It is indicated by the sample respondents that these two villages are located adjacent to the tank and benefits more, so more labour participation is voluntarily available for this inlet channel water diversion purpose from these villages.

For main / branch channel maintenance major contribution is made by well farmers in all the villages followed by tail end farmers. All the 8 sample tail end farmers in Punaithangal village did not participate since they had not received the tank supply for more than a decade previously from 2005. Contributions of hired labour in all the sample villages is reported. The maximum hired labour was from Vagai followed by Sothiyampakkam. Only under well farmers category majority of hired labour participation is noticed. For main, branch channels maintenance cash contribution by all the villages is reported. The cash contribution is the maximum from the well farmers in all the villages followed by the non-well tail end farmers (see previous page Table 5.8).

As discussed earlier, field channel maintenance was done only by farmers themselves and no hired labour was used for this purpose. In Dusi sluice villages of Vagai and Narasamangalam field channels maintenance was negligible due to non-existence of field channels in most lands. Whereas in the Mamandur sluice villages it is done by both sample villages and there has not much difference in labour contribution.

The imputed value per hectare of labour contribution indicates that ayacutdars in Narasamangalam contributed more amount (Rs. 13, 743) than those in all the remaining villages. Between the tail reach villages of Punaithangal and Sothiyampakkam the former village contributes less because a portion of ayacut did not get water, for that labour contribution was not provided. Within each village, well farmers contributed more than that of non-well farmers in Dusi villages and in Mamandur sluice villages it is more with tail segment farmers of non-well category. The overall value of labour contribution in Narasamangalam was Rs. 13, 743 followed by 12,770 in Vagai. In the other two sample villages Sothiyampakkam (Rs. 8037) contributes more than twice that of Punaithangal (Rs. 4629). Comparatively lesser contribution in the tail end village of Sothiyampakkam is mainly due to non-participation in inlet channel maintenance for which absolutely no labour contribution was made.

The above account shows the extent of labour participation in the maintenance of main/branch/field channels in the latest and the preceding supply years in KPT and DMT. The survey indicates that within each tank, across villages there is large variation in the number of times of participation in the maintenance, ranging from zero to three times in both tanks. The survey also captured the fact that in some villages, especially in Sothiyampakkam, if the farmers did not contribute the allotted share of labour man-days, they were fined. This clearly shows that the local village level organisation is effectively functioning in those villages where the tank water is the primary supply source. As against this, in a few other villages, eg., Alapakkam in KPT and Vagai and Narasamangalam in DMT, there was no penalty for the defaulters. This is mostly due to the less active functioning of local organisation. However, the fact remains that the reach factor and the availability of other source of supply (especially wells) also play a significant role in the extent of farmers' participation in maintenance.

5.4 Summary

This chapter described the maintenance of channels under KPT and DMT in the latest two supply years. The analysis shows that maintenance of channels is not evenly done across villages. It was found that tail end villages contributed more labour for maintenance than the middle and head reach villages to get tank water. In KPT the deep sluice head reach village/ tail end villages did not actively take part in maintenance work due to non-availability of supply to them. The survey also indicates that in both tanks in 2015-16, enforcement of rules and regulations, and the maintenance work were more vigorous in the tail end/ elevated sluice villages than in head reach villages. The extent of encroachment was more in the main channels than in branch channels. In the tail end villages encroachment problem was more severe in KPT, than in DMT. This affected the supply available to the villages considerably. Villages served by separate main/branch channel did not have much of an encroachment problem.

It is also found that in all sample villages a large labour force was involved to desilt and deweed the main/branch channels at the time of initial maintenance; whereas subsequent maintenance required only limited labour input. For the use of machine power for maintenance money contribution was also reported in the sample villages. Tail end villages contributed more labour than the head reach/elevated sluice villages. There are wide variations in frequency of maintenance across villages and between the tanks. While the head reach villages under KPT contributed labour once in every season, those under DMT contributed labour only once in a supply year. In the villages under KPT, the elevated sluice/tail end villages contributed labour once in every season, whereas in DMT it was once in every month. The survey reveals that wherever the village level organisation is effectively functioning, detailed records pertaining to the maintenance of channels are kept by the irrigation functionaries, whereas in villages where the organisation is weak they are not maintained. In all the villages the organisation is functioning but its effectiveness varies a great deal across reaches and between the tanks.

It is important to note that in both tanks labour contribution for maintenance of channels (including water diversion to DMT through Rajakkal channel) is impressive. The elevated sluice/tail end villages relatively contributed more than the head reach villages. The amount of contribution per annum for maintenance in 2015-16 ranged from Rs. 311 per hectare in the tail end village of Alapakkam to Rs. 1278 per hectare in the elevated sluice village of Kadaperi in KPT and Rs. 146 per hectare in the tail reach village of Punaithangal to Rs. 500 per hectare in the head reach village of Vagai served by the elevated Dusi sluice in DMT.

Notes

1. However, this point, "maintenance of tanks was not satisfactory, *Kudimaramath* so valuable an institution was practically dead" could not be understood, especially in the context of Tamil Nadu, where one may observe that in the 1960s the maximum net irrigated area by tanks that was 9.1 lakh ha which is about 91% of the total registered ayacut under tanks in Tamil Nadu attained (see table 1.1 in page 25 of chapter 1). If the *Kudimaramath* did not take place in that decade how that much area could be irrigated by tanks in Tamil Nadu? From this it follows that only from 1970s onwards *Kudimaramath* system gradually weakened and tank irrigated area steadily declined.

Chapter - 6

Water Allocation and Management in KPT and DMT Villages

In the preceding chapter, the maintenance of the inlet channels and distribution networks in the selected tanks, and the nature and extent of user involvement in itare discussed. The other important aspect of tank irrigation management is allocation of water from the tank to different villages and its distribution within the *ayacut* in normal and deficit supply periods. This aspect is dealt with in this chapter with reference to two full tank storage years (1991-92 and 2015-16) and the latest precedingsupply year 2005-06. For the 1991-92 supply year, a survey was conducted in 1992 in the same sample villages that data are used wherever required for proper understanding of the supply management in these tanks. Allocation of water across villages is discussed in the context of the procedures of *Mamulnamas* and the experiences of 1991-92 and 2015-16. Distribution of water within the villages is dealt with in terms of the ayacutdars' experience in 2005-06 and 2015-16.

6.1 Operational Rules for the Palar Anicut System

At the system level water regulation is governed by certain rules and regulations framed by Mr. Pears (Executive Engineer, North Arcot division) in 1899. A lot of changes have occurred since 1899. These include siltation in the inlet channels and tanks, encroachment in the distribution networks, cultivation in the catchment areas, increase in the number of wells in the *ayacut*, and changes in cropping pattern and in the functioning of irrigation institutions at both the tank and village levels.

6.1.1 Water Regulation at the Anicut Site

The Superintendent in charge of the Palar Anicut and head works is required to operate the scour vents in the Anicut to keep full supply in all the main channels, and prevent water going over the crest of the Anicut unnecessarily. The level of full supply in the main channels is clearly specified in terms of levels in front of the various diversion sluices andattempt to balance the area of opening in the left and right flank scour vents. In case of continuous rain or wind the Anicut under-vents are to be opened at once and kept open until the storm is over; soon after the flood starts receding, the falling shutters are to be erected and all leakages through the scour vents are to be effectively sealed using waste gunny bags, puddle or clay.

As a rule, the under-vents on the south side of the Anicut are to be opened first. When full supply is not required down Sakkaramallur channel, the vents of the sand sluice are to be regulated to let the surplus water into the river. The Superintendent is expected to act according to the weekly reports received from the channel Superintendent stating how much water is required in each channel. A water register is to be maintained at Anicut site and its daily extract is to be sent direct to the higher official. During the months of October, November and December post cards are to be sent daily to the E.E. and S.D.O. giving an abstract of the water register for the Anicut.

6.1.2 Distribution of Water between Channels

One Superintendent oversees the southern section. He has five *lascars* for help and his head quarter is at *Randam*. His functions include regulating water at the commencement of the cultivation season to give each tank under his jurisdiction approximately one month's supply; utilising specified vented weirs and regulation in specified tanks for supply purposes only and never to discharge surplus floods.

The Superintendent in charge of northern section has to inform the Kaveripakkam tank Superintendent whenever full supply is not available from the Anicut. The KPT Superintendent is supposed to open the Ponnappanthangal sluices if water can be spared from Kaveripakkam tank. He is also responsible to place the shutters in the Kachendi odai outlet when the river supply is passing down the Mahendravadi channel or when the KPT is full. If full supply is not available at *Goon Madai*, the division sluice Superintendent of this channel has to distribute the flow in the ratio of 0.452 to 0.548 from **1st May to 30th November** and 0.533 to 0.467 from **1st December to 30th April** between the Kaveripakkam and Govindavadi channels respectively.

6.1.3 Water Regulation at KPT and DMT

In both the tanks there is a Superintendent assisted by many *lascars* - all appointed by the PWD - to ensure that certain stipulations regarding management of tank storage are observed. In the case of KPT, the water level should not be allowed to rise more than 2.5 feet above the sill of the surplus weir before 15th December. The Superintendent is to report every week to their counterpart on the Govindavadi Channel about KPT requirements; to open the vents of the surplus weirs and the waste weirs above their sill levels; and to regulate Ponnappanthangal sluices only for supply regulation based on the report from the Superintendent of Mahendravadi.

The Superintendent in charge of DMT is to look after the Rajakkal sluice and channel and regulate water flows to fill the tank as soon as possible. The head sluice of the Rajakkal channel is to be closed when there is heavy and continuous rain. The low-level sluices of DMT are never to be used when there is water on the sills of the high-level sluices, and the screw gearing shutters are to be kept clear and in good working condition. The Superintendent is thus primarily responsible to ensure that water supply reaches fairly from the source to the tanks.

No shutter will be put on top of the surplus weirs of any tank for raising the water storage **until 1st December or until the lower tanks are full**, except in the case of weirs surplusing direct into the river and for regulating supply to lower tanks. Shutters in exceptional circumstances are to be removed after 1st December under orders of the PWD officer and the vents are to be kept open so long as the river water diminishes so that there may not be decrease of supply to the tank whose vent is opened.

The Superintendents are responsible to ensure that every weir is provided with vents, and every sluice whether in channel or tank has shutters in good order. If any shutter is missing or damaged, they must send a report to the SDO stating clearly the extent of loss or damage.

The Superintendents are expected to see that all trees on channel and tank bunds are cut to form a path of two metres wide with a clear headway of three metres and to extract work from their *lascars*. The work consists of clearing silt, removing prickly pear and jungle and other petty repairs.

The connecting banks at drops and other masonry works must be kept tidy and neat. Apart from these rules and regulations there are certain procedures to be followed at tank level while supplying water to the *ayacut*.

6.2 Regulation of Supply Within Ayacut

6.2.1 Inter-Channel Allocation

The allocation of water between different segments of the command in each tank is in the first instance determined by the opening and closing of the sluice. In both tanks, as pointed out earlier, up to early 2000s, this decision is made by the PWD in accordance with the desire of the Irrigation Board in KPT and the Hissa Tank Farmers Association in DMT. The opening and closing of sluices during the season is determined based on requirements from the *ayacutdars* through the IB delegates. From the mid-2000s, the territorial constituency members (TCMs) of the WUAs decided these functions in both the tanks.

6.2.2 Allocation between Channels

Within the *ayacut*, the fact that sluices are situated at different levels provide an automatic mechanism for rationing available supply. Normally, in larger tanks (like KPT) sluices are arranged at three levels, viz., deep, middle, and the elevated. Naturally, the deep sluices get earlier supply followed by middle and the elevated sluices. In some cases, the elevated sluices cannot draw any supply till the tank reaches near FTL. As the tank storage falls from FTL, the depth of supply available at the sills of the sluices also decreases automatically. Consequently, the elevated sluices face water shortage first. Any further decrease at the storage leads to no supply to the elevated sluices (see also Table 3.7). The higher the sluice the lower is the volume available for its *ayacut*, shorter the duration of supply and greater the uncertainty. The deep sluices on the other hand will begin getting supplies as soon as the tank gets water; it is able to draw water for a longer period and with great certainty.

In DMT the arrangement of sluices varies considerably from the KPT. It has two sets (one deep and one elevated in each set) of sluices and each set irrigates nine villages. Whenever the tank gets adequate supply the elevated sluices (1 and 4) are opened first. The deep sluices (2 and 3) are opened only when the supply available from the elevated sluices is inadequate to feed the *ayacut*. Then the low-level sluices are gradually opened according to the irrigation requirements. In KPT there has not been any specific procedure to open any of the sluices on a priority basis, as the arrangement of sluices and the channels taking off from them do not permit such a functioning.

6.2.3 Intra-Channel Allocation

As a rule, once the sluice is opened for irrigation it is not closed unless the *ayacut* gets heavy rainfall or the *ayacutdars* request for its closure for maintenance purposes. Once water is let out, it continuously flows along the main and branch channels from the head reach village to the tail end of the *ayacut*. However, to facilitate irrigation of *ayacut*, several villages divert supply from the channels through such arrangements as *kondam* (a temporary earthen embankment to raise the water level to irrigate elevated lands); *mattu* (a masonry structure from which a specified dimension of two or more channels take off to supply water to the *ayacut*); and *diversion vents* (vents constructed in appropriate places with specified dimensions to divert water from the main/branch channels). The existing practices are: the location, size and the *ayacut* to be irrigated of these devices are said to be roughly in consonance with the specifications laid down in the *Mamulnamas*.

The KPT and DMT supplies were augmented by construction of Palar Anicut. Consequently, there should have been modification in the principles indicated in the *Mamulnamas*. But no such modification was made. On this W. Wynne, the Collector of North Arcot district, wrote in 1895:

> The government did not amend the water *Mamulnamas* at the time when the Palar Anicut System (1858) was introduced. We are bound by them now. Under the present system the supply goes under a system of arrangement and compromise among the villagers themselves. The slavish adherence to an archaic and not very accurate document was a mistake but there is no remedy for it without legislation (GO 270 I, PWD, 24-08-1910, p. 7).

Even a century after these observations, the same procedures is followed in all the villages in both the tanks.

The way supplies in a channel are allocated to different villages is brought out in figures 6.1 through 6.6. These figures show the division points in each of the channels serving selected villages (5 in KPT and 4 in DMT), the proportions in which the flow at each point is divided and whether the regulation is automatic or supervised can be seen from these figures.

There are two types of channel level rationing: Automatic regulation; and Supervised regulation. The former method is adopted at the main/branch channel level to distribute water across channels/villages. The main channel supply is rationed by constructing a permanent masonry structure to divert water in appropriate places and in appropriate quantity so that the volume of water supplied is proportional to the regulated *ayacut* under different branches/villages served by the channel. By this method, whenever tank supply is released in the main channel, water automatically flows through the vent ways in a predetermined proportion. This method is found in the Maganipattu and Thuraiperumpakkam sluices channel and Fox sluice under KPT, and in all the main channels under DMT (see also figures 6.1 to 6.4).

As may be seen in figure 6.1 in Mulai sluice channel under KPT, there is an automatic regulation point, which is constructed by five block rounded stones to divide water equally (50:50) between the two main channels stating from the sluice cistern (thoti).In the next case, it may be seen in figure 6.2 in KPT under Maganipattu and Thuraiperumpakkam channel the automatic regulation vents are constructed roughly in 14 places in the main/branch channels. At each point the vent is constructed at the dimensions specified in the *Mamulnamas*. Likewise, in figure 6.3 under Fox sluice there are six automatic regulation points. In DMT, under both sets of Dusi and Mamandur sluices the automatic regulation points are constructed in appropriate places at specified dimensions (see figure 6.4).

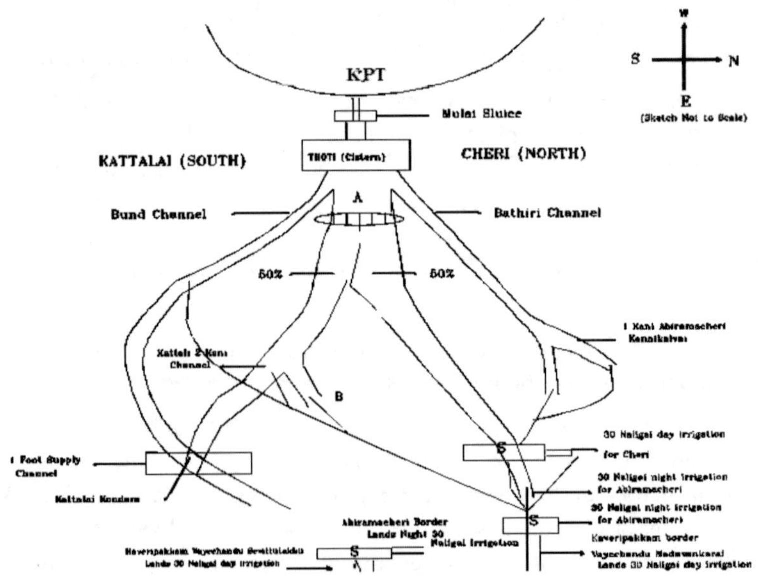

Figure 6.1: Water Distribution as per Mamulnamas in Kattalai Village under Mulai Sluice Channel in KPT

Under supervised regulation, supply from main/branch channels is regulated by the *Neerkatties/Kammukutties* of the respective channels/ villages (see figures 6.1, 6.3 and 6.6). For instance, under Moolai sluice there are three supervised regulation points denoted by "S" in figure 6.1. At each point there are two channels. One channel under each vent irrigates lands for 30 *naligai* (12 hours) in day time and the remaining one channel irrigates 30 *naligai* at night. Most probably the specified time for the diversion of supply from the respective vents are 6 a.m. and 6 p.m. During this time either the *Kammukutti* or*Thotti* has to be present for the diversion purpose. One of the important points is that irrespective of whether the *ayacut* specified in the channels are fully irrigated or not, only 30 *naligai* supply is to be permitted and the *Kammukutti* or *Thotti* of respective channels block or open the supply according to the specification.

However, this pattern has discontinued from the mid-2000s in both the tanks, where the *Neerkatties/Kammukutties* functioning has been fully

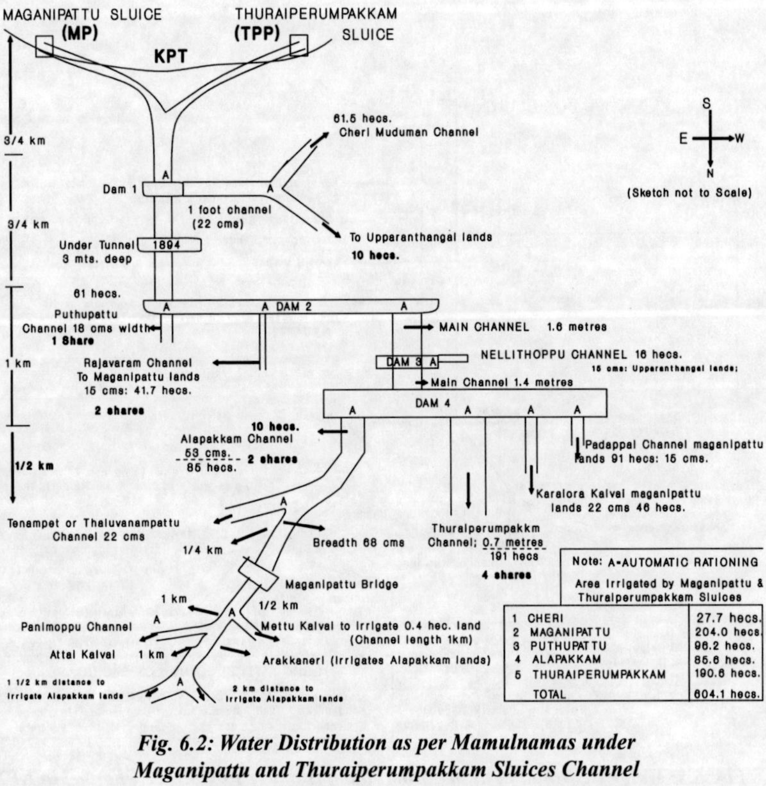

Fig. 6.2: Water Distribution as per Mamulnamas under Maganipattu and Thuraiperumpakkam Sluices Channel

vanished and only farmers themselves distribute water as per their need / requirements. This situation has happened because of changes in local water management pattern from the traditional system to government initiated WUAs where the latter's Territorial Constituency Members (TCMs) with the help of farmers managing the supply below the tank sluices. But by the farmers' point of view this pattern has not been working satisfactorily since all the villages -especially the tail end villages - serving in a sluice is not getting supply fairly, due to encroachment of channels an improper distribution of water across channels and in the ayacut by the ayacutdars.

As per the above traditional method, as per the *Mamulnamas,* there are some other time bound methods of supervised rationing as reported in the Fox sluice channel. As may be seen from figure 6.3 there are two supervised regulation points marked as S1 and S2. At point S1 the main channel supply is regulated in two ways: (i) two hours full supply (from 4 p.m. to 6 p.m.) is diverted to irrigate the lands under *Sitheri Kondam* channel belonging to Kondapuram; and (ii) two hours full supply (from 6 p.m. to 8 p.m.) is

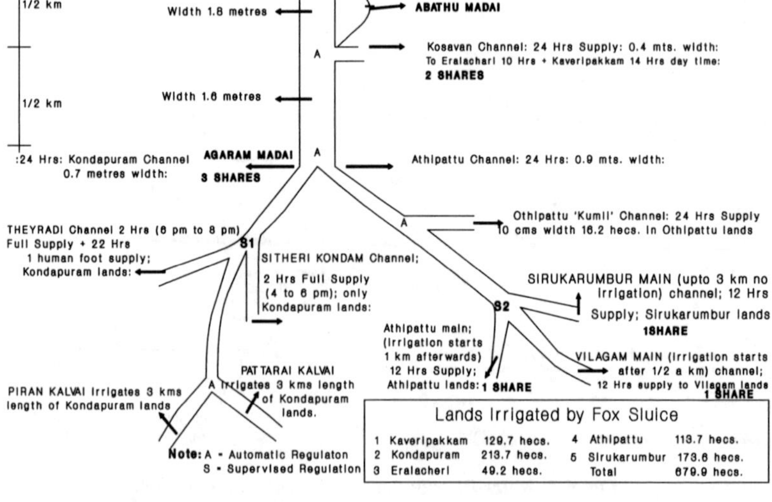

Fig. 6.3: Water Distribution as per Mamulnamas under Fox Sluice Channel in KPT.

diverted to irrigate lands under *Theyradi* channel and 20 hours (from 8 p.m. to 4 p.m. of the next day) one-foot supply is diverted to Kondapuram lands.

At point S2 the Athipattu main channel supply is regulated in the following manner: (i) 12 hours full supply (from 6 a.m. to 6 p.m.) is diverted by the *Kammukutties* of Athipattu village to irrigate the lands of that village; (ii) 12 hours full supply (from 6 p.m. to 6 a.m. of the next day) is diverted by the Sirukarumbur *Kammukutties* to irrigate the lands of that village; and (iii) 12 hours full supply (from 6 a.m. to 6 p.m.) is diverted to Vilaham lands. Hence each channel gets supply on alternate days for about 12 hours.

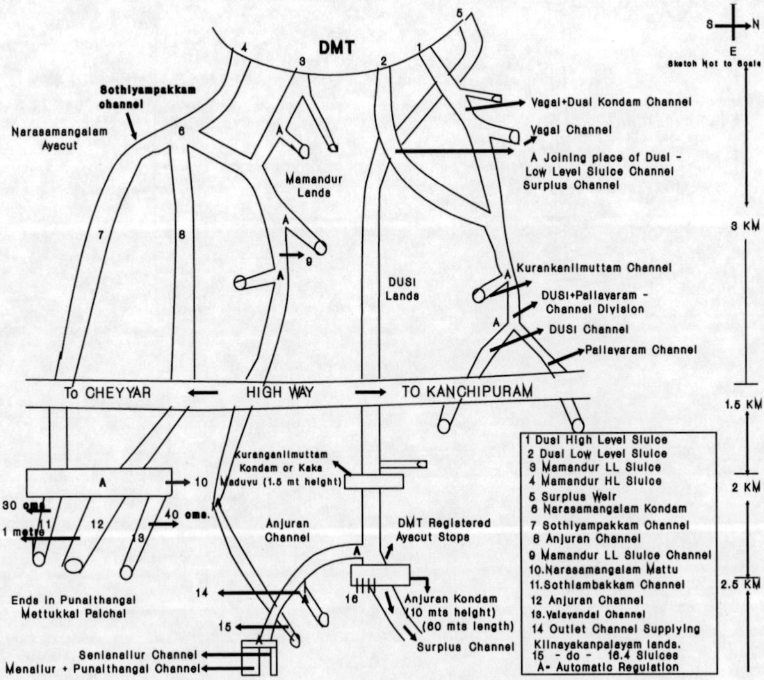

Fig. 6.4: Traditional Rights of Villages Served by DMT as per Mamulnamas.

The important point is at each diversion point only the irrigation workers regulate the supply at specified time. Personal visits along these channels in 1991-92 supply year confirmed that these irrigation workers effectively discharge their duties of regulation of supply and keeping the time schedule. However, personal visits again in 2015-16 supply year indicated that no irrigation worker is functioning in Athipattu to undertake the local water regulation tasks and more importantly the entire Athipattu / Sirukarumbur

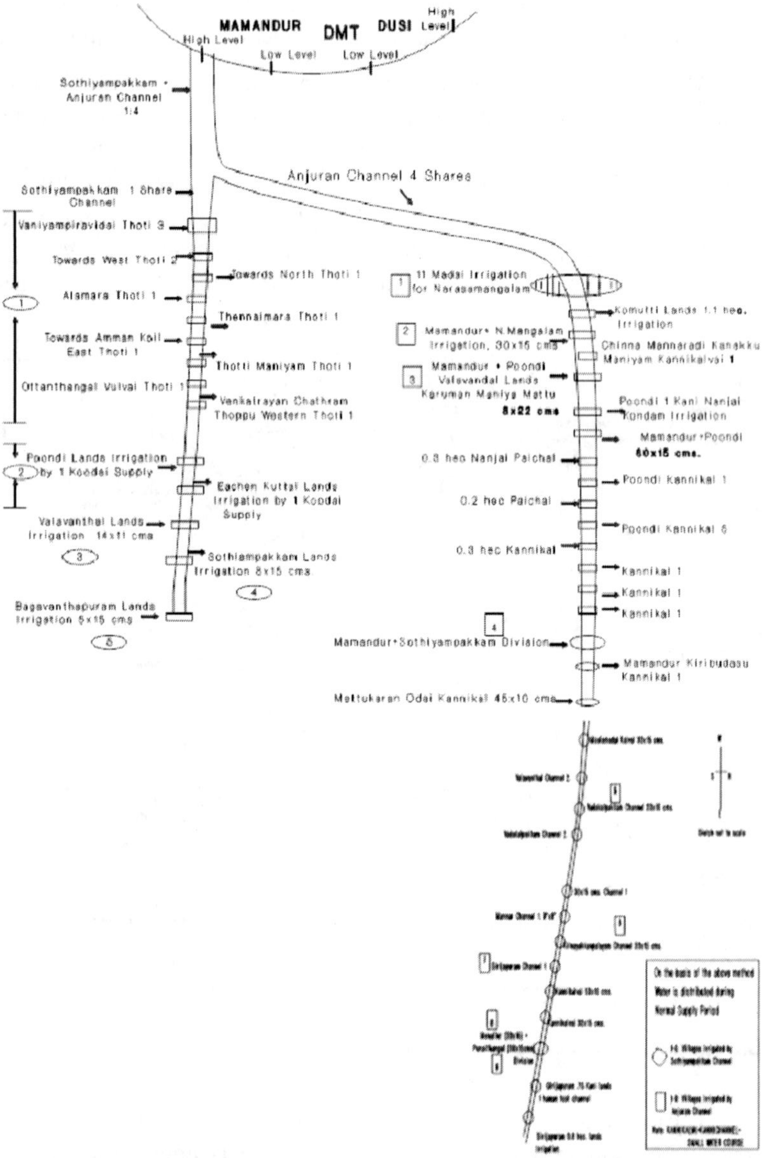

Fig. 6.5: Water Distribution under Sothiyampakkam and Anjuran Channels as per Mamulnamas in DMT during Normal Supply Period.

villages did not get any tank supply although the KPT was at full storage. Also, only farmers' themselves managing the water distribution as per their requirements in other villages where tank water was available. Also, in

2015-16, it is also reported under KPT 8 out of 14 villages had not receivedtank water for cultivation due to various reasons such as: 1. Long gap in the availability of tank supply (2005-06 and latest by 2015-16) did not permit many villagers to undertake cultivation. 2. Many of the tail end villages did not have proper main distributaries to get the tank supply (Athipattu, Sirukarumbur and Eralacheri belong to this category). 3. Main outlet channel encroachment in the head of the channel prevented supply reaches to middle and tail end villages (eg. Fox / TPP and Panniyur sluices: Thuraiperumpakkam and Puthupattu villages and also the tail end of Kilveeranam village all come under this category) and 4. Drastic change in the institutional pattern of water distribution from traditional method to elected members management through TCMs of the WUAs as per the Tamil Nadu Farmers Management of Irrigation System Act (TNFMIS Act 2001).

The methods of supervised rationing as performed, as per traditional pattern, in the channels under DMT are shown in figure 6.6. In DMT the

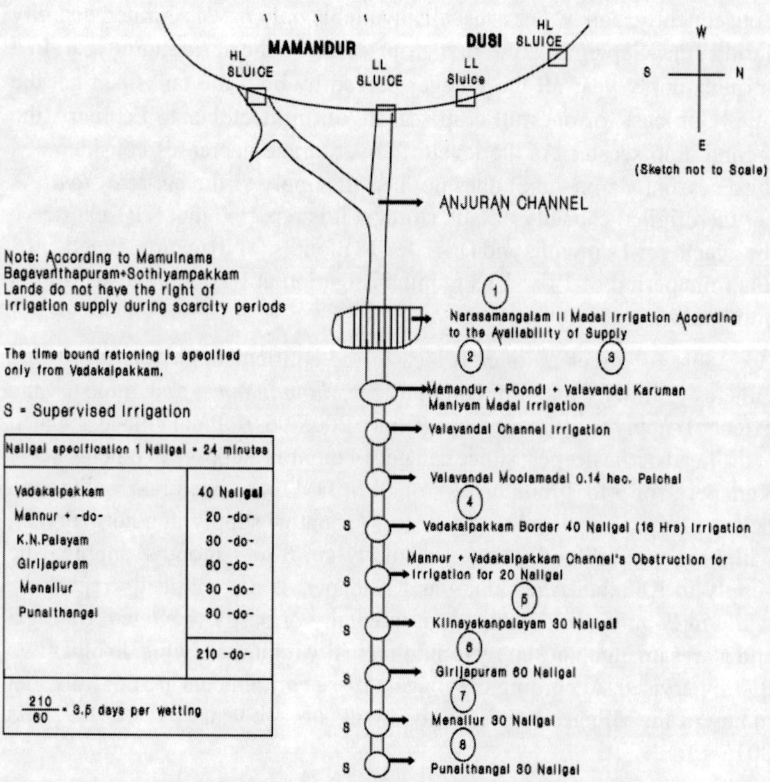

Fig. 6.6: Water Distribution as per Mamulnamas during Scarcity Period under Anjuran Channel in DMT.

supervised rationing is performed only at the scarcity supply period, and that too only in the Mamandur sluices. As may be seen in figure 6.6 there are six supervised rationing points and each one has its specified time to get the supply. Accordingly, each village gets the supply at an interval of 3.5 days. Since in the supply year 1991-92 no scarcity was reported, no supervised regulation was performed in DMT. The same is the case for the year 2015-16. However, in thelatter survey year no irrigation functionaries has been working but the required water distribution works are managed by farmers themselves. While comparing KPT and DMT, water management performance in the latter tank is better than the former one, although in both tanks the traditional irrigation management system (functioning of Kavaimaniyam / Kammukutti and Thotti) has disappeared altogether.

6.3 Supply Conditions in Selected Villages Since 2005-06

The tanks start receiving supply only by September - October. Consequently, tank water is usually available only in very limited quantity during June-September: the first crop season of the agricultural year. In a normal supply year all the villages served by both the tanks can get the supply at least for one full crop season, during October to February, the second crop season. As the level of tank storage decreases considerably, third season(March-June) does not permit supply to the entire *ayacut*. As per the original capacity of tank storage, it is reported that KPT can serve the ayacut for 14 months and DMT for 18 months for irrigation. Practically, this time period of 14 and 18 months irrigation at FTL storage serving in both tanks is not reported so far.

Table 6.1 gives an overall idea of the entitlements of the 9 selected villages (5 in KPT and 4 in DMT) as per *Mamulnamas*; and Table 6.2 the extent of supply in different seasons since 2005-06. Kattalai village, situated at the head of the deepest sluice can get 14 months' supply; all others except Kadaperi, for 6 to 7 months (2 crops). In DMT, the head reach village in both sluicesare expected to get 12 to 18 months' supply in a normal year, while the two tail end villages normally get 6 to 7 months' supply. The supply to Kattalai, Alapakkam and Kadaperi is automatically regulated; while in Panniyur and Athipattu it is wholly or partly regulated. In Vagai and Narasamangalam supply is automatically regulated; while in other two it is supervised. According to villagers these entitlements are not fully met in any of the villages even in a good tank storage years like 1991-92 and 2015-16.

During 2015-16 survey, when both the tanks got full storage, a clear picture in distribution of water to all the registered villages other than the

T able 6.1: Entitlements/Traditional Rights of the Villages for the Use of T ank Water in KPT and DMT, 2006 and 2016

Tank / Sample villages	*Entitlements/Traditional rights (According to Mamulnama)*	*Observations: Changes in water allocation, 2006*	*Observations: Changes in water allocation, 2016*
KPT Kadaperi E(H)	1. FT L storage can irrigate 66 acres of ayacut for one season. 2. Whenever the Palar river channel which is passing close to this village did not supply enough water the farmers who depend upon the channel can also get the KPT supply.	Only the tail reachers participated in the maintenance wherever the river channel failed to supply water. Encroachment and water theft resulted in poor supply to tail reachers.	Only 25-50% of sample farmers participated in maintenance. 100 days labour was used for channel maintenance. River channel lands were sold out mostly and non-cooperation of farmers in maintenance let poor supply to tail reachers.
Panniyur E(H)	Main channel water is allocated into 3 shares by the extent of acreage to be irrigated by the channel: One share goes to Gengatharanallur another pertains to Panniyur and yet another to Kilveranam.	Water theft was reported in the upper reach village and Kilveranam did not get the tank supply since 1981-82.	No change in allocation entitlements. Adequate tank supply due to breakage of sluice shutter in 2015-16 helped to get two season's good supply. Kilveranam did not take KPT supply since 1981-82.
Kattalai D(H)	This village is irrigated by the deepest sluice, hence abundant supply is available. T his village can get 14 months supply at FT L storage. Hence 3 crops of paddy or 1 Banana crop is feasible.	T he supply in 2005-06 available only for 9 months at FT L apparently due to more than 17 feet of tank siltation at the deep sluice.	T he supply in 2005-06 available only for 9 months at FT L apparently due to more than 17 feet of tank siltation at the deep sluice.
Alapakkam M (T)	Automatic rationing supply from the main channel to this village and 4 other villages (Cheri, Maganipattu, Puthupattu & T .P.P). (Refer sketch map 2 for the Entitlements of this village).	Water theft and encroachment results poor supply and 2 crops at FT L storage is not possible.	Same entitlements still persist. Water theft, failure of local institutional set up and encroachment results poor supply and 2 crops at FT L storage is not possible.
Athipattu D(T)	Automatic and Supervised rationing of tank supply from the main channel to this village and 4 other villages (KVP, Kondapuram, Eralacheri and Sirukarumbur). (Refer sketch map 3 for the Entitlements of this village).2 crops at FT L storage is possible.	Main channel encroachment affected the supply adversely; 2 crops at FT L storage is difficult. Only head reachers get a limited tank supply.	No change in entitlements. But heavy main channel encroachment affected KPT supply adversely; So, 3 villages - Athipattu, Eralacheri and Sirukarumbur - did not get KPT supply throughout the tank supply period.

(*Contd...*)

DMT			
Vagai E(H)	At FT L storage this village can get 18 months supply.	FT L supply is available for 10-12 months only.	FT L supply was available throughout the year.
Punaithangal E (T)	During scarcity period the supply available from the Anjuran Kondam is to be allocated in the following manner: 1.Vadakalpakkam 40 naligai; 2. Mannur 20 naligai; 3. K.N.P. 30 naligai; 4. Girija- puram 60 naligai; 5. Menallur 30 naligai and 6. Punaithangal 30 naligai. T otal = 3.5 days for one supply of irrigation. (note 1 naligai = 24 minutes).	T his practice is not working effectively because of water theft at Anjuran Kondam. Mettukkal Paichal lands did not get the tank supply since 1981-82.	T his practice has not been working because of water theft at Anjuran Kondam. Since village level institutions are not working effectively as before heavy encroachments in main channels are reported. Ayacut Land sales has become widespread. Metukkal Paichal lands did not get the tank supply since 1981-82.
Narasamangala m D(H)	T his village can get the supply through the Anjuran Channel and the Sothiyambakkam Channel. Water can be taken by way of 'T hoti Murai' by using human lift. From Sothiyambakkam channel. 13 'T hoti Murai' is eligible. From Anjuran channel. 11 'Madai Murai' (i.e., by direct channel) irrigation is eligible.	Instead of taking water by human lift farmers are using oil engines to lift water which affect the supply to the tail reachers.	No change in entitlements are reported. However, instead of taking water by human lift, farmers are using oil engines and even electric motors to lift water from the channel, which affect the supply to the tail enders.
Sothiyampakka m D(T)	Mamandur High Level sluice water is divided in two ways: (i) 4 shares are diverted into Anjuran channel (ii) 1 share is diverted into Sothiyambakkam chl. From the latter channel supply water must further be divided into two portions, (i) 1.5 shares to Sothiyambakkam and (ii) 1 share to Bagavanthapuram village.	Water theft and the use of oil-engines and electric pumpsets at the upper reach of the main channel adversely affected the available supply to the village.	Water theft and the use of oil-engines and electric pumpsets at the upper reach of the main channel adversely affected the available supply to the village. Also heavy encroachment in the main supply channel and poor maintenance let only limited supply.

Note: E (H) = Elevated sluice head reach village; E (T) = Elevated sluice tail reach village.
D (H) = Deep sluice head reach village; D (T) = Deep sluice tail reach village. M (T) = Middle sluice tail reach village.
Source: Mamulnamas of KPT and DMT , and the interviews with the farmers and the irrigation functionaries, 2017.

T able 6.2: Availability of T ank Water and Its Allocation to Sample Villages, 2005-06-

Name of tank and Sample villages	No. of times water was available in the past 11 years		Number of years tank supply was available in the last 11 years					Action taken when the village did not get its entitlements
			According to the entitlements		Short of entitlements		No supply years	
	Season 2	Season 3	Season 2	Season 3	Season 2	Season 3		
1	2	3	4	5	6	7	8	9
KPT								
Kadaperi E (H)	2	0	1	0	1	0	9	A few farmers used own/purchased well water. Murai system was not practised.
Panniyur E (H)	2	0	1	0	1	0	9	Murai system was followed in 2005-06.
Kattalai D (H)	4	2	1	1	3	1	7	Ayacut was divided into two halves each half got a day's supply, followed in 2006.
Alapakkam M (T)	2	1	1	0	1	1	9	A few farmers used own/purchased well water. Murai system was followed.in 2006.
Athipattu D (T)	1	1	1	0	0	1	10	Own/purchased well water was used and no murai system followed.
DMT								
Vagai E (H)	7	3	4	2	3	1	4	No action was taken.
Punaithangal E (T)	3	2	2	1	1	1	8	Steps were taken to remove unauthorised use of water at Anjuran Kondam,
Narasamangalam D (H)	5	3	4	2	1	1	6	Farmers used oil engines / electric pumpsets to avert scarcity in tank supply.
Sothiyampakkam D (T)	2	0	1	0	1	0	9	No action was taken.

Note: E (H) = Elevated sluice head reach village; E (T) = Elevated sluice tail reach village; D (H) = Deep sluice head reach village; D (T) = Deep sluice tail reach village; M (T) = Middle sluice tail reach village.

Column 8 refers to no supply years. Supply in KPT was below 6 metres storage in 7 years and in DMT the storage was below 4.5 metres in 4 years out of 11 (2006-2016). In those years in both the tanks not even a single crop was raised by using tank supply in the deep sluice villages.

Source: Interview with the farmers and irrigation functionaries, 2017.

aforesaid methods was not able to be captured. This is because, many villages in KPT and a portion of ayacut in a couple of villages in DMT did not use the tank water. This problem was due to long gap -10 years - in getting the tank supply and ineffective functioning of local irrigations institutions and the government's ignorance to strengthen the WUAs as per the regulations made in the TNFMIS ACT – 2000 to manage the tank supply appropriately in both the tanks. Consequently, in both the tanks, FTL supply lasted more than a year – an unusual occurrence noticed – and that supply was mostly utilised by the head / middle reach villages depriving the right – as a natural process - to use it by the tail end and elevated sluice villages. During 1991-92 FTL supply year, while KPT storage lasted for 7-8 months and DMT for 9 months only (in DMT a large quantum of water was wasted due to unattended nature of one of the broken shutters in the Dusi sluice.

The eighties and 2006-2015 were marked by unusually high incidence of drought: in the eighties, eight out of the 11 years the tanks were completely dry or the available storage was grossly inadequate to feed the *ayacut* even in the head reach villages served by the deepest sluice. The same trend was also observed during 2006-2015. The villages served by the elevated sluices, Kadaperi in KPT and Sothiyampakkam in DMT, received only a single season supply and that too in three years out of 11 (Table 6.2). The head reach village served by the deeper sluice of KPT (Kattalai) received tank supply for five years in the second season and three years in the third season. The remaining three villages got some supply in both seasons in three years but not to the full extent of entitlements in all these years. In DMT, Vagai is reported to have got adequate supplies in four years during November-March and three years during April-July, and Mamandur for three years in each season. Punaithangal got supplies in the first season for three years and the second season only two. In most cases the supplies were assured to have satisfied the entitlements.

An evidence shows that in the earlier survey year (1991-92) - a normal supply year - all the head reach villages served by the deep sluices in both tanks got two full seasons supply without problem. Villages served by the elevated sluices in KPT and the tail end village of Sothiyampakkam in DMT got only a single season supply (Sivasubramaniyan 1995).

However, in 2015-16 supply year, the deep sluice village of Kattalai and the tail end village of Athipattu in KPT and a portion of ayacut (tail end ayacut) in Punaithangal in DMT received no supply for cultivation (Table 6.3). Although the tail end village of Alapakkam is supposedly eligible for two seasons supply, it received supply only for a single season that too only for head and middle reach ayacutdars, because of heavy encroachment in the supply channel and also reluctance of tail end ayacutdars to take part

Table 6.3: Availability of Tank Water According to Entitlements in the Latest (Normal) Supply Year, 2015-16.

Tank & Sample villages	Supply of water in		Entitlements when water is short from the beginning of a season	Entitlements as per 2015-16 survey
	A normal year (FTL)	Scarcity periods		
KPT				
Kadaperi E(H)	5-6 Months supply was possible.	OMP system was practised, but not in 2015-16.	No step would be taken to get the tank supply.	One full crop supply was available.
Panniyur E(H)	6-7 months supply was possible	Murai system was followed in 2005-06.	No step would be taken to get the tank supply.	8 Months supply was available. No scarcity in supply was reported. Kilveeranam did not get the tank supply.
Kattalai D(H)	14 months supply was possible. But no supply received in 2015-16.	Ayacut was divided into two halves (upper & lower portions) and each one could get a day's supply. Tank supply was not received fully in 2015-16	Ayacut was divided into two halves (upper & lower portions) each one could get a day's supply.	Due to collapse of the deepest Mulai sluice, this village did not get any supply throughout the supply year.
Alapakkam M(T)	2 season supply was possible	Murai system was followed in 2005-06 (24 minutes per acre).	No step would be taken to get the tank supply.	Main channel obstruction led to inadequate supply. No murai system was followed due to defunct of local irrigation institutions.
Athipattu D(T)	2 season supply was possible. But no supply received.	40 Mts. per acre murai was followed long back and KPT supply was not fully received in 2015-16.	No step would be taken to get the tank supply.	Due to full encroachment of main channel, KPT supply was not received fully. Action taken to rectify the channel did not yield any result.
DMT				
Vagai E(H)	18 months supply was possible	OMP system was followed at the last 40 days of tank supply. But not followed in 2015-16	OMP system was followed only at the last 40 days of tank supply.	10 months adequate supply was available.
P. Thangal E(T)	12 months supply was possible.	OMP system was followed for 3 months tank supply. But not followed in 2015-16. A portion of ayacut not get tank supply.	No step would be taken to get the tank supply.	8 months adequate supply was available to Pallakkal Paichal lands but no supply to Metukkal lands.
N.Mangalam D(H)	12 months supply was possible	Only Direct Irrigation was followed.	Only Direct Irrigation was followed.	8 months full supply was available.
S. Pakkam D(T)	6-7 months supply was possible	Murai system was followed (16 minutes/acre) but not in 2015-16	No step would be taken to get the tank supply.	Adequate supply was available for one full crop season; further supply was not used by a few farmers.

Note: E (H) = Elevated sluice head reach village; E (T) = Elevated sluice tail reach village.D (H) = Deep sluice head reach village; D (T) = Deep sluice tail reach village.M (T) = Middle sluice tail reach village; OMP = Oru Madai Paichal.

Source: Interviews with the farmers, irrigation functionaries, and direct visits, 2017.

in the maintenance of main/branch channels. While the elevated sluice of Panniyur received more than 6-7 months' supply, the Kadaperi got only 5-6 months' supply but both these elevated villages reaped one full crop without any problem. In DMT, the head reach villages of Vagai and Narasamangalam and the tail end villages of Pallavaram received two seasons adequate supply, the tail end village of Sothiyampakkam got only one full crop supply due to its locational disadvantage.

There appear to have been significant changes in the availability of tank water in practically all villages: actual supplies when the tanks reached FTL being considerably less than in the past. A few decades ago the PWD had undertaken some works such as construction of vents/diversion dams with masonry structures in the main supply channel taking off from the tank sluice. These structural improvements were reported to be based on the rules and regulations codified in the *Mamulnamas*. The PWD and the Irrigation Board jointly do allocation of water across villages.

But the important matter as reported in most of the tail reach villages was the encroachment problem in the main/branch supply channels, the water theft/unauthorised use of tank water in the upper reach villages and non-cooperation of 'well farmers' in maintaining channels on a regular basis as before, which adversely affect the supply available to the tail end villages.

In recent decades especially after 1960s there have been a lot of changes in agricultural practices all over the *ayacut*. These include introduction of HYV seeds, short duration crops, changes in crop patterns, increased use of chemical fertilizers, machinery and wells with electric/diesel pumps. All of this has enhanced the demand for water; even as supplies were adversely affected. Scarcity in supply leads to competition to avail it. This competition has both favourable and unfavourable effects. Unfavourable effects include theft of water from the common source of supply and disobeying the rules and regulations in the use of the common supply of water. These tend to aggravate the tail enders disadvantage in getting tank water. Favourable effect comes from the development of wells to overcome the shortage of water - which as we shall see later - has been sizeable. These effects are widely seen in the villages served by both the tanks. However, one important point is that although many changes have occurred, there has not been any shift in the allocation/distribution rules in most of the villages surveyed (see below the Table 6.4). Only Narasamangalam in DMT, reported changes in the water lift method. In Narasamangalam, farmers do not lift water asper the entitlement, but through electric motors, which reduce the supply of tank water to tail enders.

While all the sample villages under KPT reported significant shift in cropping pattern from paddy to sugarcane and banana and the tail end village shifted the crop from long duration to short duration variety, the villages under DMT have not reported any change in the cropping pattern. DMT ayacutdars always cultivate paddy as the only crop, because of the soil conditions, which is suitable for paddy crop only. The cropping pattern has changed in KPT villages mainly because of increase in the number of wells in the ayacut. In all the elevated sluice villages and in the tail end village of Athipattu, well density is relatively high. Another tail end village of Alapakkam reported average density equal to tank average. The deep sluice village Kattalai has very low well density due to its assured tanks supply

Table 6.4: Changes in Agricultural Practices, Availability of Tank Water and Its Effect on Allocation Rules, 2015-16.

Name of tank and Sample villages	Changes in agricultural practices			Availability of tank water				Density of Wells per 100 ha	Allocation rules		
	Use of LV/HYV	Cropping pattern	Cropping decisions	Normal 2015-16		< Normal			Any change	Role played by	
				S2	S-3	S-2	S-3		Yes/No	PWD Yes/No	IB Yes/No
1	2	3	4	5	6	7	8	9	10	11	12
KPT											
Kadaperi E (H)	Shift to HYV	Significant P to B (30%)	By farmers only	A	No Supply	IA	No Supply	High 137	No	No	No
Panniyur E (H)	Shift to HYV	Significant P to SC (15%	By farmers only	A	A*	IA	No Supply	High 192	No	No	No
Kattalai D (H)	Shift to HYV	Significant P to B (70%)	By farmers only	No Supply	No Supply	A	A	Low 57	No	No	No
Alapakkam M (T)	Shift to HYV	Significant Short Term P	By farmers only	A	A#	A	IA	Average 89	No	No	No
Athipattu D (T)	Shift to HYV	Significant P to B (25%)	By farmers only	No Supply	No Supply	A	IA	High 130	Yes	No	No
DMT											
Vagai E (H)	Not Significant	Only paddy	By farmers only	A	A	A	A	Very poor 19	No	No	No
Punaithangal E (T)	Not Significant	Only paddy	By farmers only	A	A	A	IA	Significant 63	No	No	No
Narasa-D (H)mangalam	Not Significant	Only paddy	By farmers only	A	A	A	IA	Significant 96	yes	No	No
Sothiyam-pakkam D (T)	Not Significant	Only paddy	By farmers only	A	No Supply	IA	No Supply	High 141	No	No	No

Note: E (H) = Elevated sluice head reach village; E (T) = Elevated sluice tail reach village.
D (H) =Deep sluice head reach village; D (T) = Deep sluice tail reach village.
M (T) =Middle sluice tail reach village;.LV = Local Variety; HYV = High Yielding Variety
P=Paddy; B=Banana; SC=Sugarcane; A=Adequate; IA=Inadequate. S-2 = Season 2; S-3 = Season 3.
* refers that adequate supply was also available in 2015-16 in Season 3 due to non-taking of tank supply in many villages.
# refers that poor maintenance of main/branch channels led to availability of inadequate supply in season 3.
Since the deep Mulai sluice repair and main channel encroachment led to no tank supply for Kattalai and Athipattu villages.
Source: Survey, 2017.

(unfortunately in 2015-16 tank supply was not possible to this village). In DMT, well density is reported relatively high in Mamandur villages of Sothiyampakkam and Narasamangalam compared to Dusi sluice villages. However, the quantum of water available from these wells are very less as reported by the sample respondents.

Interestingly, it may be notice from table 6.4 that allocation rules of tank water supply did not change in all the villages (except one each in both tanks). Even the respondents indicated change in the allocation rules, it is not authenticated since Athipattu village in KPT did not receive the tanks supply. In DMT, in Narasamangalam lifting of water by electric motors from the traditional lifting devices certainly a development induced technology upgradation. However, the extent of ayacut receiving water by this changed technology is much limited. If the local institutions work effectively the shortages of water could be managed by other accessible ways.

6.4 Water Allocation at Village Level

In the foregoing section it is described the way water allocation between channels and between villages within each channel is regulated. Once water reaches the village, the community through its irrigation workers according to certain customary rules allocates the available supply.

Farmers are free to decide whether to sow and what to sow without any interference from the government, previously the Irrigation Board / currently the WUAs andthe village organization. The individual farmer makes his decisions based on storage available at the time of sowing, and on his judgment of the likely availability during the entire crop season. Farmers who have access to wells are clearly in an advantageous situation. But, for the non-well farmers the level of storage in the tank at the beginning of the season is a crucial determinant in their decisions to raise the crop. For lands irrigated by the elevated sluices, the storage of water level in the tank may be inadequate for supply for the entire crop season; or even under the deepest sluices supply of water may not be adequate to irrigate all reaches of the *ayacut*. Farmers in the tail end will clearly be in a disadvantageous position in such a situation. Under these circumstances the farmers located at the tail end of the sluices as well as in the elevated sluices tend to grow shorter duration crops including dry crops (Vaidyanathan and Janakarajan 1989: 76). This situation aptly applies to both KPT and DMT.

When the supplies at the village level fall below the requirements usually one of two methods of rationing is adopted: (i) *Murai* (i.e. time bound rationing); and (ii) *Oru Madai Paichal* (channelwise turn distribution).

Under *Murai* system each village *ayacut* is divided into several *pangus* (shares). These *pangus* are customary. Normally each *pangu* is 25.50hectares, but the size is not the same in all villages. Taking into consideration the number of *pangus*, the total number of days required to irrigate the *ayacut* in a village is fixed, according to a well-established convention. The time allotment per hectare so decided does not change under any circumstances.

To illustrate the water rationing under the *murai* system: assume that the number of customary *pangus* in a village is ten. The total *ayacut* in that village is 255 hectares. Then each *pangu* is equal to 25.5 hectares. The allotment of time to irrigate the entire village per wetting according to convention is, say, three days. So, 255 hectares should be irrigated within 72 hours and on an average one hectare takes roughly 17 minutes. Irrespective of the level of flow in the channel water will be supplied to lands constituting different *pangus* in rotation for the ratio of 7.2 hours per *pangu* and 17 minutes per hectare to all the *ayacutdars*. This means that whether or not the allotted time to the *ayacutdars'* land is adequate to fully irrigate it is not important. What is important is that each hectare under the *ayacut* gets a fixed time and duration of supply in each turn. The *murai* system is practiced in most of the tail reach villages. Since the allotted time per acre is fixed, disputes are rare in sharing water. When *murai* is in force the *ayacutdars* themselves irrigate their lands. The *murai* system is widely practiced in KPT till the 1990s. From the mid-2000s it is reported by the respondents that this system was not working well due to changes in the local institutional set up and farmers themselves irrigate their lands as per the availability of tank supplies.

The *Oru Madai Paichal* (OMP) permits one complete irrigation in the entire village *ayacut*. Under this system the *ayacutdars* are not allowed to irrigate their lands by themselves. Only the irrigation workers like Kammukutties can regulate water supply. When there is deficiency of water from the source, it is compensated by an increase in the number of days to irrigate the entire *ayacut*. That is to say, even during the lean supply period, the entire village *ayacut* is sought to be irrigated at least once. This system is now followed by a majority of the villages in DMT. Although irrigation functionaries are not functioning now in DMT, farmers themselves follow this pattern of OMP during scarcity supply periods.

A rapid survey (1990 and 2016) in all the 32 villages of the two tanks provides a broad picture of the water rationing followed in the *ayacut*. *Murai* was followed in 13 villages (10 in KPT and 3 in DMT); Oru *Madai*

Paichal was followed in another 13 (1 in KPT and 12 in DMT) and *Direct Irrigation* in five villages (3 in KPT and 2 in DMT). Only in Alingilpattu village in DMT, *Eravai Paichal* (lifting water by using human power) was followed, however this system is replaced now by electric motor. Similarly, the rapid survey conducted in 2016 shows that in KPT, only direct irrigation is followed in all the seven villages out of 14 currently served and in DMT almost in all villages direct irrigation is served by farmers themselves. During 2015-16 supply year, both the tanks got full storage level and inadequacy was not reported in any of the serving villages in both the tanks as per their entitlements.

Farmers in the villages irrigated by elevated sluices/tail end villages practice some form of *murai* during scarcity period in both the tanks. Villages served by the deepest sluices do not follow any *murai*. *The villagers as and when they face scarcity also not immediately follow Murai*. In the following circumstances *murai* is not to be adopted: (1) Scarcity situation faced at the final stage of a crop season when a majority of farmers harvest the crop without expecting further tank supply. (2) Inadequate tank supply to feed the entire *ayacut* from the beginning of a season. When majority of the farmers who do not have access to water other than the tank supply do not cultivate.

6.5 Detailed Survey of Selected Villages, 2015-16

6.5.1 Allocation of Water in the Sample Villages

In the supply years 2015-16 all the sample villages started receiving tank supply in the month of November/December except the tail reach village of Sothiyampakkam in DMT, which reported receiving the supply in January 2016. This is due to the fact that the *ayacutdars* of this village do not start the channel maintenance till the tank receives more than 25 feet supply. In the supply year 2015, DMT reached this level of storage at the beginning of December 2015. Hence the villagers started the maintenance of channels in the middle of December, which delayed getting the tank supply. It was pointed out earlier that once opened, sluices should not be closed even at the end of a season. This rule facilitates all villages to get a continuous supply in the third season. But the availability of supply in the third season varies across villages because of the location of villages from the tank and the sluice elevation. In KPT, as already indicated two sample villages, the tail end of Athipattu and the deep sluice serving Kattalai village did not get the tank supply.

In the second season all the sample villages (except Alapakkam in KPT) reported that they got regular and adequate supply throughout the season (Table 6.5). There were attempts to regulate water distribution. Alapakkam villagers experienced it. In the 2015-16 supply year the tail end villagers of Thuraiperumpakkam and Puthupattu (they are two among the five villages served by Maganipattu and Thuraiperumpakkam sluices) did not take tank water even from the beginning due to severe encroachment of supply channels. As a result, these two villagers did not participate in the main channel maintenance which requires huge man-power. Consequently, the maintenance burden was shifted to Alapakkam ayacutdars. In Alapakkam, the tail end lands are fully owned by the Scheduled Castes famers and they are not interested in maintenance, since the local institutions was already failed in managing the system. So, only a few head and middle reach farmers of Alapakkam village participated in channel maintenance and got tank supply to some extent to meet their requirements. Also, some of the well farmers who had adequate well supply in Alapakkam did not participate in channel maintenance activity.

In the third season only two villages, Panniyur fed by an elevated sluice and Alapakkam at the tail end of a middle level sluice, received tanksupply. It was reported Alapakkam experienced inadequate supply but it is enough for Panniyur villagers. To avoid deficit supply. Farmers have also attempted to supplement tank water with purchased well water. In the year 2015-16, due to breaking of the sluice shutter in Panniyur sluice, which was not repaired till receding the tank supply at that sluice, copious supply was received by Panniyur ayacutdars. One more added advantage to this Panniyur farmers is that Kilveeranam villagers did not take their share of Panniyur sluice water in the past 3-4 decades that supply was also utilised by Panniyur villagers.

In DMT, Vagai and Punaithangal had enough supply for two full crop seasons in 2015-16. Although Narasamangalam and Sothiyampakkam villages are situated under Mamandur sluices the former village got adequate supply but the latter got a short supply of last one month. Actually Sothiyampakkam generally gets one crop supply only. However, 2015-16 good storage in the tank helped to get some more supply to this tail end village. Further, it was reported that in a few tail end villages like Punaithangal, ayacutdars' sold their land and some farmers grown up prosopis juliflora in their ayacut land that prevented taking water from the DMT. So, adequate supply was reported by some farmers who cultivated under DMT ayacut in the tail end location.

Table 6.5: Allocation of Water in the Sample Villages of KPT and DMT, 2015-16.

Name of tank and Sample villages	Month of tank water release		Availability of supply		At what stage supply inadequate		Action taken what result		Did IA supply affect yield		Water distribution within village		Why no regulation?
	S-2	S-3	S-2	S-3	S-2	S-3	S-2	S-3	S-2	S-3	S-2	S-3	
1	2	3	4	5	6	7	8	9	10	11	12	13	14
KPT													
Kadaperi E (H)	November	Nil	R & A	No supply	-	Nil	-	Nil	-	Nil	a	Nil	Nil
Panniyur E (H)	December	Conti. supply	R & A	R & A	-	Nil	-	Nil	-	Nil	a	a	-
Kattalai D (H)	November	Conti. supply	No Tank Supply		-	-	-	-	-	-	No Tank Supply		-
Alapakkam M (T)	December	Nil	R & A	R & IA	-	Last 2 months	-	WWU	-	Yes considerably	a	a	Institutional falure
Athipattu D (T)	December	Conti. supply	No Tank Supply		-	-	-	-	-	-	No Tank Supply		-
DMT													
Vagai E (H)	November	Conti. supply	R & A	R & A	-	-	-	-	-	-	a	a	-
Punaithangal E (T)	December	Conti. supply	R & A	R & A	-	-	-	-	-	-	a	a	-
Narasamangalam D(H)	November	Conti. supply	R & A	R & A	-	-	-	-	-	-	a	a	-
Sothiyampakkam D (T)	January	Conti. supply	R & A	R & IA	-	Last 1 month	-	WWU	-	Nil	a	a	-

Note: R & A = Regular and Adequate; R & IA = Regular and Inadequate; Nil indicates supply was not used.
Dash indicates nil. MU.AD+WWU = Murai adopted and well water used. Conti. = Continuous. a = No regulation at any time.
S2, S3 = Season 2 and Season 3. E (H) = Elevated sluice head reach village; E (T) = Elevated sluice tail reach village.
D (H) = Deep sluice head reach village; D (T) = Deep sluice tail reach village. '-' indicates nil.
Source: Survey, 2017.

Table 6.6 shows that in the past all the sample villages (except the two head reach villages of Kattalai and Narasamangalam) had followed the *murai* to distribute water during scarcity period. However, in the supply year (1991-92) in KPT only Panniyur and Alapakkam followed the *murai*. In 2015-16, none of the sample villages in both tanks followed *murai* (this is because no scarcity in supply was faced by any of these villages). Understanding why other villages did not follow the *murai* calls for understanding the water situation and farmers' response to it in each of the villages in 1991-92 and 2015-16.

Kadaperi received enough supply for more than five months in both 1991-92 and 2015-16. There are two types of farmers in this village: (i) those who get supply directly from the KPT, known as *Nanjai Kalani* (wet land) farmers, a direct *ayacut* of 26.32 hectares fed by Gaffoor sluice; and (ii) those who normally get supply from *Athu Kalvai* is referred to river channel farmers, and its *ayacut*is about 16.19 hectares. In the latter case, whenever the river channel failed to supply adequate water, farmers are entitled to get KPT supply through Gaffoor sluice (as a traditional right). The *Nanjai Kalani* farmers are located at the head segment of the tank

Table 6.6: Murai in Sample Villages, 1991-92 and 2015-16.

Name of tank and sample villages	*No. of months when tank supply was available*			*No. of months of deficiency in*		*Whether Murai was practiced*		
	At FTL	*In 1991-92*	*In 2015-16*	*1991-92*	*2015-16*	*In the past*	*In 1991-92*	*In 2015-16*
KPT								
Kadaperi	5	5	6	0	0	Yes	No	No
Panniyur	6	6	6	1*	0	Yes	Yes	No
Kattalai	9	9	0	0	#1	No	No	No
Alapakkam	8	4	5	1	1	Yes	Yes	No
Athipattu	8	8	0	2*	#2	Yes	No	No
DMT								
Vagai	12	8	9	0	0	Yes	No	No
Punaithangal	12	8	9	0	0	Yes	No	No
Narasamangalam	12	8	9	1*	0	No	No	No
Sothiyampakkam	8	5	6	0	0	Yes	No	No

Note: * Deficiency was reported in the second tank supply season.

1 Due to breakdown of deep sluice and its closure, no tank supply was possible throughout the tank supply period.

2 Due to full encroachment of main supply channel Athipattu did not get KPT supply.

Source: Field Survey, 2017 and Sivasubramaniyan 1995.

sluice and the *Athu Kalvai* farmers are located at the tail segment of the tank. As a result of this locational factor, farmers in the latter case who assert their riparian right to get the tank supply had to make more efforts, both for maintenance of channels and distribution of water within the *ayacut* whenever the river supply failed. In such situations the head segment farmers (Nanjai Kalani) normally do not cooperate with them.

Since the river channel failed to supply water in 1991-92 about 32 farmers maintained the channels and also distributed the supply under *Oru Madai Paichal*. But, most of the head segment farmers did not participate in the maintenance work. Since the tail segment farmers are entitled to get the tank supply only at night that too only after satisfying the requirements of head segment farmers, there was no need for the latter group to follow *Oru Madai Paichal*. However, when the river channel provides enough supply, the tail segment farmers do not take KPT supply, maintenance of the channels and distribution of water is done exclusively by the head segment farmers. This had happened only in 1977-78, after which, no *Oru Madai Paichal* was followed by the head segment farmers.

During the supply year 2015-16, a major change in both types of farmers was reported. Due to non-availability of regular KPT supply (between 2005 and 2015) and expansion of Kaveripakkam village as semi-urban centre and these lands are located adjacent to the Kaveripakkam village, almost all the *Athu Kalvai* lands of 16.19 hectares were sold out for housing construction purpose. During the visit in 2017 it was witnessed that all these lands are occupied by pucca houses. Hence these lands are fully removed now by getting supply from either *Athu Kalvai* or Gaffoor sluice. So, the entire responsibility of channel maintenance of Gaffoor sluice main channel is vested with *Nanjai Kalani* farmers. Consequently, the total ayacut under Gaffoor sluice has reduced from 42.51 ha to 26.32 ha. Due to this change, maintenance was done by a few *Nanjai Kalani* farmers only apart from the maintenance undertaken in this channel through 100-day employment guarantee program.

Panniyur received adequate supply for about six months (a little more than Kadaperi village) in the 1991-92 supply year. After 1977-78 so much supply was available to this village only in 1991-92. However, in the third crop season tank water was inadequate for the last two months to feed the entire *ayacut*. As a result, the *ayacutdars* followed the *murai* in two ways. The main channel which feeds three villages, namely, Gangatharanellur, Kilveeranam and Panniyur, is divided in proportion to the extent of registered *ayacut* available in each village. This method laid down in the *Mamulnamas* is still followed. Presently, water is not being taken by the Kilveeranam villagers due to internal caste feuds. Their share is, therefore, utilised by the other two villages. As a result, there has not been much difficulty in sharing the available tank supply within the village.

Within the village the entire *ayacut* of 29.15hectares is divided into four *pangus* (named after eminent local persons), namely, (i) Pinchan *pangu*; (ii) Maran *pangu*; (iii) Chithooran *pangu*; and (iv) Appavu *pangu*. Each *pangu* is entitled to get daily one *Jamam* (three hours) water supply in the day time. During night, first two *pangus* can get supply one night and the second two *pangus* on the next night. Accordingly, in a week, each *pangu* is entitled to get 21 hours supply in the day time and 21 hours in the night (6+6+6+3=21). Though this method was actually followed in Panniyur in 1991-92 supply year, respondents reported that it was not as effective as it was some two to three decades ago.

During the supply year 2015-16, fortunately this village has got adequate supply for more than one crop period due to breakage of Panniyur sluice shutter. Once the KPT got its FTL storage the breakagein the shutter made continuous flow of water to this village. Due to this, no murai system was followed and as usual Kilveeranam villagers also did not take KPT supply hence that share was also utilised by Panniyur villagers. Since, initially, the KPT storage lasted a couple of months at full level, the Panniyur villagers got adequate supply for 2 crop seasons. Only a few tail end farmers affected by water scarcity in the last one month of crop period. To some extent this was managed with supplemental wells.

Kattalai being supplied by the deepest sluice; has little problem of water supply compared to other villages. In the 1991 supply year this village received the tank water for two full crop seasons. Usually, if the *ayacutdars* face any water scarcity at the final stages of crop period they also follow the *murai*. The entire village *ayacut* is divided into two halves as *Mel murai* and *Keel murai* (upper half and lower half of the village *ayacut*). Each one is entitled to get a day's supply alternately. According to the *ayacutdars* this system is followed only occasionally. There was no need for it in 1991-92.

Unfortunately, during the supply year 2015-16, as mentioned already, the deep Moolai Sluice shutter in KPT broken that led to full closure of this sluice for irrigation by the PWD. The repair / replacement of shutter was not possible throughout the supply period. Consequently, Kattalai villagers had forgone their full tank supply at least for two-crop seasons fully in 2015-16.

Alapakkam is a tail end village served by the middle sluices of Maganipattu and Thuraiperumpakkam (both sluices joined together after a 100 metre distance). It is also called Erettai Kal Madagu. These sluices also supply water to four other villages. Alapakkam is entitled to get three shares of water out of eleven from these sluices. The *Mamulnamas* rules and regulations are still followed in sharing water across villages. However, the available quantity of supply was not effectively regulated among the villages. This was due to encroachment on the main channel which reduces tank supply to Alapakkam.

Generally, when KPT gets normal supply this village receives water for two full crop seasons. Contrary to this in 1991-92, although the tank had got FTL storage this village got inadequate supply even for one crop season due to a municipal water supply pipe line was erected one foot above the sill of the main channel by the Block Development Officials. That prevented adequate supply to this village.Normally, to distribute the supply equitably among *ayacutdars* the following customary method was adopted. According to the *murai* system 1/4 *kani* (33 cents) gets eight minutes supply. Further if an *ayacutdar* is paying a compound land tax (one and a half times more than the ordinary wet assessment) he receives supply equal to 16 minutes per 1/4 *kani* (i.e., a "double *murai*" water). Most of the well farmers in this village are paying the compound land tax and are entitled to get "double *murai*" water. This method adversely affects the supply available to tail reach farmers who mostly pay the ordinary wet assessment land tax. A few tail segment of non-well respondents informed that they did not get *murai* water in 1991-92. Some others told that others "*murai* water" was purchased from those farmers who own wells and did not require tank supply (selling their "*murai* water" which is available from the tank is also widely reported in this village). The non-availability of *murai* water to all the tail reachers was also, to some extent, due to encroachment on the channels.

In 1991-92 supply year only about half the respondents in Alapakkam reported the precise magnitude of *murai* followed and the actual duration of tank supply. The remaining respondents reported that they did not get *murai* water. The total *ayacut* in this village is 85 hectares. According to *murai* each 0.13 hectare is entitled to get eight minutes supply. And within 84 hours (3.5 days) the entire *ayacut* might be irrigated once. Hence in a week two *murai* was available. Table 6.7 shows the method of *murai* adopted in Alapakkam village during 1991-92.

The major drawback in this *murai* system is that most of the non-well farmers, especially in the tail reach of the village reported that they had not

Table 6.7: Method of Functioning Murai in Alapakkam, 1991-92.

Type & No. of house-holds	*Total operational holdings (Ha)*	*Time allotment /ha according to murai (minutes)*	*No. of minutes supply required*	*No. of hours supply required*	*Frequency of supply in a week*
Well - 8	15.60	59.28	924.96	15.41	Twice
Head - 8	2.35	59.28	139.2	2.32	
Tail - 8	7.16	59.28	424.56	7.08	
Total -24	25.11	59.28	1488.72	24.81	3.5 days/ supply

Note: Operational holdings relate to each of 8 sample farmers in the three farmers' category.
Source: Sivasubramaniyan, 1995.

received adequate supply even from the beginning when the *murai* started. When the scarcity reached its peak at the end of the second week after introducing the *murai* most of the tail segment farmers did not get their *murai* share. This was mainly due to the fact that the supply from the main channel was inadequate to cater to the needs of even the head reachers. As a result, the tail reachers purchased the *murai* water (most of the well farmers did not use the *murai* water but sold it to the non-well farmers). Some farmers supplemented *murai* water with purchased well water. A few others in a disadvantageous position in all these arrangements allowed their crop to wither away.

Non-well farmers in Alapakkam reported that during scarcity period the *murai* followed was unjust, not suitable for the tail segment *ayacutdars* who received no *murai* water at all. According to them agriculture in the tail end villages served by the tank is not at all remunerative if one has to do it by using tank water alone, and even their survival itself is not feasible under this system. It is important to note that in the tail segment of the village most of the farmers belong to the Scheduled Castes.

During the supply year 2015-16, exactly the same situation has persisted in this village. Consequently, all the tail end farmers and a few middle reach ayacutdars did not receive the KPT supply fully. The channel encroachment in many places and almost disappearance of channels in the tail end areas prevented to receive any tank supply during 2015-16. As a result, all the non-well farmers in the tail end areas and majority in the middle reach did not do cultivation by using tank water. The local irrigation institution has become defunct and the representative of WUA, viz., the territorial constituency member (TCM) was practically not doing any fruitful work to get the tank supply. The caste factions between head (Reddis) and tail end (Scheduled Castes) also fueling for the defunct of local institutions. Only well farmers, who have adequate supply from their wells and farmers located close to the main supply channels able to crop their lands only for one crop season.

Athipattu is a tail end village supplied from the deeper Fox sluice, which also supplies water to four other villages. This village is entitled to get one share of tank water out of twelve from the Fox sluice. In a normal supply year this village gets tank supply for two full crop seasons (eight months). The Fox sluice supply in 1991-92 was normal. However, the supply available from the main channel was inadequate to feed all the villages served by the channel, mainly due to encroachment on the channel. Athipattu farmers reported that they had not received adequate supply because the head reach farmers tapped the maximum supply. This was because most of the well farmers were not willing to cooperate to maintain the channels and share the tank supply. This is because the village is located close to the Palar river and groundwater is abundant; and free electricity for pumpsets provided by the government enables the rich farmers to use the groundwater liberally.

As a result of non-cooperation of well farmers for maintenance, the non-well farmers alone were not able to undertake the entire maintenance work by using their own labour. Further, their fragmented holdings did not give them incentive to participate in the maintenance. Even if the tail reachers alone maintained the channels, they were not entitled to get the tank supply exclusively, denying the supply to the well farmers who did not actually participate in the maintenance. As a result, only the head reach farmers (including a few well-farmers located at the head reach) maintained the channels and took the entire supply available to the village and no *murai* was followed. The abundant availability of well water supply and the dual caste structure between the tail (majority of the farmers are Scheduled Castes) and the head reach (majority of the farmers are Mudaliars) affected the functioning of the village organisation adversely. Due to this change, the worst sufferers are the tail enders.

A few respondents recalled that some two to three decades ago they had practised the *murai*. According to the *murai* 1.8 hectares were irrigated in one *Jamam* (three hours), or 0.40 ha got 40 minutes supply. This supply was available once a week and it continued for about two months at the final period of the crop season. One of the respondents explained another version of the *murai* that was followed long back. According to this system, the entire village *ayacut* was divided into five *Karai Murai* (otherwise known as *Pangu Murai*) such as Puthuran *Veetu Karai*; Chinna *Pangu Karai*; Rayar *Veetu Karai*, Mudaliar *Veetu Karai* and *Papan Veetu Karai*. Each *Karai Murai* was having an *ayacut* of about 23 hectares (hence a total of 113 hectares) and they had to send 12 persons per *Karai* for the maintenance of channels. In 1991-92 only *Puthuran Veetu Karai* was able to get the tank water (because of its head reach location) and the remaining *Karais* did not receive tank water at all. The respondents were not able to clearly explain when and why the *murai* was adopted initially and how water was distributed among the *Karai Murai*.

However, most of the respondents specifically told that only after the 1970s the *murai* got disintegrated. This was mainly due to the introduction of new agricultural technology and extensive development of well irrigation in this village. These affected the marginal land holders. They were not able to get the tank supply due to non-cooperation of well-farmers for channel maintenance. Most of them also were not able to purchase well water for their crop production.

During the supply year 2015-16, as mentioned earlier, due to full encroachment of tank supply channel at Kaveripakkam village limit, tank water supply was fully closed for irrigation in the past two decades. As a result, two of the tail end villages viz., Athipattu and Sirukarumbur did not receive any tank supply in 2015-16. However, these two villages are situated adjacent to the river Palar that helps to get more groundwater resource for cultivation for these villagers.

In 1991-92 supply year, **Vagai** got adequate supply from DMT for two full crop seasons. In this village *Oru Madai Paichal* is to be followed only a month before the harvest when the tank supply is inadequate to feed the entire *ayacut*. This happened last in the supply year 1977-78, after which no *Oru Madai Paichal* was followed.

In the supply year 2015-16, DMT has got FTL storage. Consequently, Vagai also got adequate supply for more than two crops and no scarcity in supply was reported by any of the farmers.

The *ayacut* of **Punaithangal** is divided into two halves. The first is called *Pallakkal Paichal* (lands irrigated by lower level supply channel), about 32.40 hectares getting supply from the Dusi sluice. The second is *Mettukkal Paichal* (lands irrigated by higher level supply channel), about 26.32 hectares getting supply through the Mamandur sluice. As a result of non-cooperation by the Mudaliar well-farmers (whose lands are located mostly at the tail end of the village *ayacut*) for the maintenance of channels, the *Mettukkal Paichal* lands did not receive supply for more than ten years. Hence in the latest supply year only the *Pallakkal Paichal* lands received adequate tank supply for two full crop seasons. Since there was no water scarcity *Oru Madai Paichal* was not followed this year because of the damaged sluice shutter in the Dusi low level sluice which allowed a large quantity of water to flow through during the supply season. The problem was "excess" supply rather than deficit in 1991-92 supply year.

During the supply year 2015-16, no deficiency in supply was reported by the *Pallakkal Paichal* farmers. Whereas, in the *Mettukkal Paichal* lands part of the lands were sold out and the remaining ayacut was grown with prosopis juliflora, hence no cultivation was undertaken by the farmers.

Narasamangalam is a head reach village of the Mamandur deep sluice which received enough supply for two full crop seasons. Even if there is any scarcity of supply, the villagers follow only direct irrigation; and no rationing system was ever in vogue.

The same situation has been prevailing even in 2015-16 supply year in Narasamangalam.

In a normal supply year water is available for about six to seven months to the tail end village of **Sothiyampakkam** served by the Mamandur elevated sluice. In 1991-92, supply was available for five to six months and it was enough for one full crop. Normally, when the tank supply was inadequate for the last one or two months during the crop season, *murai* is followed. According to the *murai*, this village is entitled to get 1.5 *pangu* water from the Sothiyampakkam main supply channel. The entire *ayacut* is divided into three parts, namely, *Adi murai* (bottom rationing), *Nadu murai* (middle rationing) and *Thulir murai* (top rationing). Each part is entitled to get a supply of four *Jamum* (12 hours) period with an *ayacut* of 18.2 hectares. Hence one 0.40 hectare got 16 minutes supply per every 36 hours or 1.5 days. So in a week, 72 minutes supply is available per 0.40 hectare which was distributed in 4.5 times. According to respondents this system was practiced in this village until a decade ago.

In the supply year 2015-16, all the well and non-well farmers able to crop their lands at least for one full crop season. In the second season also, water was available but it was utilised by well farmers only. Apart from people's labour contribution for maintenance of channels, this villagers also used machine (pocklain) to undertake channel maintenance during the initial maintenance period and the amount is shared by ayacutdars themselves.

7.1 Detailed Survey of Farmers, 2015-16 and 2005-06

The above account of water supply and allocation in sample villages gives an overall picture obtained from village functionaries, which gives a broad idea of the management of tank supply. A somewhat more accurate picture on the way water is managed by the *ayacutdars* within the village and whether or not they actually followed the rationing system during the tank supply period is obtained from the farm level survey.

Data have been collected on (i) the conditions of tank water supply in terms of adequacy and reliability in two crop seasons, namely, the second (October to January) and the third (February to May) in the latest two supply years in each survey period; (ii) how inadequacies were managed; (iii) the irrigation practices at field level both in normal and scarcity periods; and (iv) specific problems faced by the farmers to irrigate lands.

7.1.1 Conditions of Tank Water Supply in KPT and DMT - 2015-16

Data codified and presented in Table 6.8. show that in most of the villages in both tanks in the second season majority of the respondents reported getting adequate and regular supply. Inadequate but regular supply was available to two villages each in both the tanks, mostly to the tail end *ayacutdars*. Among them, the most affected farmers were in Alapakkam and Sothiyampakkam (irrespective of types of farmers). "No supply" responses were mostly from the tail end villages; and as indicated already two of the KPT villages altogether did not get the tank supply. Respondents in the fourth group (water was available but not used) were negligible.

In the third season the pattern had changed significantly. Majority of farmers in Narasamangalam and Vagaiin DMT and in Panniyur village in KPT reported adequate/regular supply. Inadequate but regular supply was reported in Panniyur, and Sothiyampakkam. All the farmers in Kadaperi and Alapakkam, and a sizeable proportion in Punaithangal and Sothiyampakkam did not receive tank supply. Some farmers in Panniyur and Vagai and majority of tail enders in Narasamangalam reported that supply was available but not used. It is clear that the worst affected farmers were in the tail/elevated sluice villages; and within the village, the less fortunate farmers are the tail segment farmers without wells.

Cultivators reporting "inadequate/regular" supply have to make extra efforts to sustain cultivation. Three types of responses (as given, in Table 6.8) about the "period of inadequacy" have been received. In the second season none of the village (except Alapakkam and Narasamangalam) faced any scarcity. In these two villages, inadequacy was mainly towards the end of the season; though a few also report inadequacy in the beginning.

In the third season, except Panniyurin KPT; Vagai and Narasamangalam, all the remaining villages faced inadequacy of supply (note that Kattalai, Athipattu, Kadaperi and Alapakkam did not receive the tank water). The

T able 6.8: Conditions of Water Supply in KPT and DMT, 2015-16.

Tank and sample villages	No.of sample HHs by type	Reliability of supply								Period of Inadequacy					
		Season 2				Season 3				Season			Season		
		AR	IR	NS	AU	AR	IR	NS	AU	OT	LT	TS	OT	LT	TS
KPT															
Kadaperi	Well 8	5		1	2			8							
E (H)	Head 8	5	1	1	1	1	1	6			1			1	
	Tail 8	4			4		1	7						1	
Panniyur	Well 8	5	1	1	1	1	4	2	1			1		3	1
E (H)	Head 8	5			3	1		5	2						
	Tail 8	6	1		1		1	7				1			1
Kattalai	Well 8			8				8							
D (H)	Head 8			8				8							
	Tail 8			8				8							
Alapakkam	Well 8	1	5	2				8				5			
M (T)	Head 8	2	4	2				8				4			
*1	Tail 8	1	4	3				8				4			
Athipattu	Well 8			8				8							
D (T)	Head 8			8				8							
*1	Tail 8			8				8							
DMT															
Vagai	Well 5	3		2		4	1							1	
	Head 8	6			2	2			6						
	Tail 8	7			1				8						
P.Thangal.	Well 8	3	1	4		1	1	6			1		1		
	Head 8	6		2				8							
	Tail 8	2		6		2		6							
N.Mangalar	Well 8	4	3	1		3		3	2		1				
	Head 8	7	1			2		6		1					
*1	Tail 8	6			2			2	6						
S.Pakkam.	Well 8		6	1	1	7		1		6		1			
	Head 8	2	4	2		2	2	4		3	1			2	
*3	Tail 8	4	3	1		2	2	4		3	1			1	

(*Contd...*)

Methods of resolving inadequacy of supply																		Specific problem faced							
Season 2									Season 3									Season 2				Season 3			
1	2	3	4	5	6	7	8	9	1	2	3	4	5	6	7	8	9	NP	DS	DG	MP	NP	DS	DG	MP
2		1		5					4		4							5			3	4			4
		1		6			1				7					1		3		1	4	4			4
		4		4						1	7							4			4	3	1		4
3				5					6		1		1					3	1		4	3	1		4
		3		5							7		1					4			4	4			4
				7	1				1		7							6	1		1	6	1		1
8									8									8				8			
	8									7	1							8				8			
	5	3								5	3							8				8			
7				1					5		3							2	5		1	6	2		
2	2	1		2	1					1	7							3	4		1	8			
2		3		1	1			1			8							4	4			8			
8									7		1							8				8			
3		4		1					2		6							8				8			
3	1	4							3	1	4							8				8			
1		1		3					1				4					2		3		2		3	
		2		6							6		2					3		5		6		2	
		1		7							8							1		7		8			
4		1		3					4		3		1					6	1	1		7	1		
		2		6							6		2					7		1		7		1	
		6		2							6		2					7		1		8			
3		1		4							5		3					6		2		7		1	
1				7							2		6					7	1			8			
1		1		6					1		7							6		2		6		2	
7		1									1		7					3	5			8			
5				3							4		2	2				4	4			6	2		
3				5					2		4		2					5	3			6	2		

Note: RELIABILITY OF SUPPLY: AR = Adequate/Regular; IR = Inadequate/Regular; NS = No supply; AU = Water was available, but not used.

PERIOD OF INADEQUACY: OT = One to T wo months supply only was available at the beginning; LT = Last one to two months supply was inadequate.

T S = T hroughout season inadequate supply only was available, tank water alone was not possible for cropping.

MET HODS OF RESOLVING INADEQUACY OF SUPPLY: (T he numbers in heading denote the possible reasons attributed by the sample farmers).

1 = Using own well water; 2 = By purchased water; 3 = Not cropped fully; 4 = Ragi was cropped against paddy. 5 = Problem was not faced; 6 = Poor supply/drainage problem leads low yield/ crop failed; 7 = As a result of pest attack water was not used after 2 months cropping; 8 = Rainfall + Short duration crops + by using eravai paichal, all those helped to get some yield. 9. Shortage was adjusted by the local Arakkaneri tank supply + local T hamarai Kulam pond water + Purchased Murai water at Rs. 15/hour.

SPECIFIC PROBLEMS FACED: NP = No problem; DS = Delay in supply; DG = Drainage.

MP = With regard to specific problem we have taken only the major problem that the respondents faced: 1. Well water becomes saline, poor yield. 2. Supply channel is

*1 and *3 refer to the joint ownership of wells as reported by one, three respondents respectively. 1* refers sample farmer uses his own well water.

E (H) = Elevated sluice head reach village; D (H) = Deep sluice head reach village; D (T) = Deep sluice tail reach village; M (T) = Middle sluice tail reach village.

Source: Field survey data, 2017.

ayacutdars of Narasamangalam also faced water shortage during the last one month of third season. The most affected village was apparently Sothiyampakkam where only two month's supply was available, which was utilised by the well farmers alone (note that the two farmers each as shown in head and tail reaches also have own wells).

In order to resolve inadequacy problem, farmers adopt some mechanisms in all the villages. These are classified into nine types as shown in Table 6.8. Farmers who own wells are generally better placed and used their well supply to supplement tank water at least once in the second season(apparently the well farmers in the head/elevated villages did not face inadequacy). However, in the third season, the head (Narasamangalam) and the elevated sluice (Panniyur) villages also joined this category (note, here we are talking about the supply conditions from the tank alone but not in the wells).

In the second season purchased well water was reported mainly in Kattalai and Alapakkam. It should be mentioned, although Kattalai village had not received tank supply altogether, but all categories of farmers are able to cultivate in the first two seasons by using purchased well water. Since the village is located adjacent to the tank, well water is adequately available. In the third season, the incidence of own well water was more andpurchased well water is negligible. In this season, almost in majority of villages, there were more respondents in the "not cropped" category. In all the villages, in both the second and third seasons, a majority of respondents reported that they did not face any problem but a few reported that the problems were not resolved.

Apart from the problem of inadequacy of supply a considerable proportion of respondents faced some specific problems, which were mainly related to agronomic and locational factors. These include lack of drainage, delay in receiving supply, illegal lifting of water for irrigation, land getting submerged when the local tank reached FTL, and encroachment on the entire channel resulting in no tank supply. These problems are represented under four sub-headings in "specific problems faced" as shown in Table 6.8. The problem free farmers are more in DMT than in KPT in the second season, while the reverse is the case in third season. This was due to the high incidence of "no supply" in Kadaperi and Alapakkam in the third season. Delay in supply was reported by mostly the tail end villages in both the tanks in the second season; whereas in the third season, except Sothiyampakkam, no other village reported more on this. Within the village tail segment of non-well farmers were mostly affected by delay in receiving supply. Regarding drainage (water logging), the head reach villagesin DMT, Vagai and Punaithangal were affected in both the second and third seasons, but Narasamangalam was affected only in the second season. Only in drainage different types of farmers in a village were equally affected. Mainly Kadaperi and Panniyur faced the specific problem in the second season and only in Panniyur in the third season. This was due to locational factor and the channel encroachment.

By and large, tank water supply was favourable in the second season in all the villages (except Alapakkam and Punaithangal). But in the third season, all the villages under DMT (except Sothiyampakkam) and Panniyur in KPT reported adequate supply. Even in Punaithangal a majority of well farmers and tail segment of non-well farmers did not get the tank supply. Alapakkam was severely affected even in the second season. The reliability of supply was less favourable to the tail segment farmers within the village (eg. Panniyur, Punaithangal and Sothiyampakkam). Regarding inadequacy of supply Alapakkam and Punaithangal farmers were worst affected even in the second season (note that in Punaithangal, only the farmers served by Mamandur sluice were affected). But in third season all the tail/elevated sluice villages were affected considerably. However, we could not find any disparity between the head and tail segment farmers within the village who were affected by this problem.

Regarding methods of resolving the problems, in the second season using own well water was significant in all the tail end villages. Use of purchased well water was mainly in Kattalai and was negligible in the remaining villages. Farmers not facing any problem was widely reported in all the villages (except in Alapakkam) in all types of farmers. This was due to good availability of tank water supply. In the third season also farmers not facing problem was reported in majority of the villages. This was due to two reasons:

(1) Non-availability of tank supply to Kadaperi and Alapakkam; Also in Kattalai and Athipattu; and

(2) Availability of enough tank supply to Panniyur in KPT and all the villages in DMT except Sothiyampakkam.

Regarding specific problems, especially drainage, a few head reach villages (irrespective of type) were worst affected. The problem was more intense in the second season than in the third. The problem of drainage was attributed mainly to the non-existence of field channels and the widespread practice of "surface-flooding" irrigation. The scrutiny of the village maps in regard to the magnitude of the availability of field channels clearly shows the non-existence of field channels in two of the head reach villages (eg. Vagai and Narasamangalam) in DMT, and existence of field channels in most of the KPT villages. This was corroborated by the direct visits to the villages.

7.1.2 Conditions of Tank Water Supply in KPT and DMT - 2005-06

Data codified in Table 6.9. show that in most of the villages in both the tanks in the second season majority of the respondents reported getting adequate and regular supply. Inadequate but regular supply was available to three villages for most farmers irrespective of types of *ayacutdars* in KPT. Among them, the most affected farmers were in Alapakkam (irrespective of types of farmers). "No supply" responses were mostly from the tail end villages, mainly from Athipattu in KPT and Punaithangal in DMT. Respondents in the fourth group (water was available but not used) were negligible.

In the third season the pattern had changed significantly. All farmers in Kattalai in KPT and a majority in Vagai and Narasamangalam in DMT reported adequate/regular supply. In Vagai a different response is observed, that is all the head and tail reach of non-well farmers reported that "water was available but not used." This means the entire ayacut was flooded so the crop did not require any irrigation throughout the season. Inadequate but regular supply was reported by a few well type respondents in Athipattu, and all categories in Narasamangalam and Sothiyampakkam in DMT. All the farmers in all villages in KPT (except Kattalai), and majority in Punaithangal and Sothiyampakkam did not receive tank supply in third season. In Vagai, as already indicated middle and tail reach farmers report that supply was available but not used, due to drainage problem. It is clear that the worst affected farmers were in the tail/elevated sluice villages; and within the village, the less fortunate farmers are the tail segment farmers without wells.

Cultivators reporting "inadequate/regular" supply details are given, in Table 6.9. On the "period of inadequacy" in the second season none of the village (except Athipattu and Punaithangal and Narasamangalam) faced any scarcity. In these three villages, inadequacy was mainly towards the end of the season; though a few also report inadequacy throughout the season.

In the third season, except Kattalai and Vagai, all the remaining villages faced inadequacy of supply (note all the remaining 4 villages in KPT did

Table 6.9: Conditions of Water Supply in KPT AND DMT, 2005-06.

Tank and	No.of	Reliability of supply								Period of Inadequacy						Methods of resolving inadequacy of supply																		Specific problem faced							
sample	sample HHs	Season 2				Season 3				Season 2			Season 3			Season 2									Season 3									Season 2				Season 3			
villages	by type	AR	IR	NS	AU	AR	IR	NS	AU	OT	LT	TS	OT	LT	TS	1	2	3	4	5	6	7	8	9	1	2	3	4	5	6	7	8	9	NP	DS	DG	MP	NP	DS	DG	MP
KPT																																									
Kadaperi	Well 8	6	2					8			2					2				6					5		3							7			1	6			2
E (H)	Head 8	8						8												8							8							5		1	2	6			2
	Tail 8	6			2			8										2		6							8							5			3	5			3
Panniyur	Well 8	5	2	1				8			2	1				2				6					3		5							4	1		3	5			3
E (H)	Head 8	7	1					8			1									7	1						8							7			1	7			1
	Tail 8	4	3		1		1	7			3				1					5	3				1		7							5	2		1	6	2		
Kattalai	Well 8	8				8	3							2	1					8					3				5					8				5	3		
D (H)	Head 8	8				8														8									8					8				8			
	Tail 8	8				7			1											8									8					8				8			
Alapakkam	Well 8		7	1				8			5	3				7				1					3		5							3	5			8			
M (T)	Head 8	6	2					8				2					2			5	1				1		7							5	3			7	1		
1*	Tail 8		6	2				8				5				3		3			2						8							2	6			8			
Athipattu	Well 8	2	3	3			4	4			2	1			4	6				2					8									5	3			3	5		
D (T)	Head 8	4	1	3				8			1						1	3		4							8							7	1			8			
	Tail 8	2	1	5				8			1					2	2	1		3					1		7							8				7	1		
DMT																																									
Vagai	Well 5	5				5										1				4					1				4					2		3		1		4	
	Head 8	7			1				8									1		7							8							3		5		8			
	Tail 8	8							8											8							8							2		6		8			
P.Thangal.	Well 8		3	5		2		6			3					6		1		1					6		2							7	1			7	1		
	Head 8	6	2					8			2							2		6							8							8				8			
	Tail 8	5		3				8										2		5	1						7			1				7		1		6		2	
N.Mangalam	Well 8	7	1			7	1				1				1	1				7					4		3		1					7		1		6	1	1	
	Head 8	8				4	1		3					1						8					1		1		6					7		1		8			
*1	Tail 8	7			1	5	3							3				1		7					1		5					2		7		1		5	2	1	
S.Pakkam.	Well 8	6	1	1			1	7			1			1		1		1		6					2		2		4					7	1			7	1		
	Head 8	5	2	1			2	6			2			2		3				5							6		2					6	2			8			
*3	Tail 8	7		1			2	6				1		2				1		7					2		6							8				6	2		

Note: RELIABILITY OF SUPPLY: AR = Adequate/Regular; IR = Inadequate/Regular; NS = No supply; AU = Water was available, but not used.

PERIOD OF INADEQUACY: OT = One to Two months supply only was available at the beginning; LT = Last one to two months supply was inadequate.

TS = Throughout season inadequate supply only was available, tank water alone was not possible for cropping.

METHODS OF RESOLVING INADEQUACY OF SUPPLY: (The numbers in heading denote the possible reasons attributed by the sample farmers).

1 = Using own well water; 2 = By purchased water; 3 = Not cropped fully; 4 = Ragi was cropped against paddy.

5 = Problem was not faced; 6 = Poor supply/drainage problem leads low yield/ crop failed; 7 = As a result of pest attack water was not used after 2 months cropping; 8 = Rainfall + Short duration crops + by using eravai paichal, all those helped to get some yield.

9. Shortage was adjusted by the local Arakkaneri tank supply + local Thamarai Kulam pond water + Purchased Murai water at Rs. 15/hour.

SPECIFIC PROBLEMS FACED: NP = No problem; DS = Delay in supply; DG = Drainage.

MP = With regard to specific problem we have taken only the major problem that the respondents faced: 1. Well water becomes saline, poor yield. 2. Supply channel is

*1 and *3 refer to the joint ownership of wells as reported by one, three respondents respectively. 1* refers sample farmer uses his own well water.

E (H) = Elevated sluice head reach village; D (H) = Deep sluice head reach village; D (T) = Deep sluice tail reach village; M (T) = Middle sluice tail reach village.

Source: Field survey data, 2017.

not receive tank water at all). The *ayacutdars* of Narasamangalam also faced water shortage during the last two months of third season. The most affected village was apparently Sothiyampakkam where only a month's supply was available, which was utilised by the well farmers alone (note that the two farmers each as shown in head and tail reaches also have own wells).

For resolving inadequacy problem, farmers adopt the following measures as shown in Table 6.9. The 'well' farmers are generally better placed and used their well supply to supplement tank water as and when required in the second season (apparently the well farmers in the head/elevated villages did not face inadequacy). However, in the third season, the head (Narasamangalam) and the elevated sluice (Panniyur) villages also joined this category, that is in the third season available well water supply also decreased considerably, so not cropped responses increased (note, here we are talking about the supply conditions from the tank alone but not in the wells).

In the second season purchased well water was reported mainly in Alapakkam. In the third season, the incidence of purchased well water was more, and mostly in the tail end villages. Only in the tail end village of Punaithangal there were more respondents in the "not cropped" category in the second season, while in the third season, the respondents in Sothiyampakkam village also reported more of this problem. In all the villages, in both the second and third seasons, a majority of respondents reported that they did not face any problem but a few reported that the problems were not resolved.

Regarding specific problems, which were mainly related to agronomic and locational factors, Table 6.9 gives the following details. The problem free farmers are more in DMT than in KPT in the second season, but in the in third season DMT farmers face more problems. This was due to the high incidence of "no supply" in Kadaperi and Alapakkam in the third season. Delay in supply was reported by mostly the tail end villages in both the tanks in the second season; whereas in the third season, except Athipattu, no other village reported this. Within the village tail segment of non-well farmers were mostly affected by delay in receiving supply. Regarding drainage (water logging), the head reach villages including Kadaperi were the most affected. In DMT, Vagai and Punaithangal were affected by drainage in both the second and third seasons, but Narasamangalam and Sothiyampakkam were affected only in the second season. Mainly Kadaperi and Panniyur faced the specific problem in the second season and only in

Panniyur in the third season. This was due to locational factor and the channel encroachment.

By and large, tank water supply was favourable in the second season in all the villages (except Alapakkam, Athipattu and Punaithangal). But in the third season, all the villages under DMT (except Sothiyampakkam) and Kattalai in KPT reported adequate supply. Even in Punaithangal a majority of well farmers and tail segment of non-well farmers did not get the tank supply. Alapakkam was severely affected even in the second season. The reliability of supply was less favourable to the tail segment farmers within the village (eg. Panniyur, Athipattu, Punaithangal and Sothiyampakkam). Regarding inadequacy of supply Alapakkam and Punaithangal farmers were worst affected even in the second season (note that in Punaithangal, only the farmers served by Mamandur sluice were affected). But in third season all the tail/elevated sluice villages were affected considerably. However, we could not find any disparity between the head and tail segment farmers within the village who were affected by this problem.

Regarding methods of resolving the problems, in the second season using own well water was significant in all the tail end villages. Use of purchased well water was mainly in Alapakkam and was negligible in the remaining villages. Farmers not facing any problem was widely reported in all the villages (except in Alapakkam) in all types of farmers. This was due to good availability of tank water supply. In the third season also farmers not facing problem was reported in majority of the villages.

7.2 Irrigation Practices at Field Level

7.2.1 Methods of Irrigation

In order to study the methods of irrigation information was sought from the sample households on the method that they normally followed in each season to irrigate lands and their collective efforts for sharing tank water during normal/scarcity periods. The survey indicated three types of irrigation (Table 6.10): (1) Field to field; (2) Using field channels; and (3) Eravai Paichal.

Field to field method was found in all the villages in DMT mostly among non-well farmers. The incidence of this method is the highest in Vagai followed by Punaithangal in all categories of farmers. The method is used where field channels were not available. But this practice leads to drainage problem and loss of productivity.

Table 6.10: Irrigation Methods Follow ed in the Sample Villages in KPT and DMT, 2015-16.

Name of tank and sample villages	Sample HHs by type	HHs w hich did not follow MURAI in NORMAL supply period		HHs w hich follow ed MURAI in 2015-16		HHs w hich did not follow MURAI in 2015-16 and reasons						Methods of irrigation for all seasons			
						Season 2			Season 3						
		Season 2	Season 3	Season 2	Season 3	No Scar-city	Not cro-pped	No or partial tank supply	No Scar-city	Not cro-pped	No or partial tank supply	FF	FC	EP	OS
1	2	3	4	5	6	7	8	9	10	11	12	13	14	15	16
KPT															
Kadaperi	Well 8	8	8	0	0	6	1	1	0	2	6	0	7	0	1
E (H)	Head 8	8	8	0	0	6	2	0	2	0	6	0	6	2	0
	Tail 8	8	8	0	0	4	4	0	0	5	3	0	7	1	0
Panniyur	Well 8	8	8	0	0	5	0	3	1	1	6	0	7	0	1
E (H)	Head 8	8	8	0	0	6	2	0	7	1	0	0	8	0	0
	Tail 8	8	8	0	0	6	1	1	5	2	1	0	8	0	0
Kattalai	Well 8	8	8	tank supply was not available - sluice was damaged fully											
D (H)	Head 8	8	8	tank supply was not available - sluice was damaged fully											
	Tail 8	8	8	tank supply was not available - sluice was damaged fully											
Alapakkam	Well 8	8	8	0	0	1	0	7	0	2	6	0	8	0	0
M (T)	Head 8	8	8	0	0	2	1	5	0	8	0	0	8	0	0
	Tail 8	8	8	0	0	1	3	4	0	8	0	0	8	0	0

(Contd...)

Athipattu	Well 8	8	8	tank supply was not available - main channels encroached fully											
D (T)	Head 8	8	8	tank supply was not available - main channels encroached fully											
	Tail 8	8	8	tank supply was not available - main channels encroached fully											
DMT															
Vagai	Well 5	5	5	0	0	5	0	0	5	0	0	5	0	0	0
E (H)	Head 8	8	8	0	0	6	2	0	6	2	0	5	3	0	0
	Tail 8	8	8	0	0	7	1	0	8	0	0	6	2	0	0
Punaithangal	Well 8	8	8	0	0	7	1	0	5	3	0	4	4	0	0
E (T)	Head 8	8	8	0	0	6	2	0	0	6	2	3	5	0	0
	Tail 8	8	8	0	0	2	6	0	0	6	2	2	6	0	0
Narasamangalam	Well 8	8	8	0	0	7	1	0	5	3	0	0	6	2	0
D (H)	Head 8	8	8	0	0	8	0	0	6	2	0	4	4	0	0
	Tail 8	8	8	0	0	6	1	1	7	0	1	4	3	1	0
SothiyamPakkam	Well 8	8	8	0	0	7	0	1	7	1	0	0	6	1	1
D (T)	Head 8	8	8	0	0	3	0	5	2	2	4	3	5	0	0
	Tail 8	8	8	0	0	5	0	3	2	4	2	1	7	0	0

Note: Column 12 refers that crop raised by using well water; HHs = Households; Method of irrigation: FF = Field to field.
FC = Using field channels; EP = Eravai paichal; OS = Others - Submergeable land, not cultivated and no supply/field channel.
E (H) = Elevated sluice head reach village ; D (H) = Deep sluice head reach village; D (T) = Deep sluice tail reach village.
M (T)= Middle sluice tail reach village; E (T) = Elevated sluice tail reach village.
Source: Field survey data, 2017.

The use of**field channels** is nearly universal among well farmers, and among a large majority of non-well farmers in KPT villages. The incidence is greater in the tail end farmers. In DMT, the practice is significant only in Sothiyampakkam. Well water everywhere is costly, and their supply being scarce in tail end villages, so farmers are more careful about using well water.

Eravai Paichal is a method by lifting water manually from the supply channel. It is practiced where the land is located at a higher elevation than the channel. The use of this type is negligible, but was reported by a few farmers in Kadaperi, Narasamangalam and Sothiyampakkam. In Kadaperi and Narasamangalam three sample households in each village reported using this method. It should, however, be noted that the method of irrigation as indicated was practiced in all seasons in every supply year. But, it is also reported that in recent years some farmers instead of practicing this method, set up an oil-engine or shifting motor to lift water by using mechanical power. This method affects water availability to the tail end ayacutdars in a channel since water is continuously pumped at the head reach itself. However, this practice is reported mostly in Narasamangalam villagers.

7.2.2 No Change in Normal Practice in Scarcity Period

From the respondents' point of view the methods of irrigation followed at field level is a rigid (unchangeable) one. It cannot be modified under any circumstances. These methods have been followed for a long time and still continue. This "unchangeable" nature was said to be mainly due to the location of the village from the tank, and the prevalence of unlined supply channels which pass through the lands. Seepage water creates water logging to the nearby lands (unless these channels are lined properly the method of irrigation followed may not change). The least efficient method is the field to field type. Wherever this method is practiced, the amount of manure and fertilisers applied by a farmer to his lands could not be effectively retained in his own lands due to drainage problem. As a result, productivity of these lands was comparatively low. In the sample villages no farmer household reported any change in the method of irrigation overtime.

Table 6.10 also shows that in normal supply period no farmer has followed the *murai* in both the second and third seasons. During 2015-16, Kattalai and Athipattu in KPT did not get the tank supply and in all the remaining sample villages in both tanks did not face any water shortage in the second season. But in the third season shortage of water supply was reported in many villages where tank water was available. The three tail/

elevated villages, viz., Kadaperi, Alapakkam in KPT and Punaithangal and Sothiyampakkam in DMT received little or no tank supply in the third season. As a result of poor/no tank supply, farmers reporting "not cropped" were significantly high in the third season than in the second. Consequently, no village had practiced *murai* in the second season. This was mainly due to "no scarcity" faced by the *ayacutdars*. In Athipattu, *murai* was practiced in the past it is not practiced now; and in 2015-16 KPT supply was not fully received by Athipattu. Also, it should be noted that murai has to be followed when the ayacutdars face inadequacy in tank supply by majority of the farmers in the village. Also to execute murai a robust local institution should function devotionally, but it is lacking now in all villages.

7.3 Status of Functioning of Irrigation Institutions

In any irrigation system the two important functions, viz., maintenance of irrigation infrastructure such as different kinds of channels that lead water from the source to the lands; and management of irrigation supplies, which is let out from the source to the cultivating lands. To undertake and perform these two functions effective functioning of irrigations institutions is a must. As we have elaborately discussed these aspects in chapter 4, enquiries are made to the respondents in each sample village to understand how effective the institutions are functioning in to undertake the said functions. The results of enquires are presented in Table 6.11.

Respondents were asked **three important questions**: **1.** Opinion about the functioning of irrigation institutions / WUAs on maintenance of channels and management of tank supply. Two types of responses are received on this question.

Response 1: WUA TC member was seen only once and not contacted afterwards. The WUA practically spoiled the previously functioned traditional local system. Since no tank supply available for many years the institution became defunct in the last 10-15 years. Heavy encroachment in channels itself shows that the Irrigation Institution was not functioning. Only during supply period on individuals' voluntary basis some cooperative effort is seen among users.

Response 2: Majority of the farmers responded the following: Poor functioning of II for maintenance work and not satisfactory; not functioning as before. One season supply only possible for many villages, so no one takes any initiative. Traditional institutions defunct now. Due to formation of WUA the earlier functioned good system collapsed.

Table 6.11: Status of Functioning of Irrigation Institutions in the Sample Villages in KPT and DM T, 2015-16.

Name of tank and sample villages	Sample HHs by type	Opinion about the functioning of irrigation Institutions / WUAs on water distribution		Know how about TNFMIS Act 2000. If not known why?		Opinion on: 1. Tank fisheries G.O. 2. Tree cutting in the tank vicinity. 3. Silt removal from the tank bed. 4. Any Others - Specify	
		1	2	1	2	1	2
1	2	3		4		5	
KPT							
Kadaperi	Well 8	1	7	6	2	1	7
E (H)	Head 8	2	6	7	1	3	5
	Tail 8	3	5	6	2	3	5
Panniyur	Well 8	2	6	5	3	3	5
E (H)	Head 8		8	8			8
	Tail 8	1	7	6	2	1	7
Kattalai	Well 8	tank supply was not available - sluice was damaged fully					
D (H)	Head 8	tank supply was not available - sluice was damaged fully					
	Tail 8	tank supply was not available - sluice was damaged fully					
Alapakkam	Well 8	5	3	6	2	3	5
M (T)	Head 8	2	6	7	1	5	3
	Tail 8	1	7	6	2	4	4
Athipattu	Well 8	tank supply was not available - main channels encroached fully					
D (T)	Head 8	tank supply was not available - main channels encroached fully					
	Tail 8	tank supply was not available - main channels encroached fully					
DMT							
Vagai	Well 5	3	5	5	3	2	6
E (H)	Head 8	4	4	6	2	4	4
	Tail 8	3	5	4	4	2	6
Punaithangal	Well 8	4	4	6	2	3	5
E (T)	Head 8	2	6	7	1	2	6
	Tail 8	1	7	5	3	2	5
Narasamangalam	Well 8	2	6	5	3	2	6
D (H)	Head 8	3	5	6	2	3	5
	Tail 8	3	5	8	0	3	5
SothiyamPakkam	Well 8	4	4	6	2	4	4
D (T)	Head 8	3	5	5	3	5	3
	Tail 8	2	6	4	4	3	5

Note: HHs = Households. E (H) = Elevated sluice head reach village; D (H) = Deep sluice head reach village.
M (T)= Middle sluice tail reach village; E (T) = Elevated sluice tail reach village; D (T) = Deep sluice tail reach village.

Sl. No. 3: 1. WUA TC member was seen only once and not contacted afterwards. WUA practically spoiled the previously functioned traditional local system. Since no tank supply for many years the institution has become defunct in the last 10-15 years. Heavy encroachment in channels itself shows that the II was not functioning. Only during supply peiod some cooperative effort is seen among users.

Sl. No. 3: 2. Poor functioning only for maintenance work and not satisfactory; not functioning as before. One season supply only possible so no one takes any initiative. Traditional institutions defunct now. Due to formation of WUA the earlier functioned good system collapsed.

Sl. No. 4: 1. No Act is known & nobody informed. 2. No time to ask about the Act from anyone.

Sl. No. 5: 1. All are the responsibility of the PWD, we do not know anything about these issues. 2. Not known about anything.

Source: Field survey data, 2017.

Question 2: Know how about the TNFMIS Act -2000. If not known why?

Response 1: Irrespective of tank and reaches majority of the farmers reported that they are unaware of any Act relating irrigation management and said that nobody has informed about this act so far. Response 2:A few respondents in each village reported that they had no time to ask about the Act from anyone.This is the status position of our important irrigation management ACT, either from the government side and also from people's perception.

Question 3: Opinion on 1. Tank fisheries G.O. 2. Tree cutting G.Os in the vicinity of tanks. 3. Information about silt removal from the tank bed; and 4. Any other details.

Response 1: For all these three queries some respondents replied that for these aspects the main responsibility is only with the PWD and replied that they do not know about anything of these issues. Response: 2. A large majority of the farmers simply said that they do not know anything about these issues.

8.1 Summary

In this chapter the conditions of tank water supply across villages and within each sample village served by KPT and DMT have been discussed. It is found that in both the tanks the rules and regulations prescribed by the *Mamulnamas* for sharing tank water across villages are still followed. In KPT most of the villages followed *murai* system during scarcity period in the past and now now. Whereas in DMT, OMP system was followed. Villages served by the deep sluices in both the tanks did not follow any *murai* system.

In a normal tank supply year, like 1991-92 and 2015-16, the deep sluice village of Kattalai and the tail end villages of Athipattu and Alapakkam normally get the tank water for two full crop seasons (8-9 months) and the elevated sluice villages get one crop season (5-6 months) supply. (However, due to encroachment of main supply channelAthipattu village had not received the 2015-16 tank supply. Also, due to breakage of deep sluice shutter, the Mulai sluice was altogether closed that prevented total supply to Kattalai village in 2015-16). Wherever water deficiency is faced the *ayacutdars* followed *murai* system. Accordingly, in KPT, in 1991-92 Panniyur and Alapakkam followed *murai* system and this is not followed in 2015-16. In the former village *murai* was followed in the third crop season. But in Alapakkam it was followed in the second crop season in

1991-92, since the tank supply was deficient even from the beginning in that year. In the tail end village of Athipattu the practice of *murai* system has disappeared and most farmers do not use the tank water, whereas in 2015-16, Athipattu did not get KPT supply fully. In DMT, in 1991-92 and 2015-16, all the sample villages, except the tail end of Sothiyampakkam, got adequate tank supply for two crop seasons (7-8 months) and Sothiyampakkam got supply adequately for one crop season (5 months) as per the entitlement. Hence all these DMT villages did not follow OMP system.

The conditions of tank water supply in 2015-16 indicate that the second season supply was adequate to all sample villages that receive tank supply, except the tail end village of Alapakkam in KPT and Punaithangal in DMT. But in the third season all KPT villages got inadequate supply. But all DMT villages got adequate supply. (Note that the tail segment of the tail end village of Punaithangal did not get the tank supply since 1980 and Sothiyampakkam is entitled for a single season supply). In order to resolve the inadequacy of tank supply farmers widely supplement well water in both the tanks. It should be noted that in 2015-16 all the head reach villages served by the deep sluices in DMT were severely affected by drainage problem.

Our analysis revealed that the KPT villages are likely to be better in receiving tank water supply and exploiting groundwater. This is mostly because the assured availability of Palar Anicut supply to KPT. Since KPT is located at the head reach of the system its water supply condition is relatively better than in DMT. This picture was reversed in the case of DMT, where in recent decades the available tank supply was inadequate to feed the entire *ayacut* and the availability of groundwater was also less than in KPT.

By and large, the analysis of the conditions of water supply in both the tanks shows that well farmers in all the villages are better placed than non-well farmers in using the tank supply. Between non-well farmers, head reachers were better placed than tail enders. Again, in the tail end villages, farmers located at the tail segment and the small and marginal land holders were worst affected.

Chapter - 7

Farm Level Survey

In the previous chapter the management of tank water in the latest two supply years and anotherone available for both tanks in 1991-92 that was surveyed earlier are discussed. This chapter brings out, on the basis of a detailed survey of farmers served by the two tanks conducted in 2017, the differences in the conditions of water supply (especially adequacy and reliability) within the *ayacut*; the extent to which these reflect differences in socio-economic characteristics of the farmers (such as size of holdings, caste, whether owner or tenant cultivator), location, access to well water, soil and other material conditions, and the cumulative impact of these on cropping intensity, crop patterns and productivity of land as well as the variability in these respects between the latest two supply years.

The expectations, as formulated in chapter 1, on the basis of other available studies, are the following: (a) farmers in head reach villages and in head segment within the village would be better placed than those at the tail end; (b) owners of wells would be better placed in terms of water supply and productivity than non-well farmers using purchased well water, who in turn will be better placed than those depending wholly on tank water; (c) large farmers are likely to be better placed in terms of access to well water as owners and as purchasers and possibly even in access to tank water; and (d) the upper castes, being better placed than the Scheduled Castes in terms of both resources and power are likely to have better water accessand (e) natural factors such as climatic change effected erratic rainfall pattern and man-made severe encroachments in the main, branch and field channels prevented many tail end villagers to abstain from taking tank water.

The discussion of the results are divided into the following sections: (1) characteristics of holdings; (2) conditions of tank water supply; (3) use of groundwater; (4) land use and cropping patterns; and (5) productivity of land.

7. Results of the Survey

7.1 Characteristics of Sample Holdings

The discussion in this section focuses on two aspects: (1) the material conditions (defined by soil, location and access to well water) of the farmers; and (2) the social characteristics (defined by caste, size of holdings, incidence of tenancy and educational levels).

7.1.1 Material Conditions

7.1.2 Soil and Drainage Conditions

Soil Composition: The soils in the *ayacuts* of both tanks are broadly classified into five categories, *Kaliman* (clay); *Kalimanal* (sandy clay); *Manal* (sandy); *Vandal* (alluvial) and *Sengattuman* (red soil). The resettlement of 1913 compiled detailed plot-wise data on the soil composition of lands. The proportions of area under different soil types in the selected villages as recorded in the book of resettlement are given in Table 7.1. Since soil type does not change much over time, the Table can be taken to reflect the current situation in the villages and relates to all lands in the selected villages. Separate data for the *ayacut* lands are not readily available.

The fertility of these soils varies: *Kaliman* and *Kalimanal*, which retain water well but are prone to drainage problems, are suitable only for paddy. *Manal* is a poor soil in terms of fertility, suitable to grow coconut and other related tree crops but not paddy. *Vandal* is very rich for intensive cultivation, provided it is heavily fertilised. However, it is affected by flooding and water-logging. *Sengattuman* is a poor soil, though better than *Manal*. It has capacity to absorb water quickly, and transport it through the soil strata. It also dries up quickly and does not retain moisture very well. According to our respondents, the *Sengattuman* is good to grow groundnut; and also paddy if supplemented with adequate manure and fertilizers.

There are some variations in the soil composition across villages (Table 7.1). *Kaliman* is predominant in most villages. Its coverage ranges from 85 % in Kadaperi to 40 % in Alapakkam in KPT; and 90 % in Vagai to 60 % each in Punaithangal and Sothiyampakkam in DMT. It is around

Table 7.1: Soil and Drainage Conditions in the Sample Villages, 2017.

Tank and sample villages	***Percentage of soil type in the village ayacut***					***% of area in the ayacut having***	
	Kali-man	***Kali-manal***	***Manal***	***Vandal***	***Sen-Kattu-man***	***Good <--- drainage***	***No --->***
KPT							
Kadaperi E (H)	85	-	5	10	-	80	20
Panniyur E (H)	60	-	-	5	35	85	15
Kattalai D (H)	60	20	-	20	-	50	50
Alapakkam M (T)	40	10	-	-	50	95	5
Athipattu D (T)	65	20	15	-	-	90	10
DMT							
Vagai E (H)	90	10	-	-	-	0	100
Punaithangal E (T)	60	20	-	-	20	75	25
Narasamangalam D (H)	80	20	-	-	-	55	45
Sothiyampakkam D (T)	60	-	-	-	40	90	10

Note: Soil type relates only to the tank ayacut.
E (H) = Elevated sluice head reach village. D (H) = Deep sluice head reach village.
M (T) = Middle sluice tail reach village. D (T) = Deep sluice tail reach village.
E (T) = Elevated sluice tail reach village.
Source: Resettlement registers of 1913. Drainage details are based on the information collected from the household survey, 2017.

60 % in all the remaining villages under both the tanks. *Kalimanal* is found only to a lesser extent (between 10 and 20 %) in three villages each in both the tanks. *Manal* and *Vandal* soil types are not found significant in most villages. In KPT *Manal* is found in two villages (5 and 15 %) and *Vandal* in three villages (ranging from 5 to 20 %). In DMT these two soil types are not found at all. *Sengattuman* is found considerably in the elevated/tail reach villages (two villages each) in both the tanks, ranging from 35 to 50 % in KPT and 20 to 40 % in DMT.

Drainage Conditions: There is considerable variation in drainage conditions across villages and between the tanks. It is discussed in chapter 6, wherever field channels are absent the problem of drainage is more acute. KPT is better placed in this respect than DMT. In KPT respondents of most villages reported the availability of good drainage facilities in the *ayacut*, ranging from 50 to 95 %. Only in the deepest sluice village, Kattalai, the *ayacutdars* reported having 50 % or less of drainage facilities (Table 7.1). Even in 2015-16, tank supply was not available in Kattalai, so no drainage problem was reported. In DMT the head reach of Vagai does not have any drainage facility at all. All the respondents reported that since there is no field channel, the method of irrigation practiced is only field-to-field in the entire village *ayacut*. In the deep sluice village of Narasamangalam, about 55 % reported good drainage. The two tail end villages in the tank reported around 75 % and 90 % with good drainage facilities.

7.1.3 Access to Well Water

As discussed already in chapter 4 the supplemental wells in the *ayacut* play an important role in reducing severity in the scarcity of tank water especially in the lean supply periods. Well density, that is, wells per 100 hectares, in the sample villages in both the tanks show that wells are concentrated in the tail reach villages rather than the head reach. However, between the tanks the average well density is nearly doubled (89 to 192 wells / 100 ha) in KPT (especially in the tail end villages and villages served by elevated sluices) than in DMT (53 to 113 wells / 100 ha) in the tail end villages (see tables 3.25 and 3.26 in chapter 3). In KPT two points are worthy to note: (1) between 1989 and 2017 in eight villages – five of them are located in the tail end of the sluices and three are served by elevated sluices – percentage increase in wells during this period is impressive that ranges from -114% in Athipattu to 322% in Alapakkam; (2) during 2017, in all the head and middle reach villages (including the tail end village of Eralacheri – the SC dominated village) the well density is less than 100 wells per 100 ha of ayacut. This ranges between 31 wells /100 ha in Kaveripakkam to 64 wells/ 100 ha in Cheri village. In most of the villages the increase in number of wells / bores is mainly due to rocky nature of sub-surface conditions and poor aquifer permitted tapping of less quantum of water from these wells / bores. This curtailed the interest of farmers to dig more wells in the *ayacut*. However, due to the advent of bore-well technology farmers are able to dig more no. of bores, wherever possible, than to dig wells. So, in many villages one may notice bores are more than wells (eg. Kondapuram and Thuraiperumpakkam in KPT and Senianallur in DMT. In DMT somewhat a different situation is noticeable where the difference in well density across reaches is not much although one can find an increasing trend in well density if one moves from head reach to tail reach in both set of sluices (only in Dusi sluices in the middle and tail reach a reverse picture is observed. In DMT, in the head reach village Vagai where the well density is merely 19 wells per 100 hectares. Between Dusi and Mamandur sluices, the average well density is relatively more (65/100 ha) in the latter sluices compared to the former (46/100 ha) sluices. However, the average well density may not reflect the quantum of well water tapped. The latter depends on the yields of wells, which vary widely. It is reported that sub-soil conditions are more favourable in the KPT villages and not in DMT. Consequently, cropping pattern also differs between both tanks.

7.2. Social Characteristics of Cultivators

7.2.1 Castewise Distribution of Land Ownership

Table 7.2 shows the caste-wise distribution of land and the caste of the prominent landowners in the *ayacut* of each village in both the tanks. In two KPT villages and three DMT villages, bulk of the land is held by a single caste. In those cases, though one caste owns more than half the land,

Table 7.2: Principal Land Owning Castes and the Castes of the Large Owners in the Sample Villages in KPT and DMR, 2016.

KPT and DMR, 2016 (Area in hectares)

Name of tank and sample villages	Area Owned by Sample	Castewise distribution of ayacut land holdings <------							Large land owners
		Mudaliar	Nayakar	Brahmin	Reddy	Pillai	SCs	Others	
KPT									
Kadaperi	30.6	1.33	22.07	-	-	-	7.2	-	Nayakar
E (H)		(4.3)	(72.1)				(23.5)		
Panniyur	45.9	-	-	-	45.9	-	-	3.5	Reddy
E (H)					(92.4)			(7.6)	
Kattalai	67.2	1.1	35.3	-	-	-	30.8		Nayakar
D (H)		(1.6)	(52.6)				(45.8)		
Alapakkam	97.1	5.0	-	16.0	10.0	29.8	20.3	16.0	Pillai
M (T)		(5.1)		(16.5)	(10.3)	(30.7)	(20.9)	(16.5)	
Athipattu	94	55.7	23.4	-	-	6.6	5.1	3.2	Mudaliar
D (T)		(59.3)	(24.9)			(7.0)	(5.4)	(3.4)	
DMT									
Vagai	39.4	16.8	22.2	-	-	-	-	0.4	Nayakar
E (H)		(42.6)	(56.3)					(1.0)	
Punaithangal	46.8	21.8	-	-	-	1.0	24.0	-	Mudaliar
E (T)		(46.6)				(2.1)	(51.3)		
Narasamangalam	41.8	-	37.1	-	-	-	-	4.7	Nayakar
D (H)			(88.8)					(11.2)	
Sothiyampakkam	72.4	-	52.3	-	-	2.1	18.0	-	Nayakar
D (T)			(72.2)			(2.9)	(24.9)		

Note: E (H) = Elevated sluice head reach village.
D (H) = Deep sluice head reach village. M (T) = Middle sluice tail reach village.
D (T) = Deep sluice tail reach village. E (T) = Elevated sluice tail reach village.
Figures in brackets represent percentage to total ayacut. * indicates both wet and dry lands.
Others include Mesthri, Chettiar, Naidu and Muslim, Asari, Udaiyar, Navithar.
Source: Field Survey 2017.

significant amount is also held by other castes. In some cases, ownership is much diffused.

The dominant (upper) caste in terms of land ownership also happens to be the principal large landowners in Kadaperi, Panniyur and Kattalai in KPT and in all the DMT sample villages. In this connection it is relevant to note the following observations by Vaidyanathan and Janakarajan:

> … a situation of homogeneity between leading land owners and dominant caste in terms [of] overall land ownership should be conducive to effective collective management. But this is no guarantee of smooth management because there is always the possibility of intra-caste factions (1989: 123).

There are situations in which the large landowners belong to one caste, which is dominant in terms of total area owned, and landowners for other castes are dominant merely in terms of number. The landowners of the former category usually belong to a high caste while those of the latter category are mostly the Scheduled Castes. This is the case in Alapakkam and Athipattu in KPT and Punaithangal in DMT. Certainly, the high caste, large land holders exercise greater control over tank management. Again, as Vaidyanathan and Janakarajan observed:

> … where the caste which is numerous and prominent in terms of overall land ownership differs from the caste of the large land owners, the potentials for conflict are greater and more complex arrangements for managing village and tank matters are in evidence (ibid. 123)

The implication is that when land holdings are distributed among several different castes the problem of tank management will be greater.

The caste distribution of sample households in the selected villages gives a similar picture (see Table 7.3): In most of the villages cultivation is concentrated in either one or two castes. In all the head reach villages (and in the tail end village of Sothiyampakkam) in both the tanks a single caste (Nayakars) is dominant; in the tail end villages two castes (Scheduled Castes and Mudaliars/Nayakars) are dominant. In five out of the nine villages (two in KPT and three in DMT), Nayakars are dominant in terms of both the number and area cultivated; Reddies / Grounders are dominant in one KPT village; Mudaliars are dominant in one DMT village and Scheduled Castes are dominant one each in KPT and DMT villages. In all the villages taken together, Scheduled Castes tend to have a relatively higher concentration of lands in the tail reaches which are generally less favourable in terms of access to tank water.

Table 7.3: Sample Households by Caste, 2017

Name of tank and sample villages	Sample HHs by type	Sample households by caste					
		Nayakar	Mudaliar	Reddi	Pillai	Scheduled Castes	Others
KPT							
Kadaperi E (H)	Well 8	8	0	0	0	0	0
	Head 8	7	0	0	0	1	0
	Tail 8	6	1	0	0	1	0
Panniyur E (H)	Well 8	0	0	6	0	0	2
	Head 8	0	0	7	0	0	1
	Tail 8	0	0	8	0	0	0
Kattalai D (H)	Well 8	8	0	0	0	0	0
	Head 8	6	1	0	0	1	0
	Tail 8	6	0	0	0	2	0
Alapakkam M (T)	Well 8	0	1	1	4	0	2
	Head 8	0	0	0	8	0	0
	Tail 8	0	0	0	0	8	0
Athipattu D (T)	Well 8	1	5	0	1	1	0
	Head 8	1	4	0	2	0	1
	Tail 8	3	1	0	1	3	0
DMT							
Vagai E (H)	Well 5	3	2	0	0	0	0
	Head 8	8	0	0	0	0	0
	Tail 8	6	1	0	0	0	1
Punaithangal E (T)	Well 8	0	6	0	1	1	0
	Head 8	0	1	0	0	7	0
	Tail 8	0	0	0	0	8	0
Narasamangalam D (H)	Well 8	7	0	0	0	0	1
	Head 8	8	0	0	0	0	0
	Tail 8	5	0	0	0	0	3
Sothiyampakkam D (T)	Well 8	5	0	0	0	3	0
	Head 8	7	0	0	0	1	0
	Tail 8	2	0	0	1	5	0

D (H) = Deep sluice head reach village. M (T) = Middle sluice tail reach village.
D (T) = Deep sluice tail reach village. E (T) = Elevated sluice tail reach village.
Dash indicates nil; Others include Brahmin, Mesthri, Navithar, Chettiar, Udaiyar, Naidu, Asari and Musilim
Source: Field survey data, 2017.

7.2.2 Size of Holdings

As may be seen from Table 7.4 the average size of operational holdings, both overall and of wet land alone, is higher among well farmers than among non-well farmers in all the villages in both the tanks. Though the average size of operational holdings varies across villages, there is no consistent pattern in this respect.

The average size of holdings (both total and wet land) in the non-well group tends to be higher under the elevated Panniyur sluice of KPT than in

Table 7.4: Households by Operational and Ownership Holdings, 2017.

Tank and sample villages	Sample HHs by type	Total area of operational holdings		Average size of operational holdings			Total area of owership holdings		Average size of ownership holdings		
		Wet	Dry	Wet	Dry	Total	Wet	Dry	Wet	Dry	Total
1	2	3	4	5	6	7	8	9	10	11	12
KPT											
Kadaperi	Well 8	6.19	0.30	0.77	0.04	0.81	6.19	0.74	0.77	0.09	0.87
E (H)	Head 8	1.89	0.14	0.24	0.02	0.25	1.89	0.11	0.24	0.01	0.25
	Tail 8	2.91	0.05	0.36	0.01	0.37	2.91	0.00	0.36	0.00	0.36
Panniyur	Well 8	8.21	0.73	1.03	0.09	1.12	8.21	2.42	1.03	0.30	1.33
E (H)	Head 8	3.42	1.12	0.43	0.14	0.57	3.42	1.55	0.43	0.19	0.62
	Tail 8	2.59	0.37	0.32	0.05	0.37	2.59	0.39	0.32	0.05	0.37
Kattalai	Well 8	9.19	0.01	1.15	0.00	1.15	9.19	0.04	1.15	0.01	1.15
D (H)	Head 8	2.63	0.00	0.33	0.00	0.33	2.63	0.00	0.33	0.00	0.33
	Tail 8	3.32	0.00	0.41	0.00	0.41	3.32	0.00	0.41	0.00	0.41
Alapakkam	Well 8	16.34	1.22	2.04	0.15	2.19	16.34	8.05	2.04	1.01	3.05
M (T)	Head 8	3.30	2.55	0.41	0.32	0.73	3.14	3.40	0.39	0.43	0.82
	Tail 8	4.79	1.77	0.60	0.22	0.82	4.79	3.44	0.60	0.43	1.03
Athipattu	Well 8	22.26	0.61	2.78	0.08	2.86	22.27	5.47	2.78	0.68	3.47
D (T)	Head 8	4.59	0.00	0.57	0.00	0.57	4.59	0.00	0.57	0.00	0.57
	Tail 8	6.20	0.00	0.78	0.00	0.78	4.04	0.00	0.51	0.00	0.51
DMT											
Vagai	Well 5	9.06	0.00	1.81	0.00	1.81	7.23	0.00	1.45	0.00	1.45
E (H)	Head 8	2.38	0.00	0.74	0.00	0.74	2.28	0.00	0.70	0.00	0.70
	Tail 8	5.83	0.00	1.80	0.00	1.80	5.83	0.00	1.80	0.00	1.80
Punaithangal	Well 8	10.45	0.43	1.31	0.05	1.36	8.42	1.82	1.05	0.23	1.28
E (T)	Head 8	5.12	0.00	0.64	0.00	0.64	3.91	0.00	0.49	0.00	0.49
	Tail 8	3.49	0.93	0.44	0.12	0.55	3.49	1.32	0.44	0.16	0.60
Narasamangalan	Well 8	7.19	0.00	0.90	0.00	0.90	7.19	0.00	0.90	0.00	0.90
D (H)	Head 8	2.97	2.69	0.37	0.34	0.71	2.97	3.24	0.37	0.40	0.78
	Tail 8	3.15	1.11	0.39	0.14	0.53	3.15	1.42	0.39	0.18	0.57
Sothiyampakkan	Well 8	8.70	1.15	1.09	0.14	1.23	8.70	4.06	1.09	0.51	1.60
D (T)	Head 8	7.94	1.07	0.99	0.13	1.13	7.94	3.44	0.99	0.43	1.42
	Tail 8	3.93	0.78	0.49	0.10	0.59	3.93	1.23	0.49	0.15	0.65

Note: E (H) = Elevated sluice head reach village; D (H) = Deep sluice head reach village.
M (T) = Middle sluice tail reach village; D (T) = Deep sluice tail reach village.
E (T) = Elevated sluice tail reach village.
Source: Field survey data, 2017.

the deep sluice. In DMT, the picture differs, where the average size is higher in Narasamangalam and Sothiyampakkam served by the Mamandur (deep) sluice, compared to those under Dusi (elevated) sluice; compared to those under elevated sluice, within each sluice, the average size is lesser in the head reach villages compared to the tail end. But there is no clear pattern observed in respect of holdings in both the head and tail segments of the DMT villages.

Ownership holdings also follow a similar pattern. However, the size of ownership differs from the operational holdings in most cases. The difference gives an indication of the incidence and direction of tenancy: If operational holdings are more than the land owned, land is being leased in; the contrary being the case when operated area is smaller than the owned area. There is considerable variation in this respect between different categories of land, reach, and villages.

In DMT, unlike in KPT, in Dusi sluice villages, well farmers and head reach farmers leased in land. The incidence of tenancy is almost nil among the tail reach non-well farmers of KPT (except one in Athipattu) compared to DMT. Leasing out is almost nil in all villages among well and non-well (KPT) and non-well tail segment farmers (DMT).

There are three categories of cultivators: owners, tenants, and part owners/part tenants. Pure tenant cultivators under well holdings are nil in both KPT and DMT villages (Table 7.5). In the non-well head and tail segments, pure tenancy is very little or non-existent except one each in Alapakkam and Athipattu in KPT; and it is nil in all villages in DMT. Area under pure tenancy is less than 2 hectares. Mixed type tenancy is also very little in Athipattu in KPT and Vagai and Punaithangal in DMT villages. Area under mixed tenancy ranges between 0.86 ha in the head segment of Vagai to 3.64 ha under well farmers in Punaithangal.

On the whole, it should be mentioned that cultivation under tenancy - either pure or part – has been gradually disappearing in the villages served by tanks, which is evident from the study. This disappearance is partly because of highly interrupted supply to the tanks (the latest two supply years are 2005 and 2015 in both the tanks) and poor yield coupled with high input costs, especially human labour, prevented farmers to choose tenancy in recent decades. Moreover, unremunerative prices of output frightened people to get lands under tenancy. Further, government policy on 'land to the tiller' highly prevents farmers to give their lands under tenancy.

7.2.3 Social Status

To understand the relationship between the status of farmers including their education, participation in tank management and influence with the

Table 7.5: Households by Distribution of Land Ownership in Wet Land, 2017.

(Area in hectares)

Tank and sample villages	Sample HHs by type	Opera-tional holdings	Number of Cultivators			Area Operated By			% of Area Operated		
			Owners	Tenants	Part Owners & part Tenants	Owners	Tenants	Part Owners & part Tenants	Owners	Tenants	Part Owners & part Tenants
1	2	3	4	5	6	7	8	9	10	11	12
KPT											
Kadaperi	Well 8	6.19	8	-	-	6.19	-	-	100.0	-	-
E (H)	Head 8	1.89	8	-	-	1.89	-	-	100.0	-	-
	Tail 8	2.91	8	-	-	2.91	-	-	100.0	-	-
Panniyur	Well 8	8.21	8	-	-	8.21	-	-	100.0	-	-
E (H)	Head 8	3.42	8	-	-	3.42	-	-	100.0	-	-
	Tail 8	2.59	8	-	-	2.59	-	-	100.0	-	-
Kattalai	Well 8	9.19	8	-	-	9.19	-	-	100.0	-	-
D (H)	Head 8	2.63	8	-	-	2.63	-	-	100.0	-	-
	Tail 8	3.32	8	-	-	3.32	-	-	100.0	-	-
Alapakkam	Well 8	16.34	8	-	-	16.34	-	-	100.0	-	-
M (T)	Head 8	3.30	7	1	-	3.14	0.16	-	95.1	4.9	-
	Tail 8	4.79	8	-	-	4.79	-	-	100.0	-	-
Athipattu	Well 8	22.26	8	-	-	22.26	-	-	100.0	-	-
D (T)	Head 8	4.59	8	-	-	4.59	-	-	100.0	-	-
	Tail 8	6.20	6	1	1	2.63	1.35	2.23	42.4	21.7	35.9
DMT											
Vagai	Well 5	9.06	4	-	1	7.11	-	1.94	78.5	-	21.5
E (H)	Head 8	2.38	6	-	2	1.52	-	0.86	63.9	1.0	36.1
	Tail 8	5.83	8	-	-	5.83	-	-	100.0	-	-
Punaithangal	Well 8	10.45	6	-	2	6.80	-	3.64	65.1	-	34.9
E (T)	Head 8	5.12	7	-	1	3.77	-	1.36	73.5	1.0	26.5
	Tail 8	3.49	8	-	-	3.49	-	-	100.0	-	-
Narasamangalam	Well 8	7.19	8	-	-	7.19	-	-	100.0	-	-
D (H)	Head 8	2.97	8	-	-	2.97	-	-	100.0	-	-
	Tail 8	3.15	8	-	-	3.15	-	-	100.0	-	-
Sothiyampakkam	Well 8	8.70	8	-	-	8.70	-	-	100.0	-	-
D (T)	Head 8	7.94	8	-	-	7.94	-	-	100.0	-	-
	Tail 8	3.93	8	-	-	3.93	-	-	100.0	-	-

Note: E (H) = Elevated sluice head reach village; D (H) = Deep sluice head reach village.
M (T) = Middle sluice tail reach village; D (T) = Deep sluice tail reach village.
E (T) = Elevated sluice tail reach village.
Tenancy includes leased in + Mortgaged in lands by sample households within the village.
Source: Field survey data, 2017.

bureaucracy, data were collected from the sample respondents. The analysis shows that well farmers attained the highest average status score in all the villages (except Kadaperi and Athipattu in KPT) in both the tanks. Apparently in the well farmers' category the average score attained by the tail reach villages in KPT is more than the score attained by the head/ elevated villages (Table 7.6). Incidentally, within the village head segment

Table 7.6: Households by Level of Literacy and Socio-Political Status, 2017

Name of tank and sample villages	Sample HHs by type	No. of persons Studied					Average completed years of schooling (4 to 7)	Socio-Political status score	
		No schooling	Up to Primary	Up to Middle	Up to SSLC / PUC	Up to degree, diploma,		otal Sc	Avg score per sample household
1	2	3	4	5	6	7	8	9	10
KPT									
Kadaperi E (H)	Well 8	0	1	2	4	1	50	18	2.3
	Head 8	0	5	3	0	0	30	19	2.4
	Tail 8	2	2	1	2	1	34	21	2.6
Panniyur E (H)	Well 8	0	2	5	0	1	40	21	2.6
	Head 8	1	6	1	0	0	23	16	2.0
	Tail 8	2	1	3	2	0	32	17	2.1
Kattalai D (H)	Well 8	0	1	3	4	0	46	16	2.0
	Head 8	0	4	2	2	0	36	16	2.0
	Tail 8	0	3	1	3	1	44	16	2.0
Alapakkam M (T)	Well 8	1	0	2	5	0	45	22	2.8
	Head 8	2	4	1	1	0	24	16	2.0
	Tail 8	0	5	1	2	0	34	21	2.6
Athipattu D (T)	Well 8	0	0	1	4	3	60	22	2.8
	Head 8	2	2	2	2	0	30	17	2.1
	Tail 8	1	2	0	3	2	45	31	3.9
DMT									
Vagai E (H)	Well 5	0	1	2	2	0	27	12	2.4
	Head 8	1	4	3	0	0	27	15	1.9
	Tail 8	1	4	2	0	1	31	15	1.9
Punaithangal E (T)	Well 8	0	0	1	5	2	58	16	2.0
	Head 8	0	0	2	5	1	54	16	2.0
	Tail 8	1	2	1	3	1	41	16	2.0
Narasamangalan	Well 8	0	3	4	0	1	38	33	4.1
	Head 8	1	3	3	1	0	31	17	2.1
	Tail 8	1	1	2	3	1	43	20	2.5
Sothiyampakkan	Well 8	1	1	2	2	2	45	17	2.1
	Head 8	1	0	2	5	0	45	16	2.0
	Tail 8	1	3	1	2	1	39	16	2.0

Note : E (H) = Elevated sluice head reach village; D (H) = Deep sluice head reach village.
M (T) = Middle sluice tail reach village; D (T) = Deep sluice tail reach village.
E (T) = Elevated sluice tail reach village; Dash indicates nil.
Score: Education - No schooling, 0; upto primary: 3; upto middle: 5; upto SSLC, PUC 7; degree, 9.
Socio-Political - Ex. Mainakar/Karnam: 2; Panchayat Board President: 4; Co-op. Society member: 2; Temple Trustee: 4; Party functionary, Irrigation delegate: each 5; Others: 3.
Source: Field survey data, 2017.

farmers attained less status score in four villages in KPT and one village in DMT. This is mainly due to their poor economic position, reflected partly in their possession of smaller holdings and availability of regular jobs other than agriculture. The tail segment non-well farmers attained relatively high average status score in all the villages in both the tanks than the non-well head reach farmers. This indication shows that tail reach farmers believe education as their main motto to uplift their life compared to agricultural work, since this category of farmers only affected more than well and non-well head segment farmers in agriculture operations especially getting the tank water supply in both the tanks.

The overall result indicates that there is no strong or consistent pattern relative to reach but the following points are important. Well farmers generally tend to be better educated and enjoy a high status than non-well farmers. Among non-well farmers, tail reachers are generally better or equally educated and status-wise better than head reachers in all the sample villages. Not surprisingly these differences within the village are not correlated to land holdings.

7.3. Conditions of Tank Water Supply in the Latest Two Supply Years

The overall conditions of tank water supply in the two tanks in the latest two supply years and also the one surveyed in 1991-92 have been discussed elaborately in chapter 6. It is important to recall the main findings of reliability, adequacy and methods of resolving the problems faced by the farmers to irrigate their lands.

KPT: In 2015-16, a normal supply year, the villages irrigated by deeper sluices (except Kattalai) were better placed in terms of reliability of supply for two full crop seasons. Within the village, well farmers were better placed than head reachers, who were in turn better off than tail enders.

Water availability to the villages served by elevated sluices is generally very poor: Kadaperi (fed by Gaffor sluice) and Panniyur (served by Panniyur sluice) are cases in point. However, between the two, the former got supply only for one season; the latter got supply both in the second and third crop seasons. Within the village well farmers are better placed in terms of reliability of supply followed by head and tail reachers of non-well groups (eg. Panniyur). The significant exception to this pattern is Kadaperi where no variation was found across reaches within the village. As mentioned earlier, Kattalai, the deepest sluice head reach village, had not received KPT supply fully during this supply year, due to breakage of sluice shutter, consequently closure of the sluice throughout the supply period.

Tank water supply in the two tail reach villages of KPT, (Alapakkam and Athipattu) adversely affected not only because of their location but also due to caste factions (Pillai verses Nayakar versus Reddi verses Scheduled Castes in Alapakkam; and Mudaliars verses Nayakar versus

Scheduled Castes in Athipattu), which affect the functioning of village level arrangements for tank water management. Non-cooperation of well farmers in maintenance work was widely reported in both the villages. As a result, supply of tank was adversely affected; while the tail enders (where majority of the farmers are Scheduled Castes) were worst affected, the head reachers without wells also suffered in consequence.

Although Alapakkam village received tank supply at least for one crop (second) season, Athipattu village had not received the KPT supply fully in the last one decade due to full obstruction of main / branch channels that lead to the village. No effort has been taken to correct this fault so far either by the PWD or by the elected TCMs of the WUA.

During 2005-06, also a full tank supply year (also the only preceding supply year in both tanks between 2005-06 and 2015-16), tank water was available in the aforesaid manner to all villages, During this year Kattalai had also got adequate supply for two crop seasons. During this year also, well farmers are better placed followed by head reachers. Tail enders are worst affected. Between the two elevated sluices, Panniyur was better placed than Kadaperi in getting the tank supply, but both received only one season's tank supply. In Athipattu and Alapakkam the former was better placed than the latter, mostly due to abundant supply of groundwater.

DMT: In DMT in the latest supply year (2015-16), except a portion of *ayacut* in Punaithangal, all the villages had good access to tank supply during the second season. The tail reach farmers and a majority of well farmers in Punaithangal did not get the tank supply. This was mainly due to non-cooperation between farmers belonging to the Mudaliars and the Scheduled Castes, to maintain the channel taking-off from *Anjuran Kalvai*, which gets supply from Mamandur high-level sluice. In the third season only, the head reach villages of Vagai and Narasamangalam reported receiving good tank supply compared to the tail reach villages. Between the two tail reach villages Sothiyampakkam and Punaithangal, the former did get tank supply compared to the latter. Likewise, the head reach farmers in Punaithangal received adequate supply from the Dusi sluices. Again, in DMT within village the tail enders were relatively affected more by unreliable supply than head reach and well farmers.

In the preceding supply year (2005-06) also, the DMT storage was adequate to feed the entire ayacut. Consequently, more or less the same supply pattern prevailed in the sample villages in the preceding supply year.

7.4. Use of Groundwater in the Ayacuts Served by KPT and DMT

In both the tank *ayacuts*, tank water is supplemented by groundwater, derived largely from seepage from the tank, its channels and irrigated fields. As already noted, KPT has on the average 76 wells per 100 hectares of *ayacut*, and DMT has 56 wells per 100 hectares.

7.4.1. Characteristics of Wells in KPT and DMT Villages

Table 7.7 gives some of the main features of the wells in the selected villages of the two tanks. Generally, in KPT all the sample well farmers have only one well in the *ayacut*, (but bores are also developed in recent years that were also not recorded in VAOs adangal registers) whereas in DMT, in the two tail end sample villages (Punaithangal and Sothiyampakkam) a couple of sample farmers possess two wells each in the *ayacut*(also they developed a couple of bores also). Moreover, it should be mentioned that in six sample villages (Kattalai, Alapakkam and Athipattu in KPT; Vagai, Narasamangalam, Sothiyampakkam in DMT) some of the sample non-well households in the tail segment of selected villages and in head segment holdings of two DMT villages were found to have their own wells. This shows that the details of wells have not been properly documented in the "*Adangal*" registers of the respective villages. It was found that the VAOs in those villages have not taken note of newly constructed wells in the last few years. Also joint ownership is not always recorded. In our sample of non-well farmers, there were some who benefited from wells owned by them jointly with others. The frequent "transfer" of VAOs is a major reason for defective recording of wells in the *Adangal*. An extreme case is Sothiyampakkam in DMT, where the VAO was changed thrice in two years (2015-17). It needs noting that non-well farmers found to have their own wells (sole or joint) have generally smallholdings. Their access to well water is also less than those who are sole owners. Further, in many of the bores, that are located in the tail reach, provided very little quantum of supply and a few did not yield anything at all as per respondents' point of view.

KPT has both shallow and bore wells in all the villages. The large majority of bore wells are deeper and reach a large aquifer compared to open wells. In DMT, most wells are open wells. The head reach villages in DMT do not have even a single bore well. This is for at least two reasons: (i) as the sub-soil is rocky most of the farmers do not venture to dig wells or deepen the existing ones, and (ii) the quantity of water available from the wells is limited.

Table 7.7: General Features of Wells in the Sample Villages.

(Area in ha, depth in metre)

Tank and sample villages	Number of wells		Average depth								Area benefitted (Own use + Sale)									
			Latest year				Preceding year				Latest year (2015-16)					Preceding year (2005-05)				
	Latest year	Prece- year	Open well		Bore well		Open well		Bore well		S1	S2	S3	Total	Gross area per well	S1	S2	S3	Total	Gross area per well
			No	Avg. depth	No	Avg. depth	No	Avg. depth	No	Avg. depth										
1	2	3	4	5	6	7	8	9	10	11	12	13	14	15	16	17	18	19	20	21
KPT																				
Kadaperi	8	8	8	10.5	2*	18	8	7.5	1*	15	2.68	2.78	2.41	7.86	0.98	3.13	1.70	2.91	7.74	0.97
Panniyur	10	9	8	14.1	2	93	9	11.4	0	0	1.82	1.75	4.05	7.62	0.76	0.40	2.97	1.42	4.79	0.53
Kattalai	9	7	5	7.5	4	36	5	6	2	33	9.02	12.93	12.83	34.79	3.87	7.34	10.50	9.89	27.73	3.96
Alapakkam	10	9	8	16.5	2	63	8	12	1	54	5.87	10.87	4.25	20.99	2.10	6.48	9.66	0.81	16.94	1.88
Athipattu	12	10	9	10.5	3	21	9	9	1	21	16.76	26.28	17.77	60.81	5.07	13.93	16.70	20.74	51.37	5.14
DMT																				
Vagai	5	5	5	8.1	0	0	5	6.6	0	0	4.89	5.86	6.80	17.54	3.51	5.86	4.44	6.60	16.89	3.38
Punai-thangal	9	9	9	10.2	3	18	9	9	0	0	5.26	5.26	5.67	16.19	1.80	4.45	5.79	5.99	16.23	1.80
Narasa-mangalam	8	8	8	8.7	0	0	8	7.2	0	0	1.01	3.04	0.00	4.05	0.51	1.77	1.42	3.85	7.03	0.88
Sothiyam-pakkam	9	9	9	9.3	2	20	9	8.4	0	0	3.60	7.56	0.00	11.16	1.24	3.60	0.49	4.77	8.86	0.98
HRWF	15	13	9	9.9	6	21.9	10	8.7	3	19.5	3.87	12.87	3.60	20.34	1.36	8.95	3.74	0.64	13.33	1.03
TRWF	12	11	8	10.8	3	22.5	9	9.6	2	20.4	2.81	6.83	4.29	13.94	1.16	2.43	2.89	2.76	8.08	0.73

Note: HRWF = Head reach own-well farmers in the non-well group; Latest year = 2015-16; Preceding year = 2005-06;

TRWF = Tail reach own-well farmers in the non-well group.* indicates bores developed within the wells.

S1, S2 and S3 = Season 1, Season 2 and Season 3.

Source: Field survey data, 2017.

The average depth of bore wells ranged from 18 metres in the head reach of Kadaperi to 93 metres in Panniyur in KPT villages in 2016 (note both villages are serve by elevated sluices); while in DMT the depth ranged from 18 metre in Narasamangalam to 24 metre in Sothiyampakkam. The average depth of open wells ranged from 7.5 metre to 16.5 metre in KPT, and from 8.1 metre to 10.2 metre in DMT in the latest year (2015-16). In KPT, wells of both categories are much deeper in the tail reach villages, especially in Alapakkam, than in the head reach; the average area irrigated per well is also three to four times higher in the former. This pattern is also noticed in DMT, but the head-tail differences are much less marked. The average area irrigated per well in DMT does not reach the levels reported in the tail end of KPT.

All the sample wells are used for irrigation purpose, especially when the tank supply is inadequate or not available. The average area benefiting from well water in 2015-16 ranged from 1.75 hectares per well, (Panniyur), to over 5.0hectares in Athipattu. The abundance of well water use in the latter village in KPT is due to proximity from the Palar. Also that Palar river got good floods during 2015-16, consequently the aquifer is fully recharged. (It should be mentioned that one of the tail end sample wells farmer (an SC farmer) in Athipattu, who own a well adjacent to the Palar river bed, fitted with 10 horse power motor pumpset, irrigated more than 10 ha of ayacut land, that includes his own land plus many adjacent non-well farmers land. This farmer is considered one of the prosperous farmers in the village. Further, he is open with all details). However, when the flow in the river fails for many years continuously, the quantity of water tapped from the wells also diminishes considerably and the same farmer reported that during that period irrigation is possible only to his ayacut land and no adjacent farmers' land is irrigable.

The average depth of both open and bore wells is more or less the same as in the preceding year, but in both categories a marginal increase in depth is observed. But the average irrigated area per well is significantly different: In the two villages (Panniyur and Alapakkam) of KPT, this figure was substantially lower in 2005-06, while in the remaining villages it was marginally lower. Under DMT, in three of the four villages selected the area irrigated per well in 2005-06 was lower than in 2015-16. This suggests that the availability of groundwater has a significant bearing on the accessibility of water in the *ayacut*, and the extent to which it is tapped by the head and tail reaches.

In KPT the head reach village of Kattalai gets abundant supply of ground water even in a below-normal tank supply year. This has induced the digging of more bore wells. The elevated sluice villages of Kadaperi and Panniyur and the tail end village of Alapakkam have lesser recharging capacity from the tank supply. Though Athipattu is also located at the tail reach of the tank, its wells have the benefit of recharge from the Palar river, which flows close to the village.

Although the elevated sluices serve Kadaperi and Panniyur, the latter has better recharging capacity than the former due to its locational advantage from the KPT even between the elevated sluices village.

In DMT the rocky nature of the sub-surface prevented much of the recharge from the tank storage. However, wells in the head reach villages supply more water than in the tail reach villages. In the tail end village of Punaithangal a part of the village *ayacut* is served by the Dusi high level sluice and the rest is served by the Mamandur high-level sluice. In the former *ayacut* there is hardly any well; and all the wells of the village are located in the latter *ayacut*. The point is that Mamandur sluice has not supplied water to this village since 1980. Hence these wells did not get recharged from the DMT supply. The local rainfall is also inadequate for recharging. By and large, between the tanks the effect of tank storage on recharge of wells is more favourable in KPT villages than in DMT villages.

Both KPT and DMT had good storage in 2015-16. Consequently, two crops were possible (except in the elevated sluice/tail reach villages) by using tank supply. The farmers in most of the villages also raised some crops exclusively by using well water, but the extent of land irrigated by using own wells or purchased water was limited.

The preceding supply year selected for the survey was 2005-06 (considered to be weather-wise normal). The available tank storage, was also at FTL in both tanks in that year. Like the latest supply year in both tanks all sample villages received adequate supply. More importantly, in KPT almost all the villages including tail/elevated sluice villages received tank supply for at least one crop season, including Athipattu. Also, the deep sluice serving Kattalai village received two seasons supply. However, wherever tank supplies for the second season were inadequate farmers drew more heavily on groundwater. This seems to have affected groundwater availability in the tail end villages resulting in a reduction in area irrigated per well. In DMT tank supply was available like that of the latest supply year for all the sample villages and groundwater was also utilised where the deficit in supply was increased in the second tank supply season.

7.4.2. Gross irrigated area in the villages under KPT and DMT

Apart from the analysis of the characteristics of well water use one can also analyse the extent of overall use of well water both of "sole" use as well as "supplementary" use in the sample villages in both the tanks. The relevant data are presented in Table 7.8. The following results emerge from the Table.

Among well farmers of KPT the proportion of Gross Irrigated Area (GIA) - in terms of gross season acres - using solely wells in the latest (2015-16)and the survey year (2016-17)decreased significantly in Kadaperi and Athipattu and moderately in Panniyur. The proportion of GIA using wells as supplementary source was important in Athipattu,Alapakkam; and to a lesser extent in Kadaperi and Panniyur. No supplementary use was reported in Kattalai since tank water was not available in 2015-16 and irrigation was done solely (100%) by Wells. Not surprisingly well farmers depend on groundwater much more than non-well farmers. But except in Alapakkam, the major portion of GIA of well farmers used only tank water in Kadaperi and Panniyur.

Among the non-well head segments in all villages (except Kadaperi) sole well use by farmers varies considerably. It is 5% of GIA in Alapakkam to 45% in Panniyur and in the remaining two villages it is 100% and these two villages did not get the tank supply altogether in 2015-16. Supplementary use is more in tail end, especially in Alapakkam (25 to 83% in well and non-well head reach respectively) and in Panniyur (40%in well farmers only) of GIA used groundwater as a supplement). Among non-well tail segments sole well use by farmers is greater than in head segments of Panniyur (88%) than in Alapakkam (53%). Supplementary use is both significant and more widespread than sole well irrigation only in Alapakkam. Note that Kattalai and Athipattu used only well water for cultivation in 2015-16.

Overall, the extent of groundwater use by non-well farms - measured by the proportion of GIA in the *ayacut* receiving sole or supplementary well irrigation - is much higher in the two tail end villages, and also in the head reach village. In Alapakkam the major part of crop area uses well water, while in others (Kadaperi and Panniyur) the bulk of crop area uses only tank water.

In 2005-06, tank storage was also at Full level in both tanks. The two villages – Kattalai and Athipattu also received KPT supply for two crop and one crop season respectively. Consequently, tank water use by all

Table 7.8: Parcentage of Area Irrigated by Sources, Latest, Praceding and Survey Years.

(Area in hectares)

Name of tank and sample villages	Sample HHs by type	No. of wells in the ayacut	Net irri-gable area	Latest year: Gross crop-ped area	Latest year: GIA	Latest year: % of GIA: Well only	Latest year: % of GIA: Tank only	Latest year: % of GIA: Tank-well	Preceding year: Gross crop-ped area	Preceding year: GIA	Preceding year: % of GIA: Well only	Preceding year: % of GIA: Tank only	Preceding year: % of GIA: Tank+well	Survey year 2016-17: Gross crop-ped area	Survey year 2016-17: GIA	Survey year 2016-17: % of GIA: Well only	Survey year 2016-17: % of GIA: Tank only	Survey year 2016-17: % of GIA: Tank-well
1	2	3	4	5	6	7	8	9	10	11	12	13	14	10	11	12	13	14
KPT																		
Kadaperi	Well 8	8	6.19	8.12	8.12	57	43	0	11.32	10.72	50	50	0	8.96	8.96	100	0	0
E (H)	Head 8	0	1.89	1.97	1.83	0	100	0	2.17	2.17	13	87	0	2.79	2.79	100	0	0
	Tail 8	0	2.91	2.04	1.86	0	100	0	1.97	1.97	0	100	0	3.06	3.06	100	0	0
Panniyur	Well 8	10	8.21	13.09	12.18	18	42	40	9.34	8.43	28	53	19	11.56	11.56	100	0	0
E (H)	Head 8	0	3.42	4.76	2.67	45	55	0	3.30	2.65	0	100	0	0.00	0.00	0	0	0
	Tail 8	1#	2.59	2.19	2.19	88	12	0	2.13	2.13	0	87	13	0.54	0.27	100	0	0
Kattalai	Well 8	9	9.19	22.36	22.36	100	0	0	20.83	20.83	23	58	19	17.47	17.47	100	0	0
D (H)	Head 8	0	2.63	7.08	7.08	100	0	0	6.67	6.67	21	79	0	5.26	5.26	100	0	0
	Tail 8	0	3.32	7.98	7.98	100	0	0	8.44	8.44	23	77	0	7.66	7.66	100	0	0
Alapakkam	Well 8	10	16.34	21.60	20.99	72	3	25	16.34	13.70	40	0	60	18.36	18.36	100	0	0
M (T)	Head 8	2#	3.30	3.23	3.10	5	12	83	4.92	4.92	0	34	65	2.72	2.59	100	0	0
	Tail 8	3#	4.79	2.56	2.56	53	47	0	3.09	3.09	18	50	33	1.76	1.15	100	0	0
Athipattu	Well 8	12	22.26	45.43	45.43	100	0	0	51.37	51.38	47	13	39	48.87	48.87	100	0	0
D (T)	Head 8	6#	4.59	7.10	7.10	100	0	0	3.77	3.77	22	50	28	7.76	7.76	100	0	0
	Tail 8	3#	6.20	9.52	9.52	100	0	0	6.94	6.94	50	27	24	6.70	6.70	100	0	0

(*Contd*...)

DMT																	
Vagai	Well 5	5	9.06	21.67	21.67	38	16	46	24.17	24.17	61	13	26	26.59	26.59	17	33	51
E (H)	Head 8	0	2.38	2.47	2.47	0	100	0	2.36	2.36	0	100	0	4.42	4.42	0	100	0
	Tail 8	0	5.83	5.69	5.69	0	100	0	5.69	5.69	0	100	0	11.23	11.23	0	100	0
Punaithangal	Well 8	9	10.45	12.35	12.35	87	0	13	14.62	14.62	55	13	33	16.19	16.19	60	23	18
E (T)	Head 8	0	5.12	3.10	3.10	0	100	0	3.10	3.10	0	100	0	4.11	4.11	0	100	0
	Tail 8	0	3.49	3.97	3.97	0	100	0	5.59	4.37	0	100	0	3.77	3.77	0	100	0
Narasamangalam	Well 8	8	7.19	9.28	9.28	18	54	28	10.03	10.03	42	40	18	12.88	12.88	9	70	20
D (H)	Head 8	2#	2.97	4.70	4.70	20	75	5	4.28	4.08	27	73	0	7.16	6.49	0	100	0
	Tail 8	1#	3.15	3.28	3.28	33	67	0	3.75	3.28	10	90	0	7.13	7.13	2	98	0
Sothiyampakkam	Well 8	9	8.70	19.77	19.77	26	40	35	16.83	16.83	47	46	7	20.28	20.28	25	53	22
D (T)	Head 8	5#	7.94	10.30	10.30	32	18	49	11.18	11.18	32	50	19	17.17	17.17	22	44	34
	Tail 8	4#	3.93	6.52	6.52	33	50	17	5.08	5.08	18	73	9	9.33	8.93	15	69	16

Note: E (H) = Elevated sluice head reach village; D (H) = Deep sluice HR village. M (T) = Middle sluice tail reach village; D (T) = Deep sluice TR village.
E (T) = Elevated sluice tail reach village; GIA = Gross irrigated area. # Indicates number of farmers used their own well water.
Columns 6 and 11 include own use plus purchased water. Latest year, 2015-16; Preceding year, 2005-06; Survey year, 2016-17.
Source: Field survey data, 2017.

segments of farmers is relatively higher than sole well use in Kattalai village due to its proximity to the tank, and both sole and supplemental use is more in Athipattu in all segments, this is because of inadequacy of tank supply faced by farmers in the tail end village. For the remaining villages the aforesaid same descriptions for the year 2015-16 is more or less applicable.

It should be noted that data were collected for the survey year (2016-17) also. In that year it is reported that none of the KPT villages received tank supply, but the DMT villages received tank supply in all the sample villages at least for one season in 2016-17 (this is mainly because of large storage capacity of DMT compared to KPT and in the next year (2016) catchment run-off is considerably more in DMT, that again replenished the storage position of DMT that helped all villagers to get tank supply again in 2016-17. However, the extent of tank water availability varies considerably among the sample villages.

In DMT, among well farmers sole well use in 2015-16 ranged between 38 and 87% of GIA; the dependence being greater in tail end villages. The percentage of GIA by using well water only was less in the Mamandur sluice villages of Narasamangalam and Sothiyampakkam than in the Dusi sluice villages of Vagai and Punaithangal. Supplementary use also accounted for nearly a half of GIA in Vagai and more than a third in Sothiyampakkam. In Narasamangalam it is more than a quarter and negligible (13%) in Punaithangal.

In the non-well head segment no farmer in the Dusi sluice reported well irrigation; but in the Mamandur sluice the proportion is 20% in the head reach village of Narasamangalam and 32% in the tail end village of Sothiyampakkam. In non-well tail segment, again the Dusi sluice villages use no well water. The incidence of well water use is more in the Mamandur sluice, both as sole use and also as a supplementary source. The former is 33% of GIA in both Mamandur villages and it is 17% only in Sothiyampakkam. Even head reach non-well category in Sothiyampakkam had also access to own well water, consequently head reach accounts for 49% of supplementary use. Overall, use of well water among non-well holdings is confined to the villages under the Mamandur sluice, and that too among those in the tail segment of Narasamangalam, and the whole of Sothiyampakkam. On the whole, the use of well water by non-well farmers is far more widespread in the villages under KPT and not in DMT (except Sothiyampakkam).

Also in the preceding supply year, both the tanks had got full tank storages. Hence one may expect the same trend as it is described above for both the tanks. Further, source wise net irrigated area in the sample villages in the latest two supply years are presented in Table 7.9.

7.4.3. Variations across Seasons and Years

Data given in Table 7.9 indicate that during the first season (July-October) well farmers in both the tanks and in all the sample villages exclusively use the well water. However, within the village much variation in the extent of well irrigated area between the head and tail segments is not seen. Interestingly in some villages, under both the tanks, farmers identified as non-well owners under-reported well irrigation in the first season. In most cases, this is because the sample farmers in this category in fact owned wells either themselves or jointly with others. Some non-well farmers who raised annual crops in a limited extent of their *ayacut* purchased well water during this first season. In the second season (November-February) where tank provides the maximum water supply, pure well irrigation was rare in 2015-16. Even in where it is reported, the area irrigated is only marginal. However, in two KPT villages Kattalai and Athipattu, as already mentioned, tank water was not fully available. Hence, in these two villages pure well irrigation only is possible for all three seasons.

As stated above, use of well water - solely as well as supplement - is more significant in the tail end villages compared to the head reach/elevated sluice villages. The weighted average estimates of groundwater use for each selected village as a whole -both sole and supplement - (column 17 in Table 7.9) show that the proportion of GCA (in season hectares) using well water is much higher in the tail end villages in both the tanks. In KPT, the proportion of area using well water ranged from 59.6% in Kadaperi to 68.4% in Panniyur. It should be noted that both villages are served by elevated sluices. In Alapakkam tank supply was notadequately available in 2015-16 even for one crop season - due to encroachment in the mainsupply channel. This probably led to a relatively greater use of well water in this village and the GCA using well water is 211%. It is also observed, the two villages where tank supply was not available, the percentage of GCA irrigated by well was the highest ranging from 333% in Athipattu to 865% in Katalai. In 2005-06, tank supply was more or less equal to that of 2015-16 in all the villages. As a result, the proportion of area using well water decreased significantly in all villages. It should, however, be noted that the increase in reliance on groundwater was much more marked in the tail end villages.

Table 7.9: Area Irrigated by Sources, Lat est and Preceding Supply Years.

(Area in hectares)

Tank and sample villages	Sample HHs by type	No. o wells in the ayacu	Net irri-gable area	Total area as per adangal	Area irrigated, 2015-16 in the sample farms: Rain-fed	Pure Well S1	Pure Well S2	Pure Well S3	Pure Tank S2	Pure Tank S3	Tank+Well S2	Tank+Well S3	(6to 12)	Well-Tank Well	<---- All farms ----->: Weigh-ted average total 14/4X5	GCA	% of GC irriga-ted by wells (15/16)	Area irrigated, preceding year, 2005-06: Rain-fed	Pure Well S1	Pure Well S2	Pure Well S3	Pure Tank S2	Pure Tank S3	Tank+Well S2	Tank+Well S3	Sample farms: (18 to 24	Well+ Tank+ Well	<----- All farms ----->: Weigh-ted average total 26/4X5	GCA	% of GC irriga-ted by wells (27/28)
1	2	3	4	5	5A	6	7	8	9	10	11	12	13	14	15	16	17	17A	18	19	20	21	22	23	24	25	26	27	28	29
KPT																														
Kadaperi	Well 8	8	6.19	9.73	0	2.07	0.77	1.76	3.51	0	0	0	8.1	4.6	7.2	8.12		0	2.82	0.20	2.61	5.69	0	0	0	11.3	5.6	8.9	11.3	
E (H)	Head 8	0	1.89	8.31	0.13	0	0	0	1.70	0.14	0	0	1.8	0.0	0.0	1.97		0	0.14	0	0.14	1.90	0	0	0	2.2	0.3	1.2	2.2	
	Tail 8	0	2.91	8.57	0	0	0	0	1.63	0.40	0	0	2.0	0.0	0.0	2.04		0	0	0	0	1.97	0	0	0	2.0	0.0	0.0	2.0	
															7.2	12.12	59.6											10.1	38.3	26.3
Panniyur	Well 8	10	8.21	10.99	0.91	1.62	0.54	0	4.55	0.61	2.83	2.02	12.2	7.0	9.4	13.09		0.91	0.40	0.53	1.42	4.45	0	1.62	0	8.4	4.0	5.3	9.3	
E (H)	Head 8	0	3.42	9.64	2.06	0	0.61	0.61	1.48	0	0	0	2.7	1.2	3.4	4.76		0.65	0	0	0	2.65	0	0	0	2.6	0.0	0.0	3.3	
	Tail 8	1#	2.59	8.55	0	0	0	0	1.92	0	0	0.27	2.2	0.3	0.9	2.19		0	0	0	0	1.86	0	0	0.27	2.1	0.3	0.9	2.1	
															13.7	20.04	68.4											6.2	38.7	16.0
Kattalai	Well 8	9	9.19	43.51	0	5.34	8.44	8.58	0	0	0	0	22.4	22.4	105.9	22.36		0	4.70	0	0	8.79	3.32	0	4.04	20.8	8.7	41.4	20.8	
D (H)	Head 8	0	2.63	47.90	0	1.93	2.63	2.52	0	0	0	0	7.1	7.1	129.0	7.08		0	1.42	0	0	2.63	2.63	0	0	6.7	1.4	25.8	6.7	
	Tail 8	0	3.32	36.90	0	2.26	2.86	2.86	0	0	0	0	8.0	8.0	88.8	7.98		0	1.93	0	0	3.32	3.19	0	0	8.4	1.9	21.5	8.4	
															323.7	37.42	865.1											88.7	88.8	99.9
Alapakkan	Well 8	10	16.3	22.12	0.61	5.67	5.26	4.25	0.61	0	5.20	0	21.0	20.4	27.6	21.60		2.63	3.85	0.40	1.21	0	0	8.26	0	13.7	13.7	18.6	16.4	
M (T)	Head 8	2#	3.30	25.95	0.13	0	0.16	0	0.37	0	2.56	0	3.1	2.7	21.4	3.23		0	0	0	0	1.68	0	1.62	1.62	4.9	3.2	25.5	4.9	
	Tail 8	3#	4.79	32.43	0	0	1.30	0	1.26	0	0	0	2.6	1.3	8.8	2.56		0	0.40	0.14	0	1.53	0	1.01	0	3.1	1.6	10.5	3.1	
															57.9	27.39	211.2											54.6	60.2	90.7
Athipattu	Well 8	12	22.3	59.55	0	10.1	20.8	14.5	0	0	0	0	45.4	45.4	121.5	45.43		0	11.5	7.17	5.55	6.88	0	6.68	13.6	51.3	44.5	118.9	51.3	
D (T)	Head 8	6#	4.59	24.35	0	1.11	2.99	2.99	0	0	0	0	7.1	7.1	37.7	7.10		0	0	0.81	0	1.89	0	1.07	0	3.8	1.9	10.0	3.8	
	Tail 8	3#	6.20	30.91	0	1.42	4.25	3.85	0	0	0	0	9.5	9.5	47.4	9.52		0	1.42	2.02	0	1.86	0	0.23	1.42	6.9	5.1	25.3	6.9	
DMT															206.6	62.04	333.0											154.2	153.3	100.6

(*Contd*...)

DMT															206.6	62.04	333.0											154.2	153.3	100.6
Vagai	Well 5	5	9.06	9.06	0	6.51	1.82	0	0	3.44	4.48	5.42	21.7	18.2	18.2	21.67		0	7.72	0	7.04	3.11	0	4.48	1.82	24.2	21.1	21.1	24.2	
E (H)	Head 8	0	2.38	7.74	0	0	0	0	2.12	0.35	0	0	2.5	0.0	0.0	2.47		0	0	0	0	2.36	0	0	0	2.4	0.0	0.0	2.4	
	Tail 8	0	5.83	9.30	0	0	0	0	5.69	0	0	0	5.7	0.0	0.0	5.69		0	0	0	0	5.69	0	0	0	5.7	0.0	0.0	5.7	
															18.2	29.83	61.1											21.1	79.6	26.5
Punai-	Well 8	9	10.4	28.71	0	3.04	4.45	3.24	0	0	0	1.62	12.3	12.3	33.9	12.35		0	2.23	1.82	3.97	1.82	0	4.78	0	14.6	12.8	35.2	14.6	
thangal	Head 8	0	5.12	22.18	0	0	0	0	3.10	0	0	0	3.1	0.0	0.0	3.10		0	0	0	0	3.10	0	0	0	3.1	0.0	0.0	3.1	
E(T)	Tail 8	0	3.49	7.89	0	0	0	0	1.13	2.83	0	0	4.0	0.0	0.0	3.97		1.21	0	0	0	4.37	0	0	0	4.4	0.0	0.0	5.6	
															33.9	19.41	174.8											35.2	57.6	61.1
Narasa-	Well 8	8	7.19	13.01	0	1.21	0.40	0	1.08	3.97	2.61	0	9.3	4.2	7.6	9.28		0	1.77	0	2.40	4.04	0	1.42	0.40	10.0	6.0	10.8	10.0	
mangalam	Head 8	2#	2.97	16.84	0	0.24	0.70	0	1.52	2	0.24	0	4.7	1.2	6.7	4.70		0.20	0	0.70	0.40	2.97	0	0	0	4.1	1.1	6.3	4.3	
D (H)	Tail 8	1#	3.15	14.81	0	0	1	0	1.76	0	0	0	3.3	1.1	5.2	3.28		0.47	0.16	0	0.16	2.80	0.16	0	0	3.3	0.3	1.5	3.8	
															19.5	17.26	113.0											18.7	48.6	38.4
Sothiyam-	Well 8	9	8.70	29.05	0	3.95	0.61	0.53	0	7.81	6.9	0	19.8	12.0	39.9	19.77		0	3.92	0	4.05	7.72	0	0	1.13	16.8	9.1	30.4	16.8	
pakkam	Head 8	5#	7.94	15.70	0	2.52	0.81	0	1.08	0.81	5.09	0	10.3	8.4	16.6	10.30		0	3.53	0	0	2.09	3.44	2.12	0	11.2	5.7	11.2	11.2	
D (T)	Tail 8	4#	3.93	11.16	0	0.47	1.26	0.45	1.40	1.88	1.07	0	6.5	3.2	9.2	6.52		0	0.45	0	0.45	3.72	0	0	0.47	5.1	1.4	3.9	5.1	
															65.8	36.59	179.7											45.4	81.7	55.6

Note: GIA = Gross irrigated area. S1, S2, S3 = Season 1, Season 2, Season 3; E (H) = Elevated sluice head reach village; D (H) = Deep sluice head reach village.

M (T) = Middle sluice tail reach village; D (T) = Deep sluice tail reach village; E (T) = Elevated sluice tail reach village;

Latest year, 2015-16; Preceding year, 2005-06. GCA is in terms of "season hectares". # Indicates number of farmers used their own well water.

Source: Field survey data, 2017.

Use of well water (as a proportion of GCA) is generally much less in DMT, the maximum in 2015-16 being in Sothiyampakkam (179.7%) followed by Punaithangal (174.8%) and both are the tail reach villages. The proportion of area using well water is 61% in Vagai and 113% in Narasamangalamand both are the head reach villages. Under Mamandur sluice villages, the extent of area used well water is significantly more than that of Dusi sluice villages, but again the use is more widespread in the tail end. In the preceding supply year also good quantum of tank supply was available to all villages. So, the extent of well water use was much limited and the proportion of GCA using wells was much lower than in 2005-06. The proportion of area using well water was however again significantly higher in the tail end villages than in the head reach.

It is obvious from the above that, while tail end villages are unfavourably located to get the tank supply, this disadvantage is to a considerable degree offset by the greater use of well water. Exploitation of groundwater for conjunctive use in these two tanks thus dampens the effect of unequal distribution of tank water between different segments of the *ayacut*.

7.4.4. Purchased Well Water

Data on purchase of well water during each of the three crop seasons in the latest two supply years and during the survey year are presented in Table 7.10. They show the following:

1. Between KPT and DMT water purchase is existing only in KPT and it was not reported in DMT in any of the years. In KPT, well farmers rarely purchase well water. In the tail reach and elevated sluice villages, as a proportion of GCA purchase of well water varies widely, it is less than 15% in two villages and 30% in Alapakkam. In the head reach of Kattalai, since no tank supply was possible, water purchase is between 98% and 100%in non-well head and tail reach farmers.
2. The frequency of water purchase among non-well farmers is relatively high in KPT sample villages. Water purchase is more prevalent in the tail end villages and in both head and tail segment farmers. Well water purchases are more frequent in season 2 and season 3 than in season 1 including the survey year 2016-17.
3. In all three years including the latest two supply years a substantial proportion of non-well holdings, reported using purchased well water in the second season. The number of holdings reporting purchased water in the first and third crop seasons is generally

Table 7.10: Frequency Distribution of Sample Farmers Buying well Water in th Latest, Preceding and During the Survey Years.

(Area in hectares)

Name of tank and sample villages	Sample HHs by type	<--------- Latest year (2015-16) ----------->							<------------ Preceding supply year (2005-06)------------>							<-------------- Survey year (2016-17) ---------------->						
		GCA	Number of sample HHs buying well water for			GCA receiving well water		Type of contract	GCA	Number of sample HHs buying well water for			GCA receiving well water		Type of contract	GCA	Number of sample HHs buying well water for			GCA receiving well water		Type of contract
			Sea-son 1	Sea-son 2	Sea-son 3	Area	% to total (7/3)X100	RS GP OS 1 2 3		Sea-son 1	Sea-son 2	Sea-son 3	Area	% to total (14/10)X100	RS GP OS 1 2 3		Sea-son 1	Sea-son 2	Sea-son 3	Area	% to total (21/17)X100	RS GP OS 1 2 3
1	2	3	4	5	6	7	8	9	10	11	12	13	14	15	16	17	18	19	20	21	22	23
KPT																						
Kadaperi	Well 8	8.12	-	-	-	-	-	-	11.32	-	-	-	-	-	-	8.96	-	-	-	-	-	-
E (H)	Head 8	1.97	1	-	-	0.13	6.61	1	2.17	1	-	1	0.28	12.9	1	2.79	3	5	4	2.79	99.9	1
	Tail 8	2.04	-	-	-	-	-	-	1.97	-	-	-	-	-	-	3.06	2	3	2	2.66	86.9	1
Panniyur	Well 8	13.09	-	-	-	-	-	-	9.34	-	-	-	-	-	-	11.56	1			0.54	4.7	1
E (H)	Head 8	4.76	-	-	-	-	-	-	3.30	-	-	-	-	-	-	0.00	-	-	-	-	-	-
	Tail 8	2.19	-	-	-	-	-	-	2.13	-	-	-	-	-	-	0.54	-	-	-	-	-	-
Kattalai	Well 8	22.36	-	-	-	-	-	-	20.83	-	-	-	-	-	-	17.47	-	-	-	-	-	-
D (H)	Head 8	7.08	6	7	7	6.92	97.8	1	6.67	4	2	2	2.89	43.3	2	5.26	4	6	3	4.96	94.2	1
	Tail 8	7.98	4	5	5	7.98	100.0	2	8.44	3	-	-	1.93	22.9	1	7.66	4	5	4	7.65	99.9	1
Alapakkam	Well 8	21.60	-	-	-	-	-	-	16.34	-	-	-	-	-	-	18.36	-	-	-	-	-	-
M (T)	Head 8	3.23	-	2	-	0.97	30.0	2	4.92	-	-	-	-	-	-	2.72		2		0.97	35.6	1
	Tail 8	2.56	1	-	-	0.81	31.7	1	3.09	-	-	-	-	-	-	1.76	-	-	-	-	-	-
Athipattu	Well 8	45.43	-	-	-	-	-	-	51.37	-	-	-	-	-	-	48.87	-	-	-	-	-	-
D (T)	Head 8	7.10	-	-	-	-	-	-	3.77	-	1		1.07	28.4	1	7.76	-	-	-	-	-	-
	Tail 8	9.52	-	1	1	1.21	12.7	2	6.94	-	2		0.83	12.0	1	6.70	-	1	-	0.61	9.1	2
DMT																						
Vagai	Well 5	21.67	-	-	-	-	-	-	24.17	-	-	-	-	-	-	26.59	-	-	-	-	-	-
E (H)	Head 8	2.47	-	-	-	-	-	-	2.36	-	-	-	-	-	-	4.42	-	-	-	-	-	-
	Tail 8	5.69	-	-	-	-	-	-	5.69	-	-	-	-	-	-	11.23	-	-	-	-	-	-

(Contd...)

	Tail 8	5.69	-	-	-	-	-	-	5.69	-	-	-	-	-	-	11.23	-	-	-	-	-	-
Punaithangal	Well 8	12.35	-	-	-	-	-	-	14.62	-	-	-	-	-	-	16.19	-	-	-	-	-	-
E (T)	Head 8	3.10	-	-	-	-	-	-	3.10	-	-	-	-	-	-	4.11	-	-	-	-	-	-
	Tail 8	3.97	-	-	-	-	-	-	5.59	-	-	-	-	-	-	3.77	-	-	-	-	-	-
Narasamangalam	Well 8	9.28	-	-	-	-	-	-	10.03	-	-	-	-	-	-	12.88	-	-	-	-	-	-
D (H)	Head 8	4.70	-	-	-	-	-	-	4.28	-	-	-	-	-	-	7.16	-	-	-	-	-	-
	Tail 8	3.28	-	-	-	-	-	-	3.75	-	-	-	-	-	-	7.13	-	-	-	-	-	-
Sothiyampakkan	Well 8	19.77	-	-	-	-	-	-	16.83	-	-	-	-	-	-	20.28	-	-	-	-	-	-
D (T)	Head 8	10.30	-	-	-	-	-	-	11.18	-	-	-	-	-	-	17.17	-	-	-	-	-	-
	Tail 8	6.52	-	-	-	-	-	-	5.08	-	-	-	-	-	-	9.33	-	-	-	-	-	-

Note: E (H) = Elevated sluice head reach village; D (H) = Deep sluice head reach village. GCA in columns 3, 10 and 17 is in terms of "season-acres". '-' indicates nil.

M (T) = Middle sluice tail reach village; D (T) = Deep sluice tail reach village. E (T) = Elevated sluice tail reach village; Dash indicates nil.

Type of contract: RS (1): Amount paid in cash on hourly basis of water received; GP (2): 1/3 of gross produce to be paid; and OS (3): Others include contract of (a) either money or grains; (b) lease and (c) free of cost.

Latest year, 2015-16; Preceding year, 2005-06; Survey year, 2016-17.

Source: Field survey data, 2017.

much smaller in both the years. The frequency of the use of purchased well water was higher in KPT especially in in all reaches of villages, except in Panniyur.

4. In KPT the frequency of purchase was less in 2005-06 when the tank supplies were also considered normal. No well holdings reported the use of purchased water. Non-well holdings in all the villages purchased water (except Panniyur and Alapakkam). The proportion of GCA receiving ranged between 12% and 28% in the tail reach holdings and between 23% and 43% in the head reach holdings. Dependence on purchased well water was higher both in absolute and relative terms in 2015-16 in some segments (especially, the head segment of Kadaperi, Kattalai and Alapakkam); while in others it was the same or less.

5. In KPT, in 2016-17, Panniyur (except one well farmer with a little extent of 0.54 ha) and Alapakkam did not use purchased water, since well water is much limited in these villages. In Kadaperi and Kattalai both segments of non-well farmers used purchased well water. In both village th percentage of GCA receiving well water is between 87 and 100%. In the tail end villages of Alapakkam and Athipattu the area (0.97 ha and 0.61 ha) under purchased well water use is very limited. On the whole, the GCA using purchased water was less and it is limited to only three (an elevated, deep sluice head reach and tail end) villages in KPT.

7.4.5. Types of Contracts for Payment

The amount paid for the use of well water by the respondents in different villages varies significantly. There are three types of purchases: (1) Water purchased mostly on hourly basis for cash. This is usual when purchased water is used for a part of the buyer's holdings and for part of the crop season. (2) Use of purchased water throughout the crop period by paying *Neer Vaaram* (water rent), usually one third of gross crop produce of area using well water; and (3) use of well water on a predetermined contract of payment of either money or grains, leased-out land for some period (usually two or three years on contract). Generally, a majority of purchasers fall under the first two categories. The third type of contract prevails to a very limited extent in KPT villages; and is not at all reported in DMT villages since water sale / purchase is not existing there. Generally, amount paid per hectare per crop season for the second and third types of purchase was more than for the first type.

7.4.6. Seasonwise Payments for Purchased Well Water

The use of purchased well water was more in the second and third seasons in 2015-16 reference year, whereas it is the first season in 2005-06 (Table 7.11). In both the years purchase was mostly happened in Kattalai village compared to Kadaperi and Athipattu. This is due to the fact that the tank water supply was used only in the second and third seasons, and the extent of the use of purchased well water depends on the supply of water available from the tank and the availability of sub-surface water in a particular village. In Kattalai, since it is located below the deep sluice, groundwater is abundant. As a result, this assured availability of well water induced non-well farmers of both types to grow annual crop of banana. On the whole, within the village the incidence of water purchase was more in the tail segment than in the head segment of non-well farmers.

Amount paid per hectare for purchased well water varies greatly; it was the highest in non-well head reach holdings in Kattalai (Rs. 24500) and lowest (Rs. 1100) in the tail segment of Athipattu village in KPT in the third season of latest supply year. It should be noted that in the tail segment of Kattalai for 2.26 ha ayacut land the amount paid for purchased water was Rs. 20,900, whereas in the second season for an increased area of 2.86 ha the amount paid was Rs. 17,300 and in the third season for the same extent the amount paid was Rs. 10,300. This shows that the payment for purchased water was more in the first season, moderate in the second season and the lowest in the third season. This low amount is because the rainwater during monsoon months helped farmers to use less of purchased water than in the non-monsoon / very little rainfall months of I season.Taking the year as a whole, farmers in head segment seem to pay relatively more than that of the tail segment. These variations reflect the combined effect of differences in the rate per hectare and the extent of water used.

In the preceding supply year only non-well farmers used purchased well water in Kattalai (except one well farmer) and Athipattu. Amount paid per hectare in the head reach holdings ranged from Rs. 6000 in Kattalaiin the first season to Rs. 2700 in Kadaperi in the third season. In the tail reach holdings for Rs. 4500 in Kattalai to Rs. 2850 in Athipattu. The amount paid for purchased well water in the first season was more than that of second and third seasons. Within village non-well farmers paid more amount than well farmers.

The weighted average estimates indicate that in KPT the proportion of area using purchased well water was about 100% in the deep sluice village of Kattalai and the elevated sluice village of Kadaperi it is 6.8% in

Table 7.11: Area Irrigated and Amount Paid by Sample Households for Purchased Well Water.

(Area in ha. amount in rupees)

Tank and sample villages	Sample HHs by type and no. of wells	<---------- Latest year (2015-16) ---------- Season 1 <-----> Area	Amount paid	Season 2 <-----> Area	Amount paid	Season 3 <-----> Area	Amount paid	Total <-----> Area	Amount paid	Amount / ha (10/9)	% area under purchase (9/GSA) X100	(OAS/OAT) CI	Estimated total area 12X13	OAT/OASX purchased area	OAT X CI	% of area using p.water 15/16 ---------->	<---------- preceding year (2005-06) ---------- Season 1 <-----> Area	Amount paid	Season 2 <-----> Area	Amount paid	Season 3 <-----> Area	Amount paid	Total <-----> Area	Amount paid	Amount / ha (25/24)	% area under purchase (24/GSA) X100	(OAS/OAT) CI	Estimated total area 27X28	OAT/OASX purchased area	OAT X CI	% of area using p.water 30/31 ---------->
1	2	3	4	5	6	7	8	9	10	11	12	13	14	15	16	17	18	19	20	21	22	23	24	25	26	27	28	29	30	31	32
KPT																															
Kadaperi	Well 8	-	-	-	-	-	-	-	-	-	-	-	-	-	-		-	-	-	-	-	-	-	-	-	-	-	-	-	-	
E (H)	Head 8	0.13	500	-	-	-	-	0.13	500	3742	6.8	0.24	1.6	0.6	8.6		0.14	2700	-	-	0.14	2700	0.28	5400	19286	12.9	0.26	49.3	0.6	8.7	
	Tail 8	-	-	-	-	-	-	-	-	-	-	-	-	-	-		-	-	-	-	-	-	-	-	-	-	-	-	-	-	
														0.6	8.6	6.8													0.6	8.7	7.1
Panniyur	Well 8	-	-	-	-	-	-	-	-	-	-	-	-	-	-		-	-	-	-	-	-	-	-	-	-	-	-	-	-	
E (H)	Head 8	-	-	-	-	-	-	-	-	-	-	-	-	-	-		-	-	-	-	-	-	-	-	-	-	-	-	-	-	
	Tail 8	-	-	-	-	-	-	-	-	-	-	-	-	-	-		-	-	-	-	-	-	-	-	-	-	-	-	-	-	
Kattalai	Well 8	0.20	0.0	-	-	-	-	0.20	Free of cost		-	-	-	-	-		0.32	600	-	-	0.32	225	0.65	825	1269	3.12	0.48	1.50	3	99	
D (H)	Head 8	1.93	20000	2.52	19370	2.63	24500	7.08	63870	9025	100.0	0.15	15	129	129		1.31	6000	0.45	crop	0.45	failed	1.31	6000	4580	19.6	0.14	2.74	24	122	
	Tail 8	2.26	20900	2.86	17300	2.86	10300	7.98	48500	6078	100.0	0.22	22	89	89		1.93	4500	-	-	-	-	1.93	4500	2332	22.9	0.23	5.25	21	94	
														218	218	99.9													48	314	15.4
Alapakam	Well 8	-	-	-	-	-	-	-	-	-	-	-	-	-	-		-	-	-	-	-	-	-	-	-	-	-	-	-	-	
M (T)	Head 8	-	-	0.81	10000			0.81	10000	12350	25.1	0.1	3.1	6.4	25.4		-	-	-	-	-	-	-	-	-	-	-	-	-	-	
	Tail 8	0.81	1500	-	-	-	-	0.81	1500	1853	31.6	0.1	2.5	5.5	17.2		-	-	-	-	-	-	-	-	-	-	-	-	-	-	
														12	43	27.8															
Athipattu	Well 8	-	-	-	-	-	-	-	-	-	-	-	-	-	-		-	-	-	-	-	-	-	-	-	-	-	-	-	-	
D (T)	Head 8	-	-	-	-	-	-	-	-	-	-	-	-	-	-		-	-	1.07	1500	-	-	1.07	1500	1402	28.4	0.2	4.4	5.7	20.0	
	Tail 8	-	-	0.61	750	0.61	1100	1.21	1850	1523	12.8	0.3	3.9	6.1	47.6	12.7	-	-	0.83	2850	-	-	0.83	2850	3434	12.0	0.2	2.7	4.1	34.6	
																													9.8	54.6	18.0

(Contd...)

DMT																													
Vagai	Well 5	-	-	-	-	-	-	-	-	-	-	-	-	-	-	-	-	-	-	-	-	-	-	-	-	-	-	-	-
E (H)	Head 8	-	-	-	-	-	-	-	-	-	-	-	-	-	-	-	-	-	-	-	-	-	-	-	-	-	-	-	-
	Tail 8	-	-	-	-	-	-	-	-	-	-	-	-	-	-	-	-	-	-	-	-	-	-	-	-	-	-	-	-
Punai-	Well 8	-	-	-	-	-	-	-	-	-	-	-	-	-	-	-	-	-	-	-	-	-	-	-	-	-	-	-	-
thangal	Head 8	-	-	-	-	-	-	-	-	-	-	-	-	-	-	-	-	-	-	-	-	-	-	-	-	-	-	-	-
E(T)	Tail 8	-	-	-	-	-	-	-	-	-	-	-	-	-	-	-	-	-	-	-	-	-	-	-	-	-	-	-	-
Narasa-	Well 8	-	-	-	-	-	-	-	-	-	-	-	-	-	-	-	-	-	-	-	-	-	-	-	-	-	-	-	-
mangalam	Head 8	-	-	-	-	-	-	-	-	-	-	-	-	-	-	-	-	-	-	-	-	-	-	-	-	-	-	-	-
D (H)	Tail 8	-	-	-	-	-	-	-	-	-	-	-	-	-	-	-	-	-	-	-	-	-	-	-	-	-	-	-	-
Sothiyam-	Well 8	-	-	-	-	-	-	-	-	-	-	-	-	-	-	-	-	-	-	-	-	-	-	-	-	-	-	-	-
pakkam	Head 8	-	-	-	-	-	-	-	-	-	-	-	-	-	-	-	-	-	-	-	-	-	-	-	-	-	-	-	-
D (T)	Tail 8	-	-	-	-	-	-	-	-	-	-	-	-	-	-	-	-	-	-	-	-	-	-	-	-	-	-	-	-

Note: E (H) = Elevated sluice head reach village; D (H) = Deep sluice head reach village; E (T) = Elevated sluice tail reach village; '-' indicates nil.

OAS = Operated area sample (= NSA); OAT = Operated area total (total ayacut area); CI = Cropping intensity; GSA = Gross area (season acres); Latest year = 2015-16; Preceding year = 2005.06.

Source: Field survey data, 2017

2016-17. The corresponding figure for the tail end villages of Athipattu and Alapakkam was about 13% and 28% respectively. In 2005-06, the average figures for the deep sluice village of Kattalai also used a higher proportion (15.4% of GCA) of purchased well water along with the tail end village of Athipattu (18%).

7.4.7. Water Sales

Apart from the use of purchased well water, the relation between the extent of groundwater use for own land and for sale in the *ayacut* is also important. Between the tanks water sale was significant in KPT but it is nil in DMT. Within KPT across villages the picture varies. More of water sale was reported in the deep sluice village of Kattalai and the tail end village of Athipattu followed by the elevated sluice village Kadaperi. In Kattalai and Athipattu, as indicated, tank water was not available and only well water was used in 2015-16. Consequently, majority of head and tail segment non-well farmers used only purchased water and hencemore of water sale was reported. In DMT in all the villages water sale was not at all reported. In KPT between the latest two supply years the area under sale of well water was higher in the latest supply year in all the selected villages. In Panniyur and Alapakkam (except for a very limited acreage) water sale was not at all reported (Table 7.12).

Judging from the responses by the users of well water, the extent of water sale (as reported by the sellers) is vastly under-reported. The respondents were reluctant to reveal the well water sale due to the following reasons: (1) The costs for the use of electric energy for lifting water is nil since 1990. (This is because the government has wholly subsidised the electric power tariff for pumpsets). The farmers might have thought that revealing the actual extent of water sale may have adverse effects on them. (2) Own well farmers are also entitled to use tank water; but actually they tap more water from their wells including for sale. For gaining sympathy (expecting more help to improve the tank supply by government spending) the respondents restricted their responses on sale of well water but reported the well water use for their own land as usual, that is, the well farmers expect us to think that they were using water for crop production mostly from their wells than the tank supply. Thesefacts are generally understood from the discussions with the respondents in the villages where water sale is significant. Hence the extent of water sale requires further investigation, which is not attempted in this study.

The weighted average estimates indicate that the area under water sale was significant in the latest reference year than the preceding year. In KPT

Table 7.12: Area Irrigated by Own Wells and Amount Received from Water Sales.

(Area in hectares, amount in rupees)

Tank and sample villages	Sample and total wells in the ayacut		<---------- Latest year (2015-16) ---------->														<---------- preceding year (2005-06) ---------->													
			Season 1 <----> Area <---->		Season 2 <----> Area <---->		Season 3 <----> Area <---->		Gross area benefited <---------->			% of area under sale 11/GSA X100	(OAS/ OAT) X CI	Esti-mated total area 13X 14	OAT/ OAS)X area under sale	OAT X CI	Season 1 <----> Area <---->		Season 2 <----> Area <---->		Season 3 <----> Area <---->		Gross area benefited <---------->			% of area under sale 25/GSA X100	(OAS/ OAT) X CI	Esti-mated total area 27X28	OAT/ OAS)X area under sale	OAT X CI
			Own use	Sale	Own use	Sale	Own use	Sale	Own use	Sale	Total						Own use	Sale	Own use	Sale	Own use	Sale	Own use	Sale	Total					
1	2	3	4	5	6	7	8	9	10	11	12	13	14	15	16	17	18	19	20	21	22	23	24	25	26	27	28	29	30	31
KPT																														
Kadaperi	8	43	2.07	0.6	0.77	1.4	1.76	0.6	4.6	2.6	7.24	32.5	0.83	27.1	4.15	13	2.82	0.30	0.20	0.91	2.61	0.30	5.6	1.51	7.15	13.3	1.16	15.5	11.2	18
Amount received				900		7800		1150										200		3100		200								
Panniyur	10	85	1.62	0.20	1.75	0.20	3.64	0.40	7.0	0.8	7.82	5.4	1.3	7.3	1.1	20	0.40	0.00	2.16	0.81	1.42	0.00	4.0	0.81	4.79	8.7	0.9	7.4	1.1	13
Amount received				600		free		1000												free										
Kattalai	9	84	5.34	3.68	8.44	4.49	8.58	4.49	22.4	12.7	35.02	56.6	0.5	29.1	59.9	106	4.70	2.63	8.41	2.63	6.99	2.63	20.1	7.89	28.00	37.9	0.5	18.2	37.4	99
Amount received				20700		7300		6000										11780		700		1350								
Alapakkam	10	190	6.28	0.00	11.07	0.40	4.25	0.00	21.6	0.4	22.00	1.9	1.0	1.8	0.5	29	6.48	0.00	9.45	0.20	0.81	0.00	16.7	0.20	16.94	1.2	0.7	0.9	0.3	22
Amount received						9000														6900										
Athipattu	12	262	10.08	6.07	18.79	6.07	14.53	4.05	43	16.2	59.6	36.0	0.9	32.1	43.3	120	11.50	2.43	20.74	1.62	19.13	1.62	51	5.67	57.0	11.0	0.9	9.5	15.2	138
Amount received				100500		135000		90000										21600		20000		22000								

(Contd...)

DMT																														
Vagai	5	5	6.52	0.0	6.32	0.0	5.42	0.0	18.3	0.0	18.26	0.0	2.4	0.0	0.0	22	7.73	0.0	4.49	0.0	8.87	0.0	21.1	0.0	21.09	0.0	2.7	0.0	0.0	24
Amount received																														
Punaithangal	9	41	3.04	0.0	4.45	0.0	4.86	0.0	12.3	0.0	12.35	0.0	0.4	0.0	0.0	34	2.23	0.0	6.60	0.0	3.97	0.0	12.8	0.0	12.79	0.0	0.5	0.0	0.0	40
Amount received																														
Narasamangalam	8	43	1.21	0.0	3.01	0.0	0.00	0.0	4.2	0.0	4.23	0.0	0.7	0.0	0.0	17	1.78	0.0	1.42	0.0	2.81	0.0	6.0	0.0	6.01	0.0	0.8	0.0	0.0	18
Amount received																														
Sothiyampakkam	9	76	3.92	0.0	7.48	0.0	0.53	0.0	11.9	0.0	11.93	0.0	0.7	0.0	0.0	66	3.93	0.0	0	0.0	5.18	0.0	9.1	0.0	9.1	0.0	0.6	0.0	0.0	55
Amount received																														

Note: E (H) = Elevated sluice head reach village; D (H) = Deep sluice head reach village; E (T) = Elevated sluice tail reach village; Dash indicates nil. Dnc = Data not canvassed.

OAS = Operated area sample; OAT = Operated area total; CI = Cropping intensity; GSA = Gross area (season acres); Latest year = 2015-16; Preceding year = 2005-06.

Source: Field survey data, 2017.

Table 7.12 Contd...

Tank and sample villages	Sample and total wells in the ayacut		Survey year (2016-17)							
			Season 1 Area		Season 2 Area		Season 3 Area		Gross area benefited	
			Own use	Sale	Own use	Sale	Own use	Sale	Own use	Sale
1	2	3	32	33	34	35	36	37	38	39
KPT										
Kadaperi	8	43	2.32	Dnc	3.72	Dnc	2.93	Dnc	9.0	Dnc
Amount received										
Panniyur	10	85	1.42	Dnc	7.27	Dnc	2.87	Dnc	11.6	Dnc
Amount received										
Kattalai	9	84	5.87	Dnc	8.50	Dnc	8.03	Dnc	22.4	Dnc
Amount received										
Alapakkam	10	190	7.09	Dnc	7.02	Dnc	2.02	Dnc	16.1	Dnc
Amount received										
Athipattu	12	262	15.75	Dnc	18.99	Dnc	14.13	Dnc	49	Dnc
Amount received										
DMT										
Vagai	5	5	4.49	0.0	4.49	0.0	8.87	0.0	17.9	0.0
Amount received										
Punaithangal	9	41	3.64	0.0	4.25	0.0	4.66	0.0	12.6	0.0
Amount received										
Narasamangalam	8	43	0.40	0.0	0.40	0.0	3.01	0.0	3.8	0.0
Amount received										
Sothiyampakkam	9	76	0.93	0.0	2.70	0.0	5.87	0.0	9.5	0.0
Amount received										

Note: Dnc = Data not collected.

Source: Field survey data, 2017.

in 2015-16, the incidence of water sales was more in the deep sluice village of Kattalai (59.9 ha) followed by the tail end village of Athipattu (43.3 ha) and in the elevated sluices village of Kadaperi and Alapakkam and also tail end village of Alapakkam it is less than 5 ha. In 2005-06, no water sale was reported in the elevated sluice village of Panniyur and the tail end of Alapakkam. Again in Kattalai water sale was the highest, which was 37.9 ha. In Athipattu the sale was very less (11 ha).

During the survey year 2016-17, tank water was not available in KPT villages but it was available for one crop season in DMT. Although well water use for own farms was canvassed it was not possible to enquire the water sale in KPT in this survey year.

7.4.8. Crops Irrigated by Well Water

It is interesting to note that well farmers mostly used their well water for raising paddy, sugarcane and banana crops in all the three seasons in both the supply years and water was also sold for these crops. The following points emerge from Tables 7.13 and 7.14.

(1) In KPT in the first and third seasons in both the reference years well farmers primarily used their well water for raising sugarcane and banana crops. Paddy occupies second position in getting well supply. However, in the second season in both the supply years both annual crops and paddy equally got more well water supply. Remaining (other) crops got the well water only for a limited acreage.

(2) In DMT paddy cultivation predominates and water sale is nil. As a result, farmers used their well water mostly for paddy crop. Banana and Sugarcane cultivation is almost nil in DMT. As a result, other crops such as groundnut vegetables and ragi got well water supply in the latest supply year and that also almost nil in the preceding supply year.

(3) Well water is sold for paddy and annual crops, especially in KPT. No other crop is reported getting well water. Even for paddy and annual crops, water sale is significant for annual crops in the first season and the acreage irrigated for paddy is negligible. In the second season water sale was negligible and in the third season water sale was significant for paddy crop in the latest supply year; but it is for the annual crops in the preceding supply year.

Table 7.13: Crops Irrigated by Using Own Well Water and Sale, 2015-16.

(Area in hectares)

Name of tank and sample village	Season 1						Season 2						Season 3					
	Own use			Sale			Own use			Sale			Own use			Sale		
	Paddy	SC&B	Others	Paddy	SC&B	Others	Paddy	SC&B	Others	Paddy	SC&B	Others	Paddy	SC&B	Others	Paddy	SC&B	Others
KPT																		
Kadaperi	-	1.46	0.61	-	0.61	-	0.51	0.66	-	-	3.50	-	-	0.99	0.77	-	0.61	-
Panniyur	1.62	-	-	-	-	0.20	1.75	-	-	-	-	0.20	3.64	-	-	-	-	0.40
Kattalai	-	5.34	-	-	3.64	-	1.90	6.52	-	0.40	4.09	-	1.90	6.68	-	0.40	4.09	-
Alapakkam	5.67	-	0.61	-	-	-	10.08	-	0.40	0.40	-	-	3.85	-	-	-	-	0.40
Athipattu	4.41	5.67	-	6.07	-	-	13.93	5.67	1.21	6.07	-	-	4.01	9.31	1.21	4.05	-	-
DMT																		
Vagai	6.52	-	-	-	-	-	6.30	-	-	-	-	-	5.42	-	-	-	-	-
Punaithangal	2.63	-	0.40	-	-	-	4.05	-	0.40	-	-	-	4.86	-	-	-	-	-
Narasamangalam	1.21	-	-	-	-	-	3.01	-	-	-	-	-	-	-	-	-	-	-
Sothiyampakkam	3.92	-	-	-	-	-	7.48	-	-	-	-	-	0.53	-	-	-	-	-
HRWF*	2.77	3.04	-	-	-	-	11.38	4.29	-	-	-	-	1.82	4.29	-	-	-	-
TRWF *	0.42	3.67	-	-	-	-	6.28	4.23	-	-	-	-	4.89	2.87	-	-	-	-

Note: HRWF = Head reach own-well farmers in the non-well category; TRWF = Tail reach own-well farmers in the non-well category.

Sc&B = Sugarcane and banana; others include: coconut, groundnut, gingelly, vegetables, fodder, and ragi; Dash indicates nil.

* indicates SC & Banana cultivation under this HRWF and TRWF mostly belong to KPT villages and not for DMT villages.

Source: Field survey data, 2017.

Table 7.14: Crops Irrigated by Using Own Well Water and Sale, 2005-06

(Area in hectares)

Name of tank	Season 1						Season 2						Season 3					
and sample	Own use			Sale			Own use			Sale			Own use			Sale		
village	Paddy	SC&B	Others	Paddy	SC&B	Others	Paddy	SC&B	Others	Paddy	SC&B	Others	Paddy	SC&B	Others	Paddy	SC&B	Others
KPT																		
Kadaperi	0.75	1.80	0.27	-	0.30	-	-	-	0.50	1.50	0.75	-	-	6.45	-	-	0.75	-
Panniyur	0.40	-	-	-	-	-	2.16	-	-	0.81	-	-	1.42	-	-	-	-	-
Kattalai	4.70	-	-	-	-	-	3.28	5.51	-	-	2.03	-	1.86	5.51	-	-	2.63	-
Alapakkam	3.85	-	2.63	-	-	-	8.64	-	0.81	0.20	-	-	0.81	-	-	-	-	-
Athipattu	1.17	10.32	-	0.81	1.62	-	7.25	12.75	-	-	1.62	-	5.63	13.56	-	-	1.62	-
DMT																		
Vagai	7.73	-	-	-	-	-	4.48	-	-	-	-	-	8.87	-	-	-	-	-
Punaithangal	2.23	-	-	-	-	-	6.60	-	-	-	-	-	3.97	-	-	-	-	-
Narasamangalam	1.77	-	-	-	-	-	1.42	-	-	-	-	-	2.81	-	-	-	-	-
Sothiyampakkam	3.92	-	-	-	-	-	0.00	-	-	-	-	-	5.18	-	-	-	-	-
HRWF *	3.82	-	-	-	0.14	-	5.70	-	-	-	-	-	2.02	-	-	-	0.14	-
TRWF *	2.54	1.42	0.40	-	-	-	9.56	-	-	-	-	-	28.20	1.42	-	-	-	-

Note: HRWF = Head reach own-well farmers in the non-well category; TRWF = Tail reach own-well farmers in the non-well category.

Sc&B = Sugarcane and banana; others include: groundnut, gingelly, vegetables, fodder and ragi; '-' indicates nil.

* indicates SC & Banana cultivation under this HRWF and TRWF mostly belong to KPT villages and not for DMT villages.

Source: Field survey data, 2017.

7.5. Land Use and Cropping Pattern

Cultivation of holdings primarily requires favourable water supply from surface and sub-surface sources or rainfall. While clear idea on the availability of water supply from the first two sources are collected, data available are inadequate in the case of rainfall. However, on the basis of the information received from the respondents one can also ascertain the effect of rainfall upon cropping pattern in the first season, especially in DMT villages, where a considerable proportion of area is cultivated (broadcast paddy) by using rain water alone. In 2015-16 favourable supply was available in the second season in all the villages. But, in the third season tank supply was good only for a few villages and the tail end farmers both within village and across villages were affected much by scarcity of supply. The supply from wells was better in most of the villages due to the supplemental use of tank water. On the basis of the survey data for the latest two supply years and also for the year 2016-17 an analysisof cropping intensity, cropping pattern and productivity per hectare of land in the sample villages in both the tanks is attempted and the results are presented.

7.5.1. Cropping Intensity and Cropping Pattern

7.5.2. Cropping intensity (CI) refers to the actual area cropped in the three seasons - counted four months each - in a year. For banana, sugarcane and other perennial crops only seasons when the crop was available on the land are taken into account. (The method adopted for measuring CI is given in the Appendix 7.1.)

7.5.3. Cropping Intensity in the Latest,Preceding Supply Yearsand during the Survey Year

The following points on CI emerge from Tables 7.15, 7.16 and 7.17:

Table 7.15: Distribution of Cultivated Area (Cropping Pattern) by Sample Households, 2015-16

(Area in hectares)

Tank, sample villages and HHs by type	Operational holdings	Area in Season 1					Area in Season 2					Area in Season 3					Gross area season hectare	Cropping Intensity 18/2X10
		Paddy	Banana	Sugarcane	Others	Total	Paddy	Banana	Sugarcane	Others	Total	Paddy	Banana	Sugarcane	Others	Total		
1	2	3	4	5	6	7	8	9	10	11	12	13	14	15	16	17	18	19
KPT																		
Kadaperi E (H)																		
Well 8	6.19	0	1.46	0	0.61	2.1	2.67	0.9	0	0.7	4.3	0	1	0	0.77	1.8	8.12	131.0
Head 8	1.89	0	0	0	0.13	0.1	1.70	0	0	0	1.7	0.14	0	0	0	0.1	1.97	104.3
Tail 8	2.91	0	0	0	0	0.0	1.23	0.40	0	0	1.6	0	0.4	0	0	0.4	2.04	69.9
Panniyur E (H)																		
Well 8	8.21	1.62	0	1.62	0.91	4.1	7.91	0	0	0	7.9	2.64	0	0	0	2.6	14.71	179.1
Head 8	3.42	0	0	0	1.45	1.5	1.98	0	0	0.6	2.6	0.72	0	0	0	0.7	4.76	139.2
Tail 8 *	2.59	0	0	0	0	0.0	1.92	0	0	0	1.9	0	0	0	0	0.3	2.19	84.7
Kattalai D (H)																		
Well 8	9.19	0	5.34	0	0	5.3	1.88	6.6	0	0	8.4	2.03	6.6	0	0	8.6	22.36	243.3
Head 8	2.63	0	1.93	0	0	1.9	0.40	2.1	0	0	2.5	0.51	2.1	0	0	2.6	7.08	269.3
Tail 8	3.32	0	2.26	0	0	2.3	1.42	1.45	0	0	2.9	1.42	1.4	0	0	2.9	7.98	240.7
Alapakkam M (T)																		
Well 8	16.34	4.86	0	0	1.42	6.3	10.67	0	0	0.40	11.1	3.44	0	0	0.81	4.3	21.60	132.2
Head 8 *	3.30	0	0	0	0.13	0.1	3.10	0	0	0	3.1	0	0	0	0	0.0	3.23	97.9
Tail 8 *	4.79	0	0	0	0	0.0	2.56	0	0	0	2.6	0	0	0	0	0.0	2.56	53.4
Athipattu D (T)																		
Well 8	22.26	4.41	6	0	0	10.1	13.93	5.7	0	2.43	22.0	6.03	5.7	0	1.21	12.9	45.0	202.2
Head 8 *	4.59	0	1.11	0	0	1.1	0.81	2.2	0	0	3.0	0.81	2.2	0	0	3.0	7.10	154.7
Tail 8 *	6.20	0	1.42	0	0	1.4	2.83	1.42	0	0	4.2	2.44	1.4	0	0	3.9	9.52	153.5
DMT																		
Vagai E (H)																		
Well 5	9.06	6.51	0	0	0	6.5	6.30	0	0	0	6.3	8.86	0	0	0	8.9	21.67	239.3
Head 8	2.38	0	0	0	0	0.0	2.12	0	0	0	2.1	0.35	0	0	0	0.3	2.47	103.6
Tail 8	5.83	0	0	0	0	0.0	5.69	0.0	0	0	5.7	0	0	0	0	0.0	5.69	97.6

(Contd...)

Punaithangal E (T)																		
Well 8	10.45	2.63	0	0	0.40	3.0	4.05	0	0	0.40	4.5	4.45	0	0	0.40	4.9	12.35	118.2
Head 8	5.12	0	0	0	0	0.0	3.10	0	0	0	3.1	0	0	0	0	0.0	3.10	60.5
Tail 8	3.49	0	2	0	0	1.7	1.13	0	0	0	1.1	1.13	0	0	0	1.1	3.97	113.8
Narasamangalam D(H																		
Well 8	7.19	1.21	0	0	0	1.2	3.62	0	0	0	3.6	4.44	0	0	0	4.4	9.28	129.0
Head 8 *	2.97	0.24	0	0	0	0.2	2.57	0	0	0	2.6	1.89	0	0	0	1.9	4.70	158.2
Tail 8 *	3.15	0	0	0	0	0.5	2.21	0	0	0	2.2	0.59	0	0	0	0.6	3.28	104.4
Sothiyampakkam D(T																		
Well 8	8.70	3.92	0	0	0	3.9	7.48	0	0	0	7.5	8.37	0	0	0	8.4	19.77	227.2
Head 8 *	7.94	2.52	0	0	0	2.5	6.97	0	0	0	7.0	0.81	0	0	0	0.8	10.30	129.7
Tail 8 *	3.93	0.47	0	0	0	0.5	3.72	0	0	0	3.7	2.33	0	0	0	2.3	6.52	166.0

Note: E (H) = Elevated sluice head reach village; D (H) = Deep sluice head reach village; M (T) = Middle sluice tail reach village.
D (T) = Deep sluice tail reach village; E (T) = Elevated sluice tail reach village. * Indicates a few farmers used their own well water.
Others include groundnut, ragi, vegetables, grams, fodder, tree crops, wate apple, gingelly and coconut.
Source: Field survey data, 2017.

Table 7.16: Distribution of Cultivated Area (Cropping Pattern) by Sample Households, 2005-06

(Area in hectares)

Tank, sample villages and HHs by type	Operational holdings	Area in Season 1 <------>					Area in Season 2 <------>					Area in Season 3 <------>					Gross area season hectare	Cropping Intensity 18/2X10(
		Paddy	Banana	Sugar cane	Others	Total	Paddy	Banana	Sugar cane	Others	Total	Paddy	Banana	Sugar cane	Others	Total		
1	2	3	4	5	6	7	8	9	10	11	12	13	14	15	16	17	18	19
KPT																		
Kadaperi E (H)																		
Well 8	6.19	0.77	1.82	0	0.28	2.9	3.81	1.8	0	0.20	5.8	0	2.6	0	0	2.6	11.32	182.8
Head 8	1.89	0.28	0	0	0	0.4	1.48	0.1	0	0	1.6	0	0.1	0	0	0.1	2.17	115.2
Tail 8	2.91	0	0	0	0	0.0	1.97	0	0	0	2.0	0	0	0	0	0.0	1.97	67.6
Panniyur E (H)																		
Well 8	8.21	0.40	0	0	0.91	1.3	6.61	0	0	0	6.6	1.42	0	0	0	1.4	9.34	113.8
Head 8	3.42	0	0	0	0.65	0.6	2.65	0	0	0	2.6	0	0	0	0	0.0	3.30	96.4
Tail 8 *	2.59	0	0	0	0	0.0	1.86	0	0	0	1.9	0	0	0	0	0.3	2.13	82.3

(Contd...)

Kattalai D (H)																		
Well 8	9.19	4.70	0	0	0	4.7	3.26	5.5	0	0	8.8	1.84	5.5	0	0	7.4	20.83	226.7
Head 8	2.63	1.42	0	0	0	1.4	0.59	2	0	0	2.6	0.70	1.9	0	0	2.6	6.67	253.9
Tail 8	3.32	1.93	0	0	0	1.9	1.38	1.93	0	0	3.3	1.26	1.9	0	0	3.2	8.44	254.6
Alapakkam M (T)																		
Well 8	16.34	3.85	0	0	2.63	6.5	7.63	0	0	1.01	8.6	0.20	0	0	1.01	1.2	16.34	100.0
Head 8 *	3.30	0	0	0	0	0.0	3.30	0	0	0	3.3	1.62	0	0	0	1.6	4.92	149.1
Tail 8 *	4.79	0	0	0	0	0.4	2.68	0	0	0	2.7	0	0	0	0	0.0	3.09	64.4
Athipattu D (T)																		
Well 8	22.26	1.17	6.28	4.05	0	11.5	7.12	6.3	7.29	0	20.7	5.63	6.3	7.29	0	19.2	51.4	230.7
Head 8 *	4.59	0	0	0	0	0.0	3.77	0	0	0	3.8	0	0	0	0	0.0	3.77	82.2
Tail 8 *	6.20	0	1.42	0	0	1.4	2.69	1.42	0	0	4.1	0	1.4	0	0	1.4	6.94	111.9
DMT																		
Vagai E (H)																		
Well 5	9.06	7.72	0	0	0	7.7	7.59	0	0	0	7.6	8.86	0	0	0	8.9	24.17	266.9
Head 8	2.38	0	0	0	0	0.0	2.36	0	0	0	2.4	0	0	0	0	0.0	2.36	99.1
Tail 8	5.83	0	0	0	0	0.0	5.69	0	0	0	5.7	0	0	0	0	0.0	5.69	97.6
Punaithangal E (T)																		
Well 8	10.45	1.82	0	0	0.40	2.2	7.21	0	0	0.40	7.6	4.37	0	0	0.40	4.8	14.62	139.9
Head 8	5.12	0	0	0	0	0.0	3.10	0	0	0	3.1	0	0	0	0	0.0	3.10	60.5
Tail 8	3.49	0	0	0	0	0.0	4.37	0	0	0	4.4	0	0	0	1.21	1.2	5.59	160.3
Narasamangalam D(H																		
Well 8	7.19	1.77	0	0	0	1.8	5.45	0	0	0	5.5	2.81	0	0	0	2.8	10.03	139.5
Head 8 *	2.97	0.70	0	0	0	0.7	2.97	0	0	0	3.0	0.61	0	0	0	0.6	4.28	144.1
Tail 8 *	3.15	0.16	0	0	0	0.2	2.96	0	0	0	3.0	0.16	0	0	0.47	0.6	3.75	119.3
Sothiyampakkam D(T																		
Well 8	8.70	3.92	0	0	0	3.9	8.21	0	0	0	8.2	4.69	0	0	0	4.7	16.83	193.4
Head 8 *	7.94	3.53	0	0	0	3.5	7.65	0	0	0	7.6	0	0	0	0	0.0	11.18	140.8
Tail 8 *	3.93	0.91	0	0	0	0.9	3.72	0	0	0	3.7	0.45	0	0	0	0.4	5.08	129.4

Note: E (H) = Elevated sluice head reach village; D (H) = Deep sluice head reach village; M (T) = Middle sluice tail reach village.
D (T) = Deep sluice tail reach village; E (T) = Elevated sluice tail reach village. * Indicates a few farmers used their own well water.
Others include groundnut, ragi, vegetables, grams, fodder, tree crops, wate apple, gingelly and coconut.
Source: Field survey data, 2017.

Table 7.17: Distribution of Cultivated Area (Cropping Pattern) by Sample Households, 2016-17

(Area in hectares)

Tank, sample villages and HHs by type	Operational holdings	Area in Season 1					Area in Season 2					Area in Season 3					Gross area season hectare	Cropping Intensity 18/2X10[0]
		Paddy	Banana	Sugar cane	Others	Total	Paddy	Banana	Sugar cane	Others	Total	Paddy	Banana	Sugar cane	Others	Total		
1	2	3	4	5	6	7	8	9	10	11	12	13	14	15	16	17	18	19
KPT																		
Kadaperi E (H)																		
Well 8	6.19	0	1.34	0	0.97	2.3	1.62	1.3	0	0.8	3.7	0.61	1.3	0	0.97	2.9	8.96	144.7
Head 8	1.89	0	0.81	0	0	0.8	0	0.8	0	0.3	1.1	0	0.8	0	0.13	0.9	2.79	148.1
Tail 8	2.91	0	0	0.53	0	0.9	0	0.54	0.53	0	1.1	0	0.5	0.53	0	1.1	3.06	105.0
Panniyur E (H)																		
Well 8	8.21	1.42	0	0	0	1.4	7.27	0	0	0	7.3	2.87	0	0	0	2.9	11.56	140.8
Head 8	3.42	0	0	0	0	0.0	0	0	0	0	0.0	0	0	0	0	0.0	0.00	0.0
Tail 8 *	2.59	0	0	0	0	0.0	0	0	0	0.3	0.3	0	0	0	0.27	0.3	0.54	21.0
Kattalai D (H)																		
Well 8	9.19	0.61	4	0	0	4.5	1.48	3.8	0	0	5.3	2.17	5.5	0	0	7.7	17.47	190.1
Head 8	2.63	0	2	0	0	1.9	0.61	1.4	0	0	2.0	0	1.4	0	0	1.4	5.26	200.3
Tail 8	3.32	0	2	0	0	2.3	0	2.86	0	0	2.9	0	2.5	0	0	2.5	7.66	230.9
Alapakkam M (T)																		
Well 8	16.34	5.67	0	0	2.43	8.1	6.21	0	0	1.42	7.6	1.62	0	0	1.01	2.6	18.36	112.4
Head 8 *	3.30	0	0	0	0	0.0	2.59	0	0	0.1	2.7	0	0	0	0	0.0	2.72	82.6
Tail 8 *	4.79	0	0	0	1	0.6	1.15	0	0	0	1.2	0	0	0	0	0.0	1.76	36.7
Athipattu D (T)																		
Well 8	22.26	8.46	6.07	0	1	15.7	11.7	6.1	0	1.21	19.0	5.22	4	0	4.86	14.1	48.9	219.5
Head 8 *	4.59	0	2	0	0	2.2	1.21	2.2	0	0	3.4	1.07	1.1	0	0	2.2	7.76	169.2
Tail 8 *	6.20	0.81	1.42	0	0	2.2	2.44	1.42	0	0	3.9	0.62	0	0	0	0.6	6.70	108.1
DMT																		
Vagai E (H)																		
Well 5	9.06	8.86	0	0	0	8.9	8.86	0	0	0	8.9	8.86	0	0	0	8.9	26.59	293.6
Head 8	2.38	1.94	0	0	0	1.9	1.77	0	0	0	1.8	0.71	0	0	0	0.7	4.42	185.5
Tail 8	5.83	5.69	0	0	0	5.7	5.55	0	0	0	5.5	0	0	0	0	0.0	11.23	192.7

(Contd...)

Punaithangal E (T)																	
Well 8	10.45	5.06	0	0	0.40	5.5	5.67	0	0	0.40	6.1	3.72	0	0	0.93	4.7	16.19	155.0
Head 8	5.12	3.50	0	0	0	3.5	0.61	0	0	0	0.6	0	0	0	0	0.0	4.11	80.2
Tail 8	3.49	3.77	0	0	0	3.8	0	0	0	0	0.0	0	0	0	0	0.0	3.77	108.0
Narasamangalam D(H																	
Well 8	7.19	5.45	0	0	0	5.5	2.38	0	0	0	2.4	5.05	0	0	0	5.0	12.88	179.2
Head 8 *	2.97	2.22	0	0	0	2.5	2.20	0	0	0	2.2	2.47	0	0	0	2.5	7.16	240.9
Tail 8 *	3.15	2.96	0	0	0	3.0	2.11	0	0	0	2.1	2.07	0	0	0	2.1	7.13	226.8
Sothiyampakkam D(T																	
Well 8	8.70	5.98	0	0	0	6.0	8.43	0	0	0	8.4	5.87	0	0	0	5.9	20.28	233.1
Head 8 *	7.94	6.92	0	0	0	6.9	6.49	0	0	0	6.5	3.77	0	0	0	3.8	17.17	216.2
Tail 8 *	3.93	4.74	0	0	0	4.7	2.73	0	0	0	2.7	1.86	0	0	0	1.9	9.33	237.6

Note: E (H) = Elevated sluice head reach village; D (H) = Deep sluice head reach village; M (T) = Middle sluice tail reach village.
D (T) = Deep sluice tail reach village; E (T) = Elevated sluice tail reach village. * Indicates a few farmers used their own well water.
Others include groundnut, ragi, vegetables, grams, fodder, tree crops, wate apple, gingelly and coconut.
Source: Field survey data, 2017.

(1) CI under well holdings is higher than non-well holdings in both tank supply years (except in Kattalai where non-well got a higher CI). But, in the year of poor tank supply, CI is invariably lower, and the difference being more marked in DMT than KPT.In Kattalai, the increase in cropping intensity by non-well head and tail segments is mainly due to cultivating the annual crop banana by most of the non-well farmers. Since groundwater is rich in Kattalai and water sale/purchase is widely prevalent this induced the cropping intensity above the well farmers.

Most of the well farmers prefer paddy cultivation, mainly for their self-consumption and also for sale of paddy.Further preservation of paddy is easy than banana. If farmers cultivate paddy in one or two seasons, definitely they left their land fallow at least for one season in a year. Hence the well farmers CI is lower in Kattalai than non-well farmers. This is purely due to locational advantage to reap more of groundwater from the KPT storage. Another reason is, once the KPT reaches its full storage the Kattalai lands get the seepage effect immediately in the vicinity of the tank bund (at least for 0.5 km) that is in the head reach ayacut of the village. So this is favourable for both head and tail end non-well farmers of Kattalai to crop bananathan paddy. The same reasons are also applicable for Kadaperi head reach (Nanjai Kalani) ayacutdars. But that village is served by the elevated sluice and supply is also possible for one season even at FTL, the effect is much less compared to Kattalai village.

(2) Among non-well head segment farmers, CI in deep sluice is higher than in elevated sluices and tail end villages. The difference is much more marked in KPT than in DMT. This pattern holds good in all three reference years (except in 2005-06 in DMT).

(3) In the non-well tail segments, again, this pattern prevails. The difference between head and tail reach villages in this respect is very marked in DMT; but not so marked in KPT.

(4) CI in the preceding year (2005-06) is not varied much than in 2015-16. However, between KPT and DMT, the reduction in CI is more marked in DMT villages than in KPT. In KPT, during 2005-06 the picture is more mixed because of better access to groundwater, and not much less reduction in supply relative to normal year. Howeverin DMT, in 2005-06 supply year, the non-well holdings of Dusi village recorded very low CI that is less than 100%. This low CI was reported as crop failure due to pest attack in the lands of majority of the farmers.

In 2016-17, KPT villages did not get tank supply but DMT villages received good supply, mainly due to longer period of storage capacity in DMT and also good local catchment supplies in 2006 monsoon. Consequently CI in deep sluice of Mamandur villages showed higher CI than the Dusi sluice villages almost in all segments of the ayacut (except in the head segment of non-well farms in Punaithangal. However, in KPT, except Kattalai, Kadaperi and Athipattu the remaining two villages - Panniyur and Alapakkam – showed a poor CI including in the well farms.

It is clear from the above that access to well water and location of land in the *ayacut* have a significant bearing on CI. Well farmers have higher CI than non-well farmers. Among the latter the head segments crop land intensely than the tail segments. However, the prevalence of wells is not the only factor affecting CI. In DMT in the year 2016-17, CI of all the non-well farmers was more or less equal to that of well farmers. This high CI was due to the widespread practice of broadcast sowing of paddy in the first crop season depending primarily on rainfall. The soils in the *ayacut* of these villages in DMT are suitable to this practice to a degree not possible in KPT and farmers like to take advantage of this to get an extra, even if risky, crop. However, the average yield of the broadcast paddy is relatively low.

7.5.4. Cropping Pattern

7.5.6. Crop Area by Types of Crops

It is generally recognised that access to irrigation and the sources of irrigation have a significant bearing on cropping pattern. The source of groundwater is richer and tank supply is more favourable in KPT than in DMT. As a result, in most of the KPT villages farmers grow annual crops such as banana and sugarcane to a considerable extent apart from paddy. But in DMT only paddy is grown in all the seasons irrespective of availability of tank supply.

There are also some striking differences in the seasonal pattern of cropping (see Table 7.15). In the KPT *ayacut* the second cropping season is more prominent than the first; in DMT the importance of first crop season is more pronounced, especially under well holdings). This is more so in both the supply years than in the latest year (2016-17). In the latter year first season cropping by broadcast sowing was not possible for farmers since most farmers crop were fully / partly standing in the field during the first season. Comparison of the second and third crop seasons in the year

2016-17 reveals that farmers cropped the *ayacut* more intensely in DMT than in KPT.

Banana and sugarcane are the dominant annual crops in the KPT *ayacut* (Table 7.15), especially in the deep sluice Kattalai and tail end village of Athipattu. Kadaperi farmers cultivated lesser extent. Paddy occupies the first place in cultivated area in the second season, thanks to the availability of assured tank supply. However, there are differences in cropping pattern within the village and between well and non-well farmers. In KPT well farmers grow annual crops such as banana irrespective of reaches from the tank. In DMT almost the entire *ayacut* is cropped with paddy only. This major difference in cropping pattern between the tanks is attributed to the relative non-availability of groundwater and also due to different soil conditions in DMT.

There is also considerable difference in the cropping pattern between the two tank supply years (see Table 7.16). The proportion of area under paddy in DMT, especially in the first season in Dusi sluice villages was drastically reduced since it depends more on broadcast sowing.Again, in the second season the area under paddy is normal in both supply years but it varies in the third season. In the supply year 2005-06 the area is drastically reduced in Mamandur sluice villages in the third season compared to the year 2015-16, which primarily use the tank water. Interestingly, in Dusi sluice villages, except well farmers, none of the non-well farmers grow paddy irrespective of reaches within the village and across villages from the tank in 2005-06. This points out that other than water supply other agronomic and climatic factors also play a crucial role in deciding the cropping pattern.

The impact of rainfall upon cropping pattern was more evident in the latest supply year than in the preceding one. In DMT in the latest supply year there was a drastic reduction in both area and yield on broadcast sowing of paddy (this crop was done before starting the tank supply season). Even a few *ayacutdars* who cropped the broadcast paddy by using light showers did not get the average yield. However, in the second and third seasons tank supply changed the crop pattern to all farmers positively in the DMT *ayacut*.

In KPT since tank supply was available adequately in the preceding supply year also, it was utilised for two crop seasons by the deep sluice of Kattalai and only one season in the remaining four villages located two each at the tail end and served by elevated sluices. As a result, in the third season the area under paddy shrank much, but the area under banana and sugarcane was stabilised. This makes it clear that whenever adequate tank supply is available in both the tanks cultivation of area under paddy predominates. The survey indicates that there is no systematic relation between the proportion of area under different crops and location of villages in the tanks. The only major difference is that well farmers raise paddy and annual crops to a great extent than the non-well farmers. In DMT the latter category farmers depend exclusively upon tank water to raise crops, especially in the second and third seasons.

The outcome of cropping will be judged mostly from the output of crops raised. As it is known, better cropping provides more yield. On this aspect, in the following section, the productivity of land under sample villages could be analysed in the three reference years.

7.6. Productivity of Land

Crop productivity refers to the value of output per unit of gross cropped area (counted in terms of season hectares). Productivity is an important indicator for the efficiency of irrigation management in agriculture. The primary motive of all agricultural operations including water management is to achieve the maximum yield and thereby receive higher income. It is needless to say that ceteris paribus (other things remaining constant) better irrigated land produces greater yield than poorly irrigated land.

The average yield of different crops varies across seasons and farms depending on water supply. The combined effect of this and the variations in crop pattern is reflected in the value of output per hectare of cropped area; and taken together with variations in crop intensity accounts for the differences in output per hectare of operated area. The relevant data are given in Tables 7.18 to 7.20 for the latest, preceding supply years and in the survey year 2016-17.

7.6.1. Productivity of Land in the Latest Supply Year

The mean productivity[1] per hectare of Gross Cropped Area (GCA) for the well and non-well farms in the sample villages under KPT and DMT are given in Table 7.18 which relates to the normal tank supply year. The following points emerge from the analysis:

7.6.2. Well Farms

Output per hectare of both gross cropped area (for the method of calculation of Gross Value Output (GVO) see Appendix 7.2) and Net Sown

Table 7.18: Value Productivity of Crops Cultivated by Sample Households, 2015-16.

(Area in hectares, value in rupees)

Tank, sample villages and HHs by type	Opera-tional hold-ings	Total area as per adangal	Gross area season acres	Gross value	GVO/ Gross area 5/4	GVO/ Net area 5/2	CI (4/2) X100	Imputed value: Gross area (4/2)X3	Imputed value: Gross value (5/4)X9	Weighted average GVO/area W & NW: Gross (10/9)	Weighted average GVO/area W & NW: Net (10/3)	Weighted average GVO/area Non-Well: Gross (10/9)	Weighted average GVO/area Non-Well: Net (10/3)	% of area under P+SC+E	Ratio of well to non-well: Gross	Ratio of well to non-well: Net	Ratio of NW head to NW tail: Gross	Ratio of NW head to NW tail: Net
1	2	3	4	5	6	7	8	9	10	11	12	13	14	15	16	17	18	19
KPT																		
Kadaperi E (H)																		
Well 8	6.19	9.73	8.12	486453	59927	78532	131.0	12.75	764017	59927	78532			74	1.03	1.56		
Head 8A*	1.89	8.31	1.97	113490	57679	60155	104.3	8.66	499746	57919	50392	57679	60155	93			0.99	1.48
Tail 8A*	2.91	8.57	2.04	118437	58159	40630	69.9	5.99	348236			58159	40630	100				
Panniyur E (H)																		
Well 8	8.21	10.99	14.71	843550	57351	102740	179.1	19.68	1128893	57351	102740			94	1.49	2.38		
Head 8	3.42	9.64	4.76	185380	38969	54252	139.2	13.41	522754	38510	43233	38969	54252	56			1.02	1.68
Tail 8B*	2.59	8.55	2.19	83340	38050	32214	84.7	7.24	275583			38050	32214	100				
Kattalai D (H)																		
Well 8	9.19	43.51	22.36	743325	33243	80882	243.3	105.85	3518842	33243	80882			100	0.71	0.68		
Head 8A*	2.63	47.90	7.08	332550	46991	126564	269.3	129.02	6062761	46807	119385	46991	126564	100			1.01	1.13
Tail 8B*	3.32	36.90	7.98	372050	46624	112206	240.7	88.81	4140703			46624	112206	100				
Alapakkam M (T)																		
Well 8	16.34	22.12	21.60	1042858	48282	63838	132.2	29.25	1412187	48282	63838			80	0.81	1.44		
Head 8A*	3.30	25.95	3.23	177120	54823	53679	97.9	25.41	1392837	59912	44188	54823	53679	96			0.84	1.55
Tail 8B*	4.79	32.43	2.56	166320	65002	34697	53.4	17.31	1125329			65002	34697	81				
Athipattu D (T)																		
Well 8	22.26	59.55	45.43	2618600	57647	117620	204.0	121.51	7004838	57647	117620			92	1.12	1.48		
Head 8	4.59	24.35	7.10	328745	46321	71668	154.7	37.67	1744989	51638	79536	46321	71668	100			0.81	0.82
Tail 8	6.20	30.91	9.52	542121	56956	87405	153.5	47.44	2701756			56956	87405	100				
DMT																		
Vagai E (H)																		
Well 5	9.06	9.06	21.67	1619330	74720	178800	239.3	21.67	1619330	74720	178800			100	1.37	3.27		
Head 8	2.38	7.74	2.47	112558	45652	47282	103.6	8.02	366005	54627	54670	45652	47282	100			0.72	0.76
Tail 8	5.83	9.30	5.69	361791	63603	62057	97.6	9.08	577358			63603	62057	100				

(Contd...)

Punaithangal E (T)																		
Well 8	10.45	28.71	12.35	742720	60148	71105	118.2	33.94	2041328	60148	71105			90	0.90	1.21		
Head 8	5.12	22.18	3.10	197865	63886	38635	60.5	13.41	856843	66710	58889	63886	38635	100			0.92	0.49
Tail 8	3.49	7.89	3.97	275883	69534	79144	113.8	8.98	624180			69534	79144	100				
Narasamangalam D (H)																		
Well 8	7.19	13.01	9.28	660514	71212	91862	129.0	16.78	1194951	71212	91862			100	1.19	1.16		
Head 8B*	2.97	16.84	4.70	287329	61129	96690	158.2	26.64	1628460	59908	78973	61129	96690	100			1.04	1.58
Tail 8B*	3.15	14.81	3.28	192696	58688	61256	104.4	15.45	906935			58688	61256	87				
Sothiyampakkam D (T)																		
Well 8	8.70	29.05	19.77	1426392	72152	163945	227.2	66.00	4762383	72152	163945			100	1.11	1.70		
Head 8B*	7.94	15.70	10.30	672436	65262	84654	129.7	20.37	1329449	65150	96302	65262	84654	100			1.00	0.78
Tail 8B*	3.93	11.16	6.52	423931	65038	107949	166.0	18.53	1204925			65038	107949	100				

Note : A* A few farmers used well water through purchase; B* A few farmers used well water from their own wells.
E (H) = Elevated sluice head reach village; D (H) = Deep sluice head reach village; M (T) = Middle sluice tail reach village.
D (T) = Deep sluice tail reach village; E (T) = Elevated sluice tail reach village; W = Well; NW = Non-well.
GVO = Gross value output; CI = Cropping intensity; P+SC+B = Paddy plus sugarcane plus banana.
Source: Field survey data, 2017.

Area (NSA) under well farms are generally more than the non-well farms in all the villages under both the tanks.

In KPT, across reaches, output per hectare of GCA and operated area (NSA) in well farms is relatively higher in the elevated sluice villagesfollowed by the tail end villages than in the head reach village of Kattalai. The differences being much more marked in output per hectare of operated area. The output per hectare of GCA ranges from Rs. 33243 in Kattalai to Rs. 59927 in Kadaperi; and output per acre of NSA ranges from Rs. 63838 to Rs. 117620. This difference is because of variations in cropping intensity. In DMT, the output per hectare of GCA is generally higher in well farms compared to KPT. It is higher in the deep sluicevillages than in elevated. There is no clear pattern of variation in output per hectare of operated area across reaches.

7.6.3. Well and Non-Well Differences

In KPT, the ratio of output per gross hectare between well and non-well farms is much higher in the elevated sluice villages compared to the tail end villages and deep sluice village. This difference is due to the fact that non-well farms under elevated sluice villages get assured tank supply than that of the non-well farms in the tail end villages. In DMT, the ratio of GVO/GCA between well and non-well farms is much less (compared to KPT) in all the villages, except in the tail end of Punaithangal - where the non-well farmers did not get adequate supply in the past two-three decades, that factor reduced ratio of GVO/GCA.

In KPT, the ratio of output per hectare of operated area in the tail end villages is about two times that of the elevated sluice village of Panniyur. The difference is mainly because of poor tank supply (no supply to Kattalai) combined with low CI in the non-well farms under tail end villages. In DMT, the ratio of GVO/NSA is relatively more and highest in Dusi sluice villages compared to the Mamandur sluice. However, within each sluice, between the head and tail villages, the productivity difference is much less (except in Punaithangal).

7.6.4. Non-Well Head Segment

In both tanks, output per hectare of GCA and NSA in the non-well head segment is higher than that of non-well tail segment in the deep sluice villages in both tanks and elevated sluice village of Panniyur in KPT, except in tail end villages in KPT(where the GVO/GCA is less in the head segment due to poor availability of tank supply) and in Dusi sluice villages (where the GVO/NSA is less in the head segment due to difference in cropping intensity).

Both in KPT and DMT, across reaches, the GVO/GCA in the head segment of non-well farms is relatively high in two villages each in both tanks (KPT: Panniyur and Kattalai and the two Mamandur sluice villages in DMT) - where there is no drainage problem - compared to the head reach of Vagai village. In KPT, especially in Kattalai where no tank water was available. But in GVO/NSA, the pattern is very different - it is much lower in all the villages in KPT, except (Athipattu).This is because, of higher CI reflecting very favourable situation in respect of groundwater supply. In DMT, output per hectare of operated area in the tail end village of Sothiyampakkam is relatively more than that of Narasamangalam served by the Mamandur sluice. The pattern is reversed between head and tail reach villages under Mamandur sluice. Relatively less output per hectare of NSA in the head segment of Vagai and Punaithangal compared to that of Mamandur villages is mainly due to severe drainage problem.

7.6.5. Non-Well Tail Segment

In KPT, as against the expectation across reaches, the GVO/GCA in the tail segment farms is more in the tail end villages compared to the head reach villages of both the deep and the elevated sluices. This reflects the relatively good accessibility of well waterin this segment in tail end villages; which also accounts for higher CI at least in the village of Athipattu. However, GVO/NSA is much less in Alapakkam - the tail end village – compared to the deep sluice village of Kattalai. Also, in the two elevated sluice villages, the GVO/NSA is much less in the tail segment, mainly because of the low CI. In DMT, in the tail segment farms, not much difference in the GVO/GCA is observed except in Narasamangalam. This is because of drainage problems in this head reach village; and the availability of well water for a few farms in two of the villages served by Mamandur sluice. Cropping intensity in Mamandur sluice villages is much higher than in Dusi sluice villages. Consequently, though the GVO/GCA in Narasamangalam is less than in Vagai, the GVO/NSA is nearly equal that of Vagai. In the tail end of the Mamandur sluice, the GVO/GCA in Sothiyampakkam (at the tail) is marginally less but this compensates for the high CI. Consequently, GVO/NSA in the tail reach village turns out to be higher than the head.

7.6.6. Differences between the Non-Well Head and Tail Segments

In KPT, the ratio of GVO/GCA in the head segment of non-well farms is greater in the deep sluice and elevated sluices villages compared to the tail end villages. But GVO/NSA does not vary much if we exclude the tail end village of Athipattu - where the difference is twice that of the head reach villages. This clearly indicates poor tank water supply to the tail villages, especially to non-well tail farms. In DMT, the ratio of GVO/GCA in the head segment of non-well farms in the Dusi sluice villages is lower than the Mamandur sluice villages. But between the two villages in Mamandur sluice, there has not foundmuch variation. In these Mamandur villages, between head and tail segments, better accessibility of own well water, for a few head and tail segment farms, makes a significant difference to productivity. While the ratio of GVO/NSA in Narasamangalam is much higher - about two times - than that of Vagai in Dusi sluice, and this difference between the head and tail villages under Mamandur sluice is also larger. The greater difference in Narasamangalam is mainly due to poor accessibility of tank supply resulting in poor yield combined with low CI in the tail farms.

7.6.7. Productivity of Land in the Preceding Supply Year

Like variations in productivity in the latest supply year,an analysisof difference in productivity of land in the preceding supply year is attempted to see if any pattern is emerging in the villages under KPT & DMT. Necessary data are given in Table 7.19.

7.6.8. Well Farms

In all the sample villages in both the tanks output per hectare of GCA as well as NSA under well farms is higher than that of non-well farms (except in Alapakkam in KPT, where the non-well head reach farmers have also own some wells that make difference in values). In DMT also, in the head reach village of Narasamangalam village the same situation prevails.

In KPT, across reaches, output per hectare of GCA is higher in the elevated sluice village of Kadaperi and the tail end village of Athipattu. Whereas the GVO/NSA is much higher only in Athipattu followed by Kadaperi and Kattalai and there has not been much variation in the elevated sluiceserving Panniyur and the tail end village of Alapakkam. This is because, these villages received tank supply for one season only. But for other villages, tank water supply was more favourable for the deep sluice village followed by the tail end and the elevated sluice village.

Table 7.19: Value Productivity of Crops cultivated by Sample households, 2005-06

(Area in hectares, value in rupees)

Tank, sample villages and HHs by type	Operational holdings	Total area as per adangal	Gross area season acres	Gross value	GVO/ Gross area 5/4	GVO/ Net area 5/2	CI (4/2) X100	Imputed value: Gross area (4/2)X3	Imputed value: Gross value (5/4)X9	Weighted average GVO/area W & NW: Gross (10/9)	Weighted average GVO/area W & NW: Net (10/3)	Weighted average GVO/area Non-Well: Gross (10/9)	Weighted average GVO/area Non-Well: Net (10/3)	% of area under P+SC+E	Ratio of well to non-well: Gross	Ratio of well to non-well: Net	Ratio of NW head to NW tail: Gross	Ratio of NW head to NW tail: Net
1	2	3	4	5	6	7	8	9	10	11	12	13	14	15	16	17	18	19
KPT																		
Kadaperi E (H)																		
Well 8	6.19	9.73	11.32	667215	58921	107714	182.8	17.79	1047920	58921	107714			96	1.69	3.32		
Head 8A*	1.89	8.31	2.17	81720	37588	43315	115.2	9.57	359849	34771	32464	37588	43315	100			1.18	2.00
Tail 8A*	2.91	8.57	1.97	63002	31954	21613	67.6	5.80	185243			31954	21613	100				
Panniyur E (H)																		
Well 8	8.21	10.99	9.34	373112	39930	45443	113.8	12.50	499322	39930	45443			91	1.29	1.65		
Head 8	3.42	9.64	3.30	93515	28376	27368	96.4	9.29	263703	30931	27466	28376	27368	80			0.85	0.99
Tail 8B*	2.59	8.55	2.13	71312	33487	27565	82.3	7.04	235808			33487	27565	100				
Kattalai D (H)																		
Well 8	9.19	43.51	20.83	926018	44447	100760	226.7	98.63	4383693	44447	100760			100	1.10	0.98		
Head 8A*	2.63	47.90	6.67	265003	39718	100856	253.9	121.64	4831303	40523	103034	39718	100856	100			0.96	0.96
Tail 8B*	3.32	36.90	8.44	348859	41328	105211	254.6	93.95	3882600			41328	105211	100				
Alapakkam M (T)																		
Well 8	16.34	22.12	16.34	589950	36113	36113	100.0	22.12	798881	36113	36113			70	0.98	0.88		
Head 8A*	3.30	25.95	4.92	203400	41350	61644	149.1	38.68	1599498	36896	41262	41350	61644	100			1.27	2.95
Tail 8B*	4.79	32.43	3.09	100085	32442	20879	64.4	20.87	677176			32442	20879	87				
Athipattu D (T)																		
Well 8	22.26	59.55	51.37	2862680	55728	128584	230.7	137.41	7657760	55728	128584			100	1.86	4.42		
Head 8	4.59	24.35	3.77	111678	29629	24346	82.2	20.01	592790	29961	29119	29629	24346	100			0.98	0.72
Tail 8B*	6.20	30.91	6.94	210207	30292	33891	111.9	34.58	1047605			30292	33891	100				
DMT																		
Vagai E (H)																		
Well 5	9.06	9.06	24.17	741514	30679	81875	266.9	24.17	741514	30679	81875			100	1.42	3.85		
Head 8	2.38	7.74	2.36	44939	19039	18878	99.1	7.68	146129	21661	21285	19039	18878	100			0.78	0.80
Tail 8	5.83	9.30	5.69	138125	24282	23692	97.6	9.08	220424			24282	23692	100				

(Contd...)

Punaithangal E (T)																		
Well 8	10.45	28.71	14.62	368596	25220	35288	139.9	40.17	1013068	25220	35288			92	1.10	1.44		
Head 8	5.12	22.18	3.10	75852	24491	14811	60.5	13.41	328472	22943	24552	24491	14811	100			1.14	0.43
Tail 8	3.49	7.89	5.59	119540	21396	34293	160.3	12.64	270458			21396	34293	78				
Narasamangalam D (H)																		
Well 8	7.19	13.01	10.03	197757	19712	27503	139.5	18.15	357766	19712	27503			100	0.98	1.04		
Head 8B*	2.97	16.84	4.28	87849	20509	29562	144.1	24.28	497891	20047	26464	20509	29562	100			1.05	1.27
Tail 8B*	3.15	14.81	3.75	73504	19585	23366	119.3	17.66	345952			19585	23366	88				
Sothiyampakkam D (T)																		
Well 8	8.70	29.05	16.83	477687	28390	54904	193.4	56.18	1594883	28390	54904			100	1.10	1.57		
Head 8B*	7.94	15.70	11.18	293131	26214	36903	140.8	22.11	579539	25807	34882	26214	36903	100			1.03	1.12
Tail 8B*	3.93	11.16	5.08	129052	25399	32862	129.4	14.44	366799			25399	32862	100				

Note : A* A few farmers used well water through purchase; B* A few farmers used well water from their own wells.

E (H) = Elevated sluice head reach village; D (H) = Deep sluice head reach village; M (T) = Middle sluice tail reach village.

D (T) = Deep sluice tail reach village; E (T) = Elevated sluice tail reach village; W = Well; NW = Non-well.

GVO = Gross value output; CI = Cropping intensity; P+SC+B = Paddy plus sugarcane plus banana.

Source: Field survey data, 2017.

Table 7.20: Value Productivity of Crops cultivated by Sample households, 2016-17.

(Area in hectares, value in rupees)

Tank, sample villages and HHs by type	Opera-tional hold-ings	Total area as per adangal	Gross area season acres	Gross value	GVO/ Gross area 5/4	GVO/ Net area 5/2	CI (4/2) X100	Imputed value		Weighted average GVO/area				% of area under P+SC+B	Ratio of well to non-well		Ratio of NW head to NW tail	
								Gross area (4/2)X3	Gross value (5/4)X9	W & NW		Non-Well			Gross	Net	Gross	Net
										Gross (10/9)	Net (10/3)	Gross (10/9)	Net (10/3)					
1	2	3	4	5	6	7	8	9	10	11	12	13	14	15	16	17	18	19
KPT																		
Kadaperi E (H)																		
Well 8	6.19	9.73	8.96	479250	53466	77369	144.7	14.08	752704	53466	77369			67	1.13	1.24		
Head 8A*	1.89	8.31	2.79	163442	58507	86631	148.1	12.30	719706	47387	62355	58507	86631	86			1.61	2.28
Tail 8A*	2.91	8.57	3.06	111000	36266	38079	105.0	9.00	326371			36266	38079	100				

(*Contd*...)

Panniyur E (H)																		
Well 8	8.21	10.99	11.56	737125	63772	89778	140.8	15.47	986468	63772	89778			100	7.40	49.71		
Head 8	3.42	9.64	0.00	0	0	0	0.0	0.00		8613	1806	0	0	0			0.00	0.00
Tail 8B*	2.59	8.55	0.54	11542	21275	4461	21.0	1.79	38166			21275	4461	0				
Kattalai D (H)																		
Well 8	9.19	43.51	17.47	997118	57077	108497	190.1	82.70	4720278	57077	108497			100	1.90	1.64		
Head 8A*	2.63	47.90	5.26	108175	20553	41170	200.3	95.95	1972152	30033	66200	20553	41170	100			0.52	0.45
Tail 8B*	3.32	36.90	7.66	302500	39512	91230	230.9	85.21	3366651			39512	91230	100				
Alapakkam M (T)																		
Well 8	16.34	22.12	18.36	1072375	58407	65645	112.4	24.86	1452158	58407	65645			74	1.19	1.92		
Head 8A*	3.30	25.95	2.72	192380	70606	58304	82.6	21.43	1512839	49054	34204	70606	58304	95			2.57	5.77
Tail 8B*	4.79	32.43	1.76	48435	27502	10104	36.7	11.92	327714			27502	10104	81				
Athipattu D (T)																		
Well 8	22.26	59.55	48.87	3601958	73710	161790	219.5	130.72	9635351	73710	161790			85	1.43	2.28		
Head 8	4.59	24.35	7.76	388995	50121	84803	169.2	41.20	2064798	51589	71078	50121	84803	100			0.94	1.48
Tail 8B*	6.20	30.91	6.70	355724	53058	57352	108.1	33.41	1772813			53058	57352	100				
DMT																		
Vagai E (H)																		
Well 5	9.06	9.06	26.59	1801679	67765	198934	293.6	26.59	1801679	67765	198934			100	1.27	1.97		
Head 8	2.38	7.74	4.42	203115	45985	85322	185.5	14.36	660469	53212	100896	45985	85322	100			0.76	0.73
Tail 8	5.83	9.30	11.23	679019	60439	116471	192.7	17.93	1083601			60439	116471	100				
Punaithangal E (T)																		
Well 8	10.45	28.71	16.19	993190	61329	95084	155.0	44.51	2729733	61329	95084			89	1.07	1.75		
Head 8	5.12	22.18	4.11	230615	56120	45029	80.2	17.80	998663	57538	54355	56120	45029	100			0.95	0.71
Tail 8	3.49	7.89	3.77	221980	58956	63681	108.0	8.52	502227			58956	63681	100				
Narasamangalam D (H)																		
Well 8	7.19	13.01	12.88	778613	60439	108287	179.2	23.31	1408606	60439	108287			100	1.11	0.85		
Head 8B*	2.97	16.84	7.16	366685	51228	123394	240.9	40.57	2078215	54646	127533	51228	123394	100			0.88	0.94
Tail 8B*	3.15	14.81	7.13	414205	58064	131671	226.8	33.57	1949482			58064	131671	98				
Sothiyampakkam D (T)																		
Well 8	8.70	29.05	20.28	1389201	68490	159671	233.1	67.72	4638212	68490	159671			100	1.22	1.27		
Head 8B*	7.94	15.70	17.17	1187076	69120	149443	216.2	33.95	2346925	55943	125534	69120	149443	100			1.62	1.47
Tail 8B*	3.93	11.16	9.33	399091	42766	101624	237.6	26.52	1134324			42766	101624	100				

Note : A* A few farmers used well water through purchase; B* A few farmers used well water from their own wells.
E (H) = Elevated sluice head reach village; D (H) = Deep sluice head reach village; M (T) = Middle sluice tail reach village.
D (T) = Deep sluice tail reach village; E (T) = Elevated sluice tail reach village; W = Well; NW = Non-well.
GVO = Gross value output; CI = Cropping intensity; P+SC+B = Paddy plus sugarcane plus banana.
Source: Field survey data, 2017.

In DMT, the GVO/GCA is generally low in all the villages, especially under Mamandur sluice. This difference being more marked in the output per hectare of NSA. This is due to growing of paddy routinely in DMT villages that had not made much output due to agronomic and climatic conditions. Consequently the values are lower than in KPT. The variation in the productivity between well and non-well farms is not differ much due to equal availability of tank supply to all villages, especially within each village, in 2005-06. However, difference may be observed larger under Dusi sluice villages than under Mamandur sluice villages. This is mainly because a few non-well farms under Mamandur sluice villages have access to well water which was not available for the Dusi sluice villages.

7.6.9. Well and Non-Well Differences

In KPT, the ratio of GVO/GCA between well and non-well farms is higher in the tail end village of Athipattu and the elevated sluice village of Kadaperi. This difference is due to poor tank supply to these villages. In all other villages there is not much difference in productivity. The ratio of GVO/NSA is higher in the tail end village of Athipattu and elevated sluice village of Kadaperi. This is because of low CI in these villages. In DMT, the ratio of both the GVO/GCA and GVO/NSA is less in Punaithangal compared to the head village of Vagai. But under Mamandur villages there is not much difference in it. This is mostly becauseavailable tank supply in DMT was more or favourable to both villages, and also due to the availability of well water supply and the extent of broadcast sowing of paddy in the first season.

7.6.10. Non-Well Head Segment

In both tanks, among non-well farms the GVO/GCA is higher in the head segment farms than in the tail in most of the villages.In Panniyur, Kattalaiand Athipattu in KPT and Vagai in DMT tail segments recorded marginally a higher productivity. This increase is mainly due to the use of own well water by a few tail segment farmers.

In KPT, across reaches, output per hectare of both the GCA and NSA in the head segment of non-well farms is higher in the deep sluice village of Kattalai and the tail end villages. The low productivity in the elevated sluice village of Panniyur is basically due to less cropping intensity. In DMT, between Dusi and Mamandur sluices, both the GVO/GCA and GVO/NSA are higher in the head segment of tail reach village of Sothiyampakkam compared to Punaithangal; but the pattern is reversed in the villages served by Mamandur sluice, where in the tail end village of Sothiyampakkam the output per hectare of GCA and NSA is higher than that of the head reach village of Narasamangalam. This variation is due to accessibility of well water supply and variation in CI.

7.6.11. Non-Well Tail Segment

In KPT, across reaches, in the tail segment farms, output per hectare of both the GCA and NSA is higher in the head reach village of Kattalai; lower in the elevated sluice villages followed by tail end villages. This clearly indicates that the tank water supply is much better in the deep sluice village followed by elevated sluice villages and very poor in the tail end villages. In DMT, between Dusi and Mamandur sluices, output per hectare of GCA in the tail segment farms under the tail village of Sothiyampakkam is better than that of the head reach village of Narasamangalam. The same point is also applicable to Dusi villages. This is due to better accessibility of well water to the tail segment farms under Sothiyampakkam, but this was not possible under Punaithangal. The GVO/NSA under DMT villages show, in both Dusi and Mamandur sluices the output values in the tail segment of tail end villages are much better than the head reach villages. This is mainly due to less CI compared to the tail end villages.

7.6.12. Differences between the Non-Well Head and Tail Segments

In KPT, the ratio of GVO/GCA in the non-well head segment is relatively more in the tail end villages followed by elevated sluices than that of the head reach village served by the deep sluice. These differences indicate that the tank water supply was favourable for the head reach villages compared to the tail end. Between the two tail villages - Alapakkam and Athipattu - tank water supply was favourable to the latter than the former. However, the output per hectare of operated area is much less in the elevated sluice village of Alapakkam and deep sluice village of Kattalai. The difference is much higher in the tail end village of Alapakkam followed by Kadaperi. This is because the CI in the head segments of elevated sluice villages is much less and in Alapakkam the pattern is reversed, where the CI in the tail segment is much less than in any other village.

In DMT, the ratio of GVO/GCA is higher in the tail end of Punaithangal than that of other villages in both the sluices. However, the difference is much less in the output per hectare of NSA between the head and the tail villages within Mamandur sluice but the difference is large in Dusi sluice villages. The difference is more marked in the latter sluice than the former. This is due to differences in CI. Since the non-well farmers in Mamandur sluice also had access to well water in DMT in the preceding supply year also, this comparison is useful only to indicate differences in the productivity levels between well and non-well farms.

On the whole, it might be clear that during the preceding normal tank supply year, in KPT, head reach villages served by the deep and the elevated sluices are better placed than the tail end villages. Among the latter, middle sluice serving tail village of Alapakkam is better placed than that of deep sluice serving Athipattu. Among non-well farms, head segments are better served under the deep sluice and the tail end villages than under the elevated sluices. In DMT, among non-well farms head segments are better placed than the tail segments.

7.6.13. Variations in Productivity between the Two Tank Supply Normal Years

In all villages of KPT (except the well farms under Kattalai, Kadaperi and Athipattu) productivity per hectare of operated area in 2015-16 was higher than in the preceding supply year (2005-06) when the tank was at FTL in both reference years. In the three villages in the latest supply year tanks water availability was much curtailed. Kattalai and Athipattu did not get the tank supply and part of the Kadaperi (tail reach farms) ayacut was sold out. The difference between the years, in percentage terms, was higher in non-well farms, than well farms in head reach villages. But in the tail end villages the difference between the two years was larger among well farms than the non-well farms. These variations are mainly on account of variations in CI between the two years - CI is relatively lower in 2015-16 in the elevated sluice and tail end villages in KPT and in Dusi sluice villages in DMT than in 2005-06 - rather than in output per gross season hectare.

In DMT, the differential between 2015-16 and the preceding supply year (2005-06) are evident, in relative terms, the difference between GVO/ NSA in 2005-06 and 201-16 is much higher in the non-well farms; the difference being very marked in the tail end villages of both sluices. Among non-well farms, those in the tail segment eventually suffer much more from inadequate tank supply than those in the head segment. This effect being more marked in the tail end villages. Here again differences in GVO/NSA between years is mostly on account of differences in cropping intensity. When tank supplies are short, the average yield per gross hectare (in season hectares) is of course affected but its effect on the intensity with which land is used is much greater.

The productivity differential, in relative terms, between well and non-well farms in the head reach villages of KPT is wider in 2005-06 compared to 2015-16. But in Alapakkam, Athipattu and Kadaperi, where a relatively high proportion of *ayacut* uses well water, the disparity between well and non-well farms is in fact greater in a good year, mainly because a combination of good tank supply and well water use increases cropping intensity.

Comparing non-well farms in the head and tail segment of various villages, one would expect that differentials to be considerably less in a good years like 2015 and 2005-06. This is generally borne out when one compare GVO/NSA in the two years in all villages except Panniyur. The extent of reduction is more marked in Kattalai - which is very well supplied by tank water - and Alapakkam and Athipattu where use of well water is relatively widespread. Again variations in CI are more important than those in output per gross hectare.

In DMT, the differential between head and tail segments of non-well farms is higher than in KPT even during a good years like 20015-16 and 2005-06. The ratio of land productivity between head and tail segments in 2005-06 was in all cases greater than in 2015-16. The difference being more marked in the tail end village of both sluices.

7.6.14. Factors responsible for productivity variation

The overall productivity of the sample farms depends upon CI and the value of output per hectare of GCA. These two are affected by a number of factors which can be broadly grouped into two: (1) Technical and (2) Institutional. Since we have discussed the effect of institutional factors in greater detail in chapter 4, the importance of technical factors influencing productivity is discussed here.

The most important among the technical factors are: (a) conditions of water supply; (b) drainage conditions and (c) soil quality. Among these the conditions of water supply is the most important followed by drainage conditions. Table 7.21 presents in summary form of the water supply conditions and agricultural performance of the overall *ayacut* in selected villages during 2015-16, 2005-06 and 2016-17. It brings out the considerable variations within each tank in the availability of tank supply, extent of well irrigation and incidence of drainage problems, and gives some insight with their effects on productivity.

Thus in KPT, Although Kattalai is best placed in terms of access to tank water (it got nine months supply in 2005-06 and did not get any in 2015-16); but had good well irrigation - especially during normal supply

Table 7.21: Productivity of Sample Holding and Some Related Characteristics in KPT and DMT in the Latest, Preceding and Survey Years (Value in Rupees)

Tank and sample villages	Free catch-ment sq. km	% of drai-nage faci-lity	Latest supply year (2015-16)					Preceding supply year (2005-06)					Survey year (2016-17)				
			Tank water supply months	% of area using well water	GVO/ GCA	GVO/ NSA	CI (7/ 6)	Tank water supply months	% of area using well water	GVO/ GCA	GVO/ NSA	CI (12/ 11)	Tank water supply months	% of area using well water	GVO/ GCA	GVO/ NSA	CI (17/ 16)
1	2	3	4.00	5	6	7	8	9	10.0	11	12	13	14	15	16	17	18
KPT																	
Kadaperi E (H)	0.0052	80	5	40.2	58588	59772	102	5	55.9	42821	57547	134	0	100	49413	67360	136
Panniyur E (H)	0.0052	85	6	19.5	44790	63069	141	6	30.1	33931	33459	99	0	100	28349	31413	111
Kattalai D (H)	0.0052	50	0*	88.9	42286	106550	252	9	42.6	41831	102276	244	0	100	39048	80299	206
Alapakkam M(T)	0.0052	95	4*1	72.8	56036	50738	91	4	53.5	36635	39545	108	0	100	52172	44684	86
Athipattu D (T)	0.0052	90	0*2	100.0	53641	92231	172	6	79.5	38550	62274	162	0	100	58963	101315	172
DMT																	
Vagai E (H)	0.0073	0	8	12.5	61325	96046	157	8	11.3	24667	41482	168	6	21.5	58063	133575	230
P.thangal E (T)	0.0073	75	8	45.7	64523	62961	98	8	38.6	23702	28131	119	4	48.9	58802	67931	116
N.mangalam D(H)	0.0073	55	8	22.6	63676	83269	131	8	19.1	19935	26811	134	6	26.7	56577	121117	214
S.pakkam D (T)	0.0073	90	8*3	66.2	67484	118850	176	5	54.3	26668	41556	156	2	70.1	60125	136913	228

Note: E (H) = Elevated sluice head reach village; D (H) = Deep sluice head reach village; M (T) = Middle sluice tail reach village;

D (T) = Deep sluice tail reach village; E (T) = Elevated sluice tail reach village; GVO = Gross value output.

GCA = Gross cropped area; NSA = Net sown area; CI = Cropping intensity; Latest year, 2015-16; Preceding year, 2005-06.

Column 2 represents the average extent of free catchment area per acre of ayacut.

* Due to complete break-down of Moolai Sluice, tank supply in this year was fully closed for irrigation.

*1 Only part of the ayacut i.e., in the head reach only received some supply due to channel encroachment and breakdown of one of the twin (Maganipattu & Thuraiperumpakkam) sluices that supply water to this village.

*2 Due to heavy channel encroachment and blockage, in Kaveripakkam village, water was not received totally.

*3 Usually 5 months supply was available but due to longer period of FTL storage maintained in the DMT during this supply year, 8 months supply was possible

Source: Field survey data, 2017.

period - and half of its land faced drainage problems. The latter drainage problem depresses the yield per gross hectare, but given the assured well irrigation – including good part of sale of well water-, it had the highest CI (252%) and the highest GVO/NSA (Rs. 106550). Athipattu is a tail end village, but being fed by the deep sluice, got almost eight months supply in 2005-06, but not received any supply in 2015-16, which hampered yield per gross hectare and also a relatively moderate CI resulting in a moderate GVO/NSA of Rs. 92231.

Alapakkam, being at the tail of the middle sluice is less well placed in access to tank water, but best placed in access to groundwater. Yield per hectare of GCA is relatively high but CI is low which depresses overall productivity. Clearly access to well water in these cases helps to compensate to a considerable degree the consequences of unequal access to tank supplies on account of location.

Kadaperi, despite being at the head, got only five months supply because it was fed by an elevated sluice and despite considerable well water. Cropping intensity is very low and GVO/NSA is also lowerbetween the elevated sluice villages. Panniyur is also served by an elevated sluice. But, it is served by a sluice which is located slightly deeper than that of the Kadaperi's Gaffor sluice. Therefore, it gets tank water for a longer time. Cropping intensity is relatively high and consequently GVO/NSA is higher in Panniyur (Rs. 63069).

Compared to 2015-16, water supply in 2005-06 was also a normal tank supply year in all the villages, but the reported distribution was best in Kattalai; the reduction in yield per hectare of NSA is also the least in this village. In all others, there was a decline in both GVO/GCA and CI. GVO/NSA shows a steep reduction in the Panniyur (elevated sluice, poor groundwater) and in the two tail end villages of Alapakkam and Athipattu.

In DMT, which has little well irrigation, productivity per GCA is generally lower than in KPT for comparable reaches even in good years; cropping intensities are much higher (mainly on account of the practice of growing broadcast paddy in June-September and also because three of the four villages got eight months supply) and GVO/NSA ranged between Rs. 62961 and Rs. 118850 comparable to KPT.

In DMT, the inter-village differences are entirely on account of differences in tank supply and drainage. The GVO/GCA is lower in the head reach villages because of severe drainage problems. But with more assured access to tank water compared to the tail end. They have high CI.

The two factors - tank supply and drainage - seem largely to compensate each other, thereby narrowing the difference in GVO/NSA. However, as may be expected, productivity per hectare of operated area is somewhat high in the *ayacut* of the deeper Mamandur sluice.

A comparable another normal tank supply year (2005-06) results in a marginal increase of CI, but reduction in the yield of area cropped. Overall GVO/NSA falls by marginally as a consequence. It needs noting that groundwater availability and accessibility in KPT is in very high proportion and it is very little effect in DMT villages.

In another survey year 2016-17, one can observe a different pattern of tank water supply between the tanks. Where the KPT villages use solely on well water for cultivation and tank supply is meagre for all villages, including the head reach village of Kattalai (due to full closure of deep sluice for irrigation due to shutter breakage). In DMT all the villages got at least one season tank supply – mainly because of longer storage capacity and good monsoon rains in the vicinity of local catchment of DMT. Due to this changed pattern of tank water supply in 2016-17, it is observed that productivity of almost all DMT villages is higher than that of DMT villages in comparable reaches. Likewise, the GVO/NSA in DMT also reached its highest level compared to KPT villages. This observation indicates, on the one hand tank supply helped much to enhance productivity and cropping intensity, especially in DMT and on the other hand, especially in KPT villages, even when there is no tank supply, well irrigation helped the ayacutdars for better cropping and relatively to get normal yield even without getting the tanks supply. It should further be noted that once the KPT gets full storage, its effect on groundwater can lead for better cropping in its ayacut at least for one more year even when the tank got dry in the next year and that effect is clearly observed in KPT villages.

7.7. Summary

In this chapter it is tried to get a detailed grassroots field level view (based on a survey) of the farmers knowledge of various issues relating to the land and water relationship in terms of their land holding size, caste, literacy, status score of the farmers, access to tank/ground water, and its impact on cropping intensity, cropping pattern and productivity. It is sen that the drainage and soil conditions are better in KPT villages than in DMT villages; Nayakars are the single largest caste cultivating the *ayacut* of the selected villages taken as a whole; the average size of (both operational as well as ownership) holdings of well farmers is larger than that of the non-well farmers; In the head reach villages farmers attained higher status score in both the tanks than in tail reach villages.

Analysis of conditions of tank water supply has shown the following: In both tanks, supply was more assured in the second season (November - February) than in the third. Head reach villages served by deeper sluices got two full crop seasons supply in both the tanks. The number of bore wells and the extent of use of groundwater (sole as well as supplemental) are relatively high in KPT command compared to DMT. Consequently, conjunctive use of well water would seem to be more significant in KPT villages compared to DMT villages.

The incidence of purchased well water in non-well farms was significant but variable across villages. In KPT, the proportion of area using well water (ranging from 6.6% to 100% in 2015-16) was higher in the head reach / tail end villages than in the elevated sluice villages. By contrast, water sales under the DMT villages is not existing.

Overall, the tail end villages in both tanks are unfavourably located to get the tank supply. However, this disadvantage is compensated to a considerable extent by the greater use of well water in the tail end villages. The conjunctive use certainly dampens the effect of unequal distribution of tank water across villages.

There is a systematic relation between CI and the location of villages in the tank *ayacut*; within villages significant differences exist between well and non-well farms. High CI was recorded in the villages under deep sluices because they got supplies for a longer time than tail reach/elevated sluice villages. Within villages, well farms ranked first in the order of CI followed by non-well farms of head segment and the tail. In DMT too CI in the head reach is much higher than the tail in 2015-16. However, CI in all the non-well farms was more or less equal to that of well farms. This is partly due to the widespread practice of broadcast sowing of paddy in the first season depending primarily on rainfall. But the extent of well irrigation is also an important factor affecting CI in DMT. Moreover, in the head /tail segment of non-well holdings cropping intensity is relatively more than that of head reach villages. This is mainly due to accessibility of well irrigation in those villages. This is evident in the villages under DMT.

The cropping pattern differs widely between the tanks. In KPT villages apart from paddy cultivation, banana and sugarcane are relatively important. But in DMT villages paddy is the predominant crop grown in all the seasons, and hardly any sugarcane or banana is grown here. During the first season, where there is no tank supply broadcast paddy is widely sown in DMT villages but it is not sown in any of the KPT villages. In the latter, well farmers grow banana more widely than the non-well farmers. Further, in the deep sluice of Kattalai, due to rich availability of groundwater not only the well farmers but also non-well farmers of both types grow banana even in non-tank supply years.

One of the important factors affecting the yield per hectare of GCA in the head reach villages, especially under the deep sluice, has drainage problems. This is clearly identified in our survey in both the tanks. In the head reach village of Kattalai though GVO/GCA is less, yield per acre of operated area (GVO/NSA) remains higher than in other villages. This is mainly because of high CI made possible by the longer duration of tank supply. Tail end villages in KPT get relatively more output per hectare of both GCA and NSA compared to the elevated sluice villages. Despite shorter duration of tank supply they have little drainage problem and well irrigation benefits more of the cropped area. Better access to well water raises output per hectare of both GCA and operated area.

In DMT, output per hectare of GCA and NSA in the head reach villages in both the reference years is less compared to tail end villages. This is mainly because of severe drainage problems in the head reach villages resulting to less GVO/GCA. However, the GVO/NSA did not decline much. This is mostly due to higher CI in the head reach villages. It is the maximum of 176 in the tail end village of Sothiyampakkam and the minimum of 98 in the head reach of Punaithangal. It is obvious that the condition of drainage facilities and the CI are the most important factors affecting the output per hectare of both GCA and NSA in the selected villages.

Note

1. Gross value output per hectare of GCA and NSA is arrived by finding out the weighted average estimates of each category of well, non-well head and non-well tail segments of sample farmers. Output per hectare of GCA is arrived by dividing the estimated gross value output by estimated gross area and to find out the output per hectare of NSA (GVO/net area) by dividing the estimated gross value output by total area owned (operated) by all farmers in the sample villages.

Chapter - 8

Summary and Findings

This 'tank institutions' study focuses on the organisation, working and impact of irrigation management in the two large, multivillage, and largely user managed tanks in Tamil Nadu, namely, Kaveripakkam and DusiMamandur. The ayacut served by KPT is about 2,430 hectares covering 14 villages, and ten sluices supply water to the ayacut. DusiMamandur tank has over 1,625 hectares of ayacut covering 18 revenue villages and the tank supply is controlled through four sluices. Both are part of a larger system, the Palar Anicut System, but differ significantly in terms of quantum and assurance of supply relative to ayacut villages. Kaveripakkam tank is located at the head reach of its single source supply channel. While DMT is situated at the tail end of its multisource supply channels. However, the supply position is less favourable in DMT compared to KPT. Both the tanks have a long tradition of user involvement for management of the system. But, over the years such involvement has undergone significant changes. A systematic study of them is expected to show how large, more complex systems of management by beneficiary communities are structured, how they have evolved, and what factors have shaped them.

The Palar Anicut System, which serves the tanks, has undergone significant changes during the last 160 years. The system is several centuries old, and was entirely managed by the user communities till the government came into the picture in 1857 and constructed the present Palar Anicut. This was followed by improvements in the distribution network, which increased supplies and brought more tanks under the scope of the system. The modernisation however did not benefit the DMT at the tail end as much as the KPT.

The modernisation of PAS benefited KPT more than DMT. The increase in the total supply to the ayacut as a whole did not benefit DMT much, because it is at the tail end of a long channel of about 36 kms. The construction of the anicut reduced the volume available through its other source, namely, Rajakkal channel. And the construction of the Thandarai Anicut under pressure of tail end ayacutdars did not bring any additional supply to DMT. But it should be noted that DMT has a larger storage capacity per hectare of ayacut. At present the inlet channels especially in DMT are not satisfactory due to improper and inadequate maintenance. Siltation and encroachment in the tank bed are a more severe problem in KPT than in DMT.

Analysis on water realisation from 1906 - 2016 (111 years) shows that the Palar Anicut received on the average 188 mcm of water a year of which 74 mcm (39.4%) was drawn through the four main channels. The entire available flow was diverted in 45 out of 87 years; while in another 22, less than 50 % was diverted. The volume of water drawn by the four channels relative to aggregate capacity of tanks under PAS during 19062016 shows that in 24 years no flow was realised; the flow exceeded storage capacity (275 mcm) only in six years; it was in the range of 5199% in 16 years; 2650% in 15 years and less than 25% in 47 years (about half of the supply years). Taking the no flow years and years when the flow was below 25 per cent (which accounted for about two thirds of the total years) the quantity realised at the Anicut was very low compared to storage capacity.

During the last 54 years (1964 - 2017), KPT attained full capacity in 14 years compared to nine years in DMT. In the case of KPT the maximum storage was between 20 and 30 feet in 16 years, 15 to 20 feet in five years and less than 15 feet in 19 years (a level at which no irrigation is considered possible and also the tank is considered as dry). DMT attained maximum depth of 2030 feet in 18 years, 15 to 20 feet in seven years and less than 15 feet in 20 years at which only the head reach of the ayacut under deep sluice could be irrigated. The KPT remained dry or water not usable for irrigation in 19 years, while the DMT was so in 11 years. It should be noted that the period 1980s is different from the 1960s and the 1970s. KPT was dry only once during the 1960s and the 1970s and DMT was dry only once in the 1960s. But during the 1980s both the tanks remained dry for three years mainly because of lower rainfall than in the 1960s and 1970s. In KPT, out of 50 years (1966-2016) the storage was adequate to raise three crops in 23 years, two crops in 13 years, and only one crop in 8 years.

During the same period, in DMT the tank supply permitted three crops in 25 years, two crops in 11 years, and one crop in 17 years.

There has been a big increase in the number of wells and most are energised. A rough calculation of annual water use, based on the crop pattern details during the years 1913 and 1985 suggests an increase of water use by 17.6 % in KPT and 17.5 % in DMT. A similar exercise for the years 1985 and 2015 shows that water use was increased to 13% over the first period of water use. However, it should be mentioned that the KPT storage in 1985 was 7.2 metre out of 9.3 whereas DMT attained 8.4 metre out of 9.3. Note that the capacity of DMT at FTL is larger (50.95 mcm) than in DMT (32.92 mcm). This clearly indicates that the groundwater use in KPT contributed a higher proportion of water use both solely and supplemental than in DMT.

There is a three tired structure of management for the tanks. (1) For the system as a whole; (2) the tank level and (3) the village level. The system management is entirely in the hands of the PWD which is responsible for maintenance and for regulating the flow of water through different channels from the anicut and from the channels to the tank. These are supposed to be governed by certain rules; which however are rather vague. Moreover these rules do not seem to have changed in spite of the fact that crop patterns, agricultural technology have changed; and well irrigation has developed extensively, though unevenly. The PWD is also responsible for maintenance of the tank and related structures, as well as inlet channels. But neither function has been performed satisfactorily resulting in considerable siltation and encroachment of tank beds and channels. It is widely believed that as a result the supply to both tanks has been adversely affected.

There is supposed to be a formal tank level organisation in the form of a Board consisting of elected representatives of all villages in the ayacut. This Board apprises the government official (a Junior engineer) in charge of the tank, the water requirements of different villages from time to time. Based on this the engineer in charge operates the sluices. The Irrigation Boards were set up in both the tanks during the early 1920s. But there have been no elections after 1969. The KPT Board is practically defunct and many of the delegates have died or migrated. Even so the system continues to function informally through the remaining delegates and the traditional leaders. In DMT the Board is not only defunct but has been replaced by an informal association, consisting of representatives from their ayacut villages.

The informal organisation interacts with the PWD engineer in regard to water release for the ayacut.

Based on the TNFMIS ACT – 2000, during the mid-2000s the government has established WUAs in all major tanks, including KPT and DMT. As a result, the traditional pattern of tank management came to an end and the elected representatives, called Territorial Constituency Members (TCMs) of WUAs managing the water distribution under the tank system. For KPT, there are 49 TCMs and for DMT 38 TCMs of WUAs have been elected in 2006-07. Although their 5-year term has come to an end in 2011-12, no election to select new members is done so far by the government. Consequently, already functioning TCMs have been continuing. But their work effort to manage the tanks, especially on maintenance of channels and eviction of encroachments in the channels are barely appreciable and hence th farmers are fully in a dissatisfied situation under both tanks.

The village level organisations functioned in both tanks till the TCMs were elected. After their election the irrigation functionaries has gradually withdrawn from doing tank related work. The traditional irrigation organisation has undergone significant changes in recent decades on account of several factors. These include (i) increasing population pressure on land contributing to fragmentation of holdings; (ii) shift in the pattern of land ownership from a relatively few upper castes to a large number of lower castes and weakening of the traditional power structure in the villages; and (iii) widespread development of more assured and controllable well irrigation in the tank ayacut.

As a result of transfer of a major portion of land from the landlords to the tenants, the number of owner cultivators has increased substantially. This has had its effects on the organisational setup in the villages. Upper caste dominance in land control has significantly declined in both the tanks; with the Nayakars emerging as the single largest caste, controlling about half (46%) of the total Pattas in KPT and 60.2% of those in DMT. The Scheduled Castes acquired a sizeable proportion of land ownership in a few tail end villages under both the tanks. Changes in land control, and the large increase in number of ayacutdars must have made enforcement of rules, previously vested in the dominant and powerful upper castes more difficult. This aspect of change has however proved difficult to document. What it is given in the report is a picture of the working of village organisation at present.

The weakening of the traditional organisation is most marked in the villages where land is controlled by two or more antagonistic castes or factions of the same caste and where well irrigation is extensive. The organisations are stronger in villages which depend mostly on tank water and where there is a reasonable prospect that collective effort will make significant difference to the availability of water to the tank. Though at village level the institutions are effective in both the tanks the overall analysis indicates that the functioning of institutions at tank level is better in DMT than in KPT.

The maintenance of channels by the ayacutdars below the tank sluices is discussed in the context of the latest two supply years. Village level management was previously vested with Nattanmaikar/ Kavaimaniyam/ Irrigation Board delegates with the help of Neerkatties/ Kammukutties/ Thotties to implement their directives. The latter category of personnel were paid in kind by the ayacutdars during every tank supply season in KPT and in all seasons in DMT. This practice of channel maintenance and local irrigation institutional set up has drastically changed from the 2000s, especially after induction of TCMs of WUAs in both tanks. Now, the maintenance is fully managed by the elected representatives of TCMs in both tanks.

The survey shows that participation in maintenance is widespread in both the tanks but more in DMT than in KPT; and that farmers contribute substantial amounts both for maintenance and for payment to machinery use, especially for the first time channel maintenance in both tanks. The maintenance of channels is not evenly done across villages. In KPT the head reach/elevated sluice villages did not actively take part in the maintenance work. The survey also indicates that in both the tanks in 2015-16 the enforcement of rules and regulations, and the maintenance work were more vigorous in a few villages and not in all. This is so especially in KPT. Due to long gap in the tank supply – after 10 years: 2005 and 2015 – most of the tail end villages abstained in taking tank water mostly in KPT. The six tail end villages – Athipattu, Sirukarumbur, Eralacheri, Puthupattu, Thuraiperumpakkam and Kilveeranam - served by different sluices in KPT did not take KPT supply in 2015-16 supply year. Main channel encroachment problem is the main reason for abstinence of taking supply in the first four villages. In DMT also, the tail end farmers in Punaithangal did not take water, due to sale of ayacut land to the city dwellers. The tail end of Elacheri farmers in Mamandur sluice also did not take water for decades.

A large labour force was mobilised to desilt and deweed the main/ branch channels at the time of initial maintenance; whereas subsequent maintenance required only limited labour input. Previously, tail end villages contributed more labour than the head reach/elevated sluice villages. However, this pattern reversed now, especially in KPT villages. There are wide variations in frequency of maintenance of main/branch channels across villages and between the tanks. While the head reach villages under KPT contributed labour once in every season for main/branch channels, those under DMT contributed labour only once in a supply year. In KPT, the elevated sluice/tail end villages contributed labour once in every season, whereas in DMT it was once in every month. The survey reveals that wherever the village level organisation is functioning effectively, detailed records pertaining to the maintenance of channels are kept by the irrigation functionaries, whereas in villages where the organisation is weak they are not maintained.

In both tanks labour contribution for maintenance of all channels (including water diversion to DMT through Rajakkal channel) is impressive. The elevated sluice/tail end villages contributed more than the head reach villages. The value of contribution per annum for maintenance is ranged from Rs. 763 to Rs. 1278 per hectare in the elevated sluice villages of Panniyur and Kadaperi to Rs. 311 per hectare in the tail end village of Alapakkam in KPT and Rs. 146 per hectare in the tail reach village of Punaithangal to Rs. 264 per hectare in the tail end village of Sothiyampakkam in DMT. It should be noted that the annual total payments made per hectare for maintenance of main/branch channels ranged from Rs. 631 to 1084 per hectare in KPT and from Rs. 859 to 1564 per hectare in DMT in 2015-16. Apparently DMT ayacutdars contributed more amount compared to KPT ayacutdars.

The gradual decay of local irrigation functionaries were enquired in this study. It was indicated that in a couple of decades ago, the remuneration given to irrigation workers (IWs) differs between the tanks and across villages within the tank. It depends upon both the availability of tank supply and the productivity of crops. Whenever farmers face crop failure the IWs are not paid properly. In DMT villages the ayacutdars also paid for the first crop season which mostly depends on the monsoon rains. In all the KPT villages nonwell farmers (irrespective of reaches) pay more than the wellfarmers especially in the second season. In DMT wellfarmers pay more than the nonwell farmers. This is perhaps because supply from wells is not as widespread or assured, and the tank supply is an important determinant of yields.

A detailed account of the sources of water supply and the manner in which the water was to be distributed between villages is set out in the

Mamulnamas published in 1815. This document spells out the rules, conventions and procedures which traditionally governed the inter village allocations. These conventions are still the basis on which allocation is regulated. A part of the task of rationing is done automatically by differentiation of the sluices and location in the ayacut of each sluice. In general the elevated sluice and tail end villages bear a greater part of shortage in tank supply. Within each sluice, the regulation is achieved in some cases automatically by the physical structures diverting the flows in main/branch channels; and in others there is supervised regulation. There are disputes but the survey has not been able to document them because the villagers are reluctant to talk about such matters. These disputes are evidently somehow managed (perhaps by compromise, mediation, force or a combination of these) without disrupting the system. However, this mediation system that happened two-decades ago through supervisory irrigation functionaries such as Nattanmaikars and Kavaimaniyam is presently discontinued. As a result, many tail end villages as shown above, especially in KPT, are unable to report their problems to any effective person that lead to discontinuance of tail end ayacut villages taking water for irrigation even in good tank storage years like 2015-16. The presently functioning TCMs are very poor in managing even small disputes.

The government staff who are responsible for the allocation of water at the tank level lack formal training for their specialised work. The local irrigation organisations and their functionaries have better knowledge (gained from long experience) of water management. They also have a direct interest in equitable distribution of water among the farmers. So they play a key role. Even though the traditional supervisory level functionaries are not involved in direct tank management activities, in most cases they are still consulted by the famers when the need arise for dispute settlements, etc.

In a good supply year, like 2015-16, the villages served by the deep sluice got tank water for two full crop seasons (89 months) while the elevated sluice villages got one crop season (56 months) supply. A part of the task of rationing available supply within the ayacut is accomplished automatically by the elevation of the sluice and/or location in the ayacut of each sluice. Villages (under elevated sluices or tail end) sometimes face water deficiency even in such a good year in which case they followed a system of rationing by turns (called murai). In 2015-16, which was also a good year for DMT, all the sample villages, except the tail end of Sothiyampakkam, got adequate tank supply for two crop seasons (8 months). Sothiyampakkam got adequate supply for one crop season (5 months) as per its entitlement. None of the DMT villages followed rationing system which they normally practice in times of scarcity.

Analysis of conditions of tank water supply has shown the following: In both tanks, supply was more assured in the second season (November February) than in the third. During a less than normal supply year the ayacut served by the elevated sluices/tail end villages was more severely affected in the third season by water scarcity. Head reach villages served by deeper sluices got two full crop seasons supply in both the tanks (due to deep sluice shutter breakage in KPT the Kattalai village was altogether stopped in getting tank water in 2015-16). The number of bore wells and the extent of use of groundwater (sole as well as supplemental) are relatively high in KPT command compared to DMT. Consequently, conjunctive use of well water would seem to be more significant in KPT villages compared to DMT villages.

Analysis of the use of purchased well water in 2015-16 indicates that well farmers rarely purchase well water. Use of purchased well water is more widespread in KPT villages than in DMT. The frequency of water purchase among nonwell farmers is relatively high in KPT and it is non-existing in DMT villages. In KPT, water purchase is more prevalent in the head reach tail end villages and among both head and tail segment farmers. The use of purchased well water was more in the second and the third seasons than in the first. Within village, especially in KPT, the incidence of water purchase is more in the head segment than in the tail segment of nonwell farms.

Well water purchase is made under three categories: (1) Water purchased on hourly basis for cash. (2) Water purchased throughout the crop period by paying one third of gross crop produce of the area using well water. And (3) water purchase on contract by ayacutdars, usually for two to three years. Generally, a majority of purchasers fall in the first two categories. The amount paid per hectare for purchased well water is higher only in KPT. In 2015-16, it was the highest in non-well head segment holdings in Alapakkam (Rs. 24500) and lowest (Rs. 1100) in the tail segment of non-well farm in Athipattu in KPT. This shows that even for those without wells, in the tail end villages locational disadvantage to avail of tank supply is to a considerable extent compensated by the use of purchased well water especially KPT. But this possibility does not prevail in DMT.

Overall, the tail end villages in both tanks are unfavourably located to get the tank supply. However, this disadvantage is compensated to a considerable extent by the greater use of well water in the tail end villages. The weighted average estimates of groundwater use show that the proportion of GCA using well water is much higher in the tail end villages. In KPT, in 2015-16, the proportion of area using well water ranged from 7% in Kadaperi to 100 % in Alapakkam. Whereas, in the preceding tank supply year the percentage of purchased well water use was poorer in all the villages. As a result, the proportion of area using well water decreased significantly in all villages. The well water use in KPT shows that the conjunctive use certainly dampens the effect of unequal distribution of tank water across villages in normal supply year.

The analysis indicates that water sales was significant only in KPT but not in DMT. Within KPT across villages the picture varies. More of water sale was reported in the head reach village of Kattalai (56.6%) followed by the tail end village of Athipattu (36.0%) and in the elevated sluice village Kadaperi (32.5%). In Kattalai, due to full closure of deep sluice for irrigation, the entire village includes all types of non-well farmers use the well water for cultivation. In KPT between the latest two supply years the area under sale of well water was higher in the latest supply year and it is less in the preceding supply year in all the villages. In Panniyur and Alapakkam (except a very limited hectare) water sale was not at all reported.

There is a systematic relation between CI and the location of villages in the tank ayacut; within villages significant differences exist between well and nonwell farms. In KPT, high CI was recorded in the villages under deep sluices because they got supplies for a longer time than tail reach/ elevated sluice villages. Within villages, well farms ranked first in terms of CI followed by nonwell farms of head segment and the tail. In DMT too CI in the head reach is much higher than the tail in 2015-16. However, CI in all the nonwell farms was more or less equal to that of well farms. This is partly due to the widespread practice of broadcast sowing of paddy in the first season depending primarily on rainfall. Also, the extent of well irrigation in the 'non-well farms' is also an important factor contributing high CI, especially in tail end villages of DMT.

The cropping pattern differs widely between the tanks. In KPT villages apart from paddy cultivation, banana is relatively important. But in DMT villages, paddy is the predominant crop grown in all the seasons; and hardly any sugarcane or banana is grown here. During the first season, where there is no tank supply, broadcast paddy is widely sown in DMT villages but it is not sown in any of the KPT villages. In the latter, well farmers grow banana more widely than the nonwell farmers.

One of the important factors affecting the yield per hectare of GCA in the head reach villages, especially under the deep sluice, is drainage problems. This is clearly identified in our survey in both the tanks. In the head reach village of Kattalai though GVO/GCA is less, yield per hectare of operated area (GVO/NSA) remains higher than in other villages. This is mainly because of high CI made possible by the longer duration of tank supply (in the 2015-16 assured and dependable nature of well irrigation helped to retain the high CI). Tail end villages in KPT get relatively more output per hectare of both GCA and NSA compared to the elevated sluice villages. Despite shorter duration of tank supply they have little drainage problem and well irrigation benefits more for the cropped area. Better access to well water raises output per hectare of both GCA and operated area.

In DMT, output per hectare of GCA and NSA in the head reach villages in both the reference years was less compared to tail end villages. This is mainly because of severe drainage problems in the head reach villages resulting in less GVO/GCA. However, the GVO/NSA did not decline much. This is mostly due to higher CI in the head reach villages. It is the maximum of 176 in the tail end of Sothiyampakkam (mainly due to use of well water in all segments farmers) and also the minimum of 98 in the tail end of Punaithangal (here a part of the ayacutdars, especially in the tail segment, had not cultivated lands). It is obvious that the condition of drainage facilities and the CI are the most important factors affecting the output per hectare of both GCA and NSA in the selected villages.

Issues for further study

The basic objective of this study is to evolve an integrated approach to the study of irrigation in two rather complex, user managed tanks. For this, we formulated a framework for analysing the Palar Anicut system and then applied this to the specific cases of KPT and DMT in Tamil Nadu. In this process, we have come across many interesting issues which call for further work.

In Chapter 2, it is analysed the conditions of water supply in PAS as a whole over time and found that the pattern of rainfall is far more important for the availability of river supply. A given amount of total rainfall in the catchment largely result in longer flows in the river if it is concentrated in a few days than when it is evenly distributed over the year. In order to capture this aspect, we need to take into account the number of rainy days; the number of high intensity rainy days, and the quantum of rainfall of these days. But there has been no systematic analysis of these relations and we could not attempt it for lack of data. In the same manner it is not attempted to record how much water was actually released at the Anicut to different channels; how much of it finally reached the tanks and drawal capacity and the extent of seepage, percolation and evaporation losses in different channels. These aspects require further investigation.

In Chapter 3, though we have attempted to find out the volume of storage available in the selected tanks and its utilisation in the ayacut in each year in the latest five decades, the analysis does not provide a clear picture on the extent of water stored in the tanks (gross quantum of storage) and the quantum of water released from the tank in each sluice for irrigation in every season. This aspect requires the relevant data on the volume of storage available at the tank at different heights and more importantly the data on the quantum of daily flow reached to the tank and its discharge over the ayacut. Apparently no data on these aspects are available which needs further work.

In Chapter 4 though we have analysed the factors which affect the functioning of institutions at some length further investigation is necessary on the following aspects: (a) disputes arising out of sharing of water among the farmers in each village and in different reaches of the system; (b) the problems of water management as a result of increased fragmentation and subdivision of holdings and (c) the effect of the extensive exploitation of groundwater among different categories of farmers in the ayacut.

In Chapters 5 and 6, it is analysed the maintenance of irrigation network and the control, regulation and use of water from the tank to the fields. Our findings indicate that a proper allocation of water by the local irrigation functionaries (Neerkatties, Kammukutties and Thotties) should improve the situation to a great extent. But this institution has vanished at present in both the tanks. Also, it is important to note that in most of the head reach villages, where there is no scarcity of water, there are no local irrigation functionaries and farmers themselves irrigate their lands who tend to misuse water. Even in a few villages where the irrigation functionaries existed then, they were not able to discharge their duties because of the problems within villages, especially the breakdown of the village society and poor remuneration to the functionaries. The nature of these problems, and the manner in which they affect the irrigation institutions call for further work.

In Chapter 7 we have analysed the extent of use of well water in the selected villages. The analysis indicates that wells are an important factor, not only to compensate the shortages of tank water supply but also to help increase the cropping intensity as well as productivity of land. It is also found from the survey that water sales were significant and more widespread, especially, in KPT villages. But most of the respondents were reluctant to reveal the actual extent of sale of water from their wells. Hence a detailed investigation is necessary to understand the actual sale of well water, which is not attempted in this study.

In sum, the study has attempted to bring out the irrigation system in its totality and various factors responsible for changes in the irrigation system. On the whole, it is clear from this study that a proper understanding of any irrigation system requires an approach from a historical perspective.

APPENDICES

Appendix 1.1

TNFMIS ACT-2000: WUA

GOTN PWD WRO

1. TN Farmers' Management of Irrigation Systems Act, 2000. Act No. 7/2001.
2. TN Farmers' Management of Irrigation Systems Rules 2002.
3. Sl. no 1. Amendments (January, 2004).
4. TN Protection of Tanks and Eviction of Encroachment Act and Rules 2007 (TN Act 8/2007).

TN Farmers' Management of Irrigation Systems Rules, 2002 (G.O. Ms. No. 359 PW (W2), 24-09-2002.

ABBREVIATIONS

TC = Territorial Constituency
MC = Managing Committee
FO = Farmers' Organization
WUA = Water Users Association
TCM = Territorial Constituency Member
SC = Sub-Committee
O & M = Operation and Maintenance
CA = Competent Authority
FY = Financial Year
GB = General Body

Key Points of Village WUA VS Tank WUA

1. Every WUA shall have the name of a village as its district name.
2. If there is more than one association in a village, then such association shall be called by adding numerals to the village name.
3. In tanks, the WUA may be named by the name of the tank.
4. The minimum area for a WUA is 40 ha and above – Keeping the area under each Territorial Constituency according to site conditions based on hydraulic boundaries.
5. The Collector shall have a command area map/sketch dividing each WUA area equally divided into not exceeding 10 TCs.
6. The final map/Sketch shall be displayed in the office of the village Panchayat and P. Union.
7. Landholders List is prepared based on the record of rights, which is given in form V.

   Voters List: Based on landholders list, TC-Wise voters list in form V-A is Prepared (Completed age of 18 years as a norm). The Voters elect the President, and members of the Managing Committee (TCMs and the convener of the Sub Committees).
8. Each landholder in the WUA area shall have one vote only.
9. For the Purpose of Operational viability in major of medium irrigation system, Distribution Committee shall be formed.
10. **Constitution and Functions of Apex Committee**
    (i) The Chairman nominated by the government; and
    (ii) The following members namely;
        (a) All the Chairmen of the Project Committees (Ex. Officio);
        (b) All four Regional C.Es and Engineers-in-Chief of WRO. Chief Engineer Agricultural Engineering Department and Director of Agriculture; and
        (c) Two representatives of selected NGO to be nominated by the government.

    Functions: Lay down the guidelines for implementation of the Act. Give directions to any farmers' organization as may be considered necessary.

    Meetings: Shall meet once in three months. Special meetings may be held if it is required. Seven days' notice to conduct meetings and the quorum shall be one third of the members.

11. Minutes of the meeting: Every Proceeding shall be recorded in the minute's book maintained for the purpose.

12. **Constitution and Functions of Sub – Committees**

    Managing Committee of the FO may constitute specific sub-committees to carry out specific functions as assigned by the managing committee.

    (i) Finance and Resources Sub-Committee.

    (ii) Works Sub-Committee.

    (iii) Water Management Sub-Committee.

    (iv) Monitoring Evaluation & Training Sub-Committee.

13. Each Sub-Committee shall consist of a convener and four members.

    A Convener shall be a member of the Managing Committee.

    Project Committee & Distributary Committee are only for major irrigation Projects.

    In each WUA, members not exceeding four to be drawn for each Sub-Committee from among the land holders within its jurisdiction.

    The Sub-Committee shall be constituted by the General Body of the respective Committee.

    No member shall represent more than one Sub-Committee.

14. Functions: Sub-Committees should meet once a month.

    (i) Finance and Resources Sub-Committee - to mobilize and collect resources;

    (a) To ensure collection of dues from members-Rs. 250 to Rs.500 per ha per year - from every Water User for carrying out:

        1. Operation and Maintenance of irrigation system;
        2. Provide infrastructure facilities;
        3. Provide specific services;
        4. To meet any urgent needs of association;
        5. Build up assets of farmers' organization and
        6. Improve the system.

    (b) Recommend to the MC for the use and deployment of resources.

    (c) Maintain records relating to financial matters.

    (ii) Works Sub Committee:

    (a) Recommend estimates of works for administrative approval;

(b) to supervise works and ensure quality control; and

(c) to approve payments for the works.

(iii) Water Management Sub Committee

(a) To carry out the decision of the Managing Committee on water Regulation, schedule of water release.

(b) Organize Patrolling of the Canal, Channels and regulate water use.

(c) Check the irrigation and drainage system regularly.

(d) Record the water deliveries.

(e) Report to the Managing Committee any violations in water use; and

(f) Maintain records of landowners and water users.

(iv) Monitoring, Evaluation and Training Sub-Committee

(a) Identify training needs and organize training to the water users;

(b) Educate in optimum use of water; and

(c) Monitor specific items like area irrigated, productivity, disputes settlement and resources building.

15. Meetings of Sub Committee: It shall meet as frequently as necessary.

(a) The Convener of the Sub Committee will preside over the meetings and maintain the record of discussions and decisions thereof.

(b) The Sub Committee shall function under the guidance, control, direction of the Chairman/ President of the Managing Committee.

16. General Body: of WUA shall comprise of all members of Water users.

- The General Body shall be assisted by the competent authority who shall have the right to attend the meeting and record his views, but shall have no right to vote.

17. The General Body: shall meet at least twice a year. Once before the I crop season and once before the II crop.

18. Notice for General Body:

- The Managing Committee of the farmers organization shall convene a General Body meeting within twenty days by giving seven days prior notice of the date, time and place of the meeting and also agenda. 1/3 as quorum for the General Body.

19. Minutes:
    - Every Proceeding of the General Body shall be recorded in the minute's book. A copy of the minutes shall be sent to the authority at the next higher tier.
20. Managing Committee:
    - It shall be held at least once a month at the office of the organization. Meetings shall be held within seven days of the receipt of requisition for such meetings.
21. Minutes: shall be recorded by the person chairing the meeting.
    - A copy shall be sent to the authority of the next higher-tier. Quorum 1/3 of the members.
22. Powers of General Body:
    1. Approve the operational plan for each crop season and review its implementation.
    2. Allocate water according to the operational plan approved.
    3. To decide on the manner of regulation and distribution of water.
    4. Prepare annual / long-term financial and works plan and prioritize works for Maintenance, repairs, upkeep and rehabilitation of the irrigation system.
    5. Approve annual budget and review previous year's performance
    6. Set-up sub committees for various activities.
    7. Create funds required for different activities/works.
    8. To levy a fee for carrying out any specific purpose.
23. Operational plan and water Budgeting:
    1. The Managing Committee (MC) of the FO shall, along with the assistance of the Competent Authority (CA), prepare a water budget for the area under its control.
    2. One month before the onset of the monsoon season (Oct-Dec), workout the anticipated-inflows to the tank storage and workout water allocation / distribution plan to the ayacut.
    3. Each FO, shall draw up an operational plan which shall specify the quantity of water to be drawn on a fortnightly basis.

4. Drawals of water shall be monitored daily at specified point as decided by the FO.
5. Review of the drawals and distribution shall be done fortnightly.
6. At the end of each season the FO shall prepare a report of water received and Utilized, area irrigated, and extent of crops cultivated.
7. The FO shall analyze the shortcomings and deviations in water budget and report to the next higher tier.

24. Water Regulation
    1. After Preparation of water Budget, the FO shall draw up a plan of water regulation.
    2. The dates of release and closure shall be informed to all members well in advance.
    3. Equitable distribution of water among all users is the main Principle.
    4. A FO shall draw water and monitor flows based on operational plan.
    5. A rotational water supply shall be prepared for each sluice.
    6. The FO shall, Carryout, Azmoish (Crop Survey) of the ayacut with the assistance of the Competent Authority along with the Agriculture and Revenue personnel, and
    7. A FO may, for the purpose of monitoring, install such devices to observe water flows and document, within its jurisdiction.

25. Levy and Collection of Fees
    (i) A FO may levy a fee ranging from Rs. 250 to Rs. 500 per ha per year from every water user for carrying out any specific purpose.
    (ii) The purpose of levying a fee is given in point 14 (i).
    (iii) The Managing Committee shall levy a fee proportionate to the land holding or to the number of members and through a demand notice.
    (iv) All fees collected shall be duly accounted for, through Proper receipts.
    (v) A fee collected shall be used only for the said (Specific) purpose.

(a) A member has defaulted in payment; the Managing Committee shall prepare a list of defaulters along with the amount due.

(b) The defaulters list shall be furnished to the Tahsildar, for recovery, A copy of the defaulters list shall be furnished to the Collector.

26. Grants from the Government

- Information related to share of water charges collected and any other grants to augment the financial resources of the FO should be notified by the state Government (then and there).

27. Accounts / Finance

(a) The Farmers Organization shall open an account in a Nationalized Bank / Co-operative bank / Agricultural co-operative Society in its name and shall be operated jointly by the President/ Chairman and one of the managing Committee member.

(b) Every expenditure should be supported by a receipt or voucher, which shall be passed for payment by the president or any authorized person by him.

(c) All expenditure has to be approved by the finance and resources Sub-Committee, at least once in a month.

(d) Every Farmers Organization shall maintain six accounts register. Each record shall bear the name, address and seal of the FO and shall be machine numbered.

1. Cash Book
2. Bill Register
3. Contingent Register
4. Amanath Register (Day book)
5. Receipt Book
6. Cheque Register.

28. **Social Audit of Farmers Organization**

(1) At the end of each crop season the FO shall conduct social audit through an experienced Auditor.

(2) Social Audit shall be for both water utilization against water budgeting and expenditure incurred for maintenance with reference to the funds available.

(3) The Social Audit shall cover:

(a) Equity in water distribution

(b) Increase in production

(c) Increase in productivity

(d) Crop diversification

(e) Multiple cropping

(f) Water use efficiency

(g) Utilization of resources for execution of works

(h) Improvement in the cultivated areas compared to previous season.

(i) Quality of works undertaken

(4) The social audit shall be made known to all the beneficiaries by way of displaying a list on the office board containing the benefits accrued with reference to funds spent.

(5) Whenever a work is taken up, the estimated cost of the work, item of work etc. are to be exhibited on a board at the place of the work, so that each beneficiary is aware of the details of the work.

(6) The competent Authority shall render all assistance in the conduct of the social audit. The revenue and agriculture officials shall also render the requisite assistance.

(7) The auditor shall incorporate the social audit report in his report together with his specific observation on rectification of defects, if any, notices in the social audit.

29. Financial Audit:

- At the end of each financial year, and not later than 3 months after the commencement of the new financial year, each FO shall cause its accounts to be audited. If the transactions exceed Rs.10 lakhs per annum the services of a chartered Accountant is utilized.

30. Budget:

- Each Managing Committee shall prepare a detailed budget covering all the activities for the next financial year, commencing from April 1st

31. Procedure for compounding of offences

1. The President of the FO shall give a notice of the offence to the individual.
2. The individual who has committed the offence shall be given an opportunity to explain his point of view.
3. The president of the Managing Committee may fix an amount not exceeding the amount of penalty as given below (as per section 32 of the act) as compounding fee and recover it.

Offences and Penalties

(a) Damage, alter, enlarge or obstruct any irrigation system.

(b) Being responsible for the maintenance of the irrigation system, neglects to take precautions or use of water in an unauthorized manner, or in such manner as to cause damage to the adjacent land.

(c) Corrupts or fouls water less fit for its ordinary use.

(d) Obstruct or removes any level marks fixed by the authority of a public servant.

(e) Violates the rotational water supply, shall, on conviction, be punished with imprisonment for a term which may extend to two years or with fine which may extend to Rs.5000 or with both.

32. Records to be maintained

(i) Map showing the boundaries and jurisdiction of the Association, water conveyance system within the boundaries of the Association.

(ii) Map showing the localized ayacut with survey numbers.

(iii) Map showing the area under irrigation which does not fall within notified ayacut.

(iv) Ayacut Register as per Settlement (1880s) / Resettlement (1910s) / Updated settlement (1980s).

(v) Inventory Register: Hydraulic particulars of structures, including details of canals / channels and their details.

(vi) Register of land, Buildings and so on.

(vii) Miscellaneous property Register: Minor properties like trees, grass etc.

(viii) Register of machineries.

(ix) Membership Register and Records.

(x) Landholders Register and voters Register as in form V and V-A

(xi) Rainfall and Water Flow Register- Apart from local rainfall this detail can be obtained from the nearby recording station periodically.

(xii) Area and Crop Register- Village wise extent of crops grown.

(xiii) Farmer wise Demand Register-based on area irrigated and crops grown.

(xiv) Works Register-based on Operation and Maintenance as per the procedures prescribed.

(xv) Register of Administrative Sanctions - date wise notices issued by a FO.

(xvi) Register of Technical Clearance - Contains all technical clearances given by the Competent Authority to take up works.

(xvii) Minutes Register and Records - All records of General Body, Managing Committee and Sub Committee.

(xviii) Power of Commissioner – He/she shall call for a meeting of any FO to discuss matters relating to urgent public importance.

Appendix 3.1

Water Mamulnamas for Kaveripakkam and Dusi-mamandur Tanks Written in 1815

காவேரிபாக்கம் ஏரி

காவேரிபாக்கம் வெங்குஜயங்கார், மேட்டுகுடி கிருஷ்ணயங்கார், வினாயகப்பிள்ளை, கணக்கு நடராஜப்பிள்ளை, தலையாரி நல்லமுத்தன், கருப்புமுத்தன், இராகவன், படவேட்டான், வீரன், சின்ன கந்தன் தோட்டி, தோட்டியான் கந்தன் ஆகியோர் காவேரிபாக்கம் தாலுக்கா மசுக்கூரில் உள்ள நாங்களனைவரும் எழுதிக் கொடுத்த மாமூல் நாமா யாதெனில் கூஷ தாலுக்கா தாசில்தார் இராமப்பா அவர்கள் 1815ஆம் ஆண்டு விசுவாசு வருடம் ஜனவரி மாதம் 7ம் தேதி மாமூல் நாமாவுக்காக அசுரிலிருந்து 8ம் இலக்கம் வந்து, ஜூலை மாதம் 27ம் தேதி வந்த தாக்கீதும், கிஸ்தி நாமாவையும் எங்களுக்கு படித்து காண்பித்து ஏரிதாங்கல், கால்வாய் தண்ணீர் மாமூலாய் பாய்ச்சலாகிக் கொண்டு வர்ற முறைக்கு முன்பு எந்த ஒரு பிரச்சினையுமின்றி நாங்கள் நடந்து கொள்ள வேண்டிய மாமூல் நாமாவை எழுதி எங்கள் கையையெழுத்திட்டு அக்கம் பக்கம் கிராமத்தார் கையெழுத்தும் போடவித்து குடுக்க சொல்லி கேட்கையில் நாங்களவைரும் ஒருமித்து கிராமத்தில் நஞ்சை, புஞ்சை தோட்டக்கால் நிலங்கள் பிரிச்சு பார்த்து ஏரி தாங்கல் சிறிய ஏரி (1மாதம், 2 மாதம்) கசிவுக்கால் (spring channel) பாய்ச்சலாகப்பட்டது கண்ணாலேபாத்து மாமூலாய் நடந்து கொண்டு வாற முறைபடிக்கு இதற்கு அடியில் எழுதிக் கொடுத்திருக்கிற தபசில் (List).

1. ரெவின்யூ கிராமம் 1 மதுரா (துணை) கிராமம் பூச்சியம் ஆக கிராமம் 1. இதற்கு நஞ்சையில் பாய்கிற ஏரி 1க்கு மதகு 10க்கு பாசனம் முறை. ஏரியின் தென்கரையில் மேல்புரம் இராமாபுரம் மதகு 1க்கு இராமாபுரம் முழுமைக்கும் பாய்வதற்கான கால்வாய் 1.

2. இதன் கிழக்கே நரிமதகு 1க்கு பெருங்கால்வாய் 1க்கு தண்ணிபங்கு 12க்கு காவேரிபாக்கத்திற்கு பாயுது. வீரவர்கோவில் மடை 1க்கு கால்வாய் 1க்கு பங்கு 3 இதன் கிழக்கே வரத்து மடை 1க்கு கால்வாய் 1க்கு தண்ணி பங்கு 3க்கு காவேரிபாக்கம் பாச்சல் கால்வாய் 1க்கு பங்கு 1. கிழக்கே போற கால்வாய் 1க்கு

தண்ணீர் பங்கு 2ல் காவேரிபாக்கம் கோட்டை பாகத்துக்கு மத்தியானம் 30 நாழிகையும் ஒரு ஏத்து தண்ணிவிடனும். மீதி தண்ணீரில் கோட்டையோரம் குதிரை குளத்து மானியத்தண்டை கொண்டாபுரத்துக்கு மெரிதண்ணி கால்வாய் 1. கொசவன் கால்வாயண்டை கொண்டாபுரத்துக்கு பாயுற பிரிவு மெரி தண்ணி கால்வாய் 1 மீதி தண்ணீர் காவேரிபாக்கம் பாச்சல் மத்தியானம் முழுமையும் பங்கு 1. நண்டேரி (ஈராளச்சேரி) பெரு கரும்பூருக்கு இரவு முழுமையும் பாய்கிற பங்கு 1. இந்த இரவு பங்கில் காவேரிபாக்கம் வெண்ணைப்பட்டி கழணிக்கு பக்கிரி தக்காவண்டை ஒருமெரி தண்ணி வருகிற கால்வாய் 1 இந்த நண்டேரி பெருகரும்பூர் கால்வாய் தண்ணீரை அதிகாலையில் (கிழக்கு வெளுக்க) காவேரிபாக்கத்தார் ஈராளச்சேரிக்கு போற தண்ணியை துண்டித்து, பெருகரும்பூர் கால்வாயயை துண்டிச்சுக் கொள்ளுகிறது. அகரத்து மடைக்கு தெற்கே போகிற பெருங்கால்வாயில் இராமாபுரத்து எல்லையில் பெரியதம்பி முதலியார் பங்கு கழனிக்கு வடக்கே கொண்டாபுரத்துக்கு பாய்கிற பரிவு கால்வாய் 1க்கு தண்ணீர் பங்கு 3. இதன் தெற்கே கொண்டாபுரம் சித்தேரியண்டை விளாகத்து ஒத்திப்பட்டு கழணிகளுக்கு பாய்கிற 1 அடி தண்ணி கால்வாய் 1. இதன் தெற்கே விளாகத்து எல்லை சரகத்து கொல்லை ஓரத்தில் சிறுகரும்பூருக்கு பாயுற பிரிவு கால்வாய் 1க்கு தண்ணீர் பங்கு 1. இதன் தெற்கே விளாகத்து கொல்லையண்டை அத்திப்பட்டுக்குப் பாயுற பிரிவு கால்வாய் 1க்கு தண்ணி பங்கு 1. கிழண்டை விளாகத்துக்கு பாய்கிற கால்வாய் 1க்கு தண்ணி பங்கு 1 ஆக மதகு 1க்கு தண்ணீர் பங்கு 12.

3. இதன் வடக்கு கீழண்டை கரையோரம் சிங்கமதகு 1க்கு பின் தொட்டியில் கொண்டம் (earthen weir) மூலம் காவேரிப்பாக்கம் பாச்சல் மேல்படி தொட்டிக்கு வடபுறம் மதகு 1க்கு முறந்தூக்கி வகையரு கழனிகளுக்கு பாய்கிற கால்வாய் 1 ஷ தொட்டிக்கு தென்பகுதி மதகு 1க்கு கொண்டம் கலமேடு வகையரு கழணிகளுக்கு பாச்சல் கால்வாய் 1 மேல்படி தொட்டி 1 கிழக்கே புறப்படுகிற மதகு 1 பெருங்கால்வாய் 1க்கு பிரிவு காவேரிப்பாக்கம் பாச்சல் வடபுற கரையோறம் போய் கன்னிகோவில் கால்வாய் 1. இதன் கிழக்கு 1 அடி தண்ணி குமிளி கால்வாய் 1. இதன் கிழக்கு மணப்பாந்துரை கால்வாய் 1. இதன் கிழக்கே தேவதானம் கால்வாய் 1. இதன் கிழக்கே மஞ்சினிர் தக்கையில் வடபுறம் போற ஓரடிக் கால்வாய் 1 இதன் கிழக்கு நடுக்கால்வாய் 1. இதன் தென்புறம் மாமன் கால்வாய் 1க்கு torn காவேரிப்பாக்கம் பாச்சல் மேலெழுதிய கால்வாய்களிலே ஈராளக்சேரிக்கி பாச்சல் வயணம். மாமன் கால்வாயில் போகிற தண்ணீரில் சூரியன் அஸ்தமான முதல் கிழக்கு வெளுக்கிற தருணம் விடிய 3 நாழிகை வரை காவேரிபாக்கம் உப்புமேடு பங்கு கம்பத்தக்க கழனிக்கு ஒரு மெரி தண்ணி விட்டு மற்ற நீர் ஈராளச்சேரிக்கு பாயுறது. கிழக்கு வெளுக்க துண்டித்து (intercepting the flow) காவேரிப்பாக்கம் பாய்வது. நடுக்கால்வாயில் வருகிற தண்ணீரில் உப்புமேடு பங்கு புலிப்பட்டு கால்வாயில் வருகிற தண்ணீர் 3 நாளைக்கு 1 இரவு தண்ணீர் ஈராளச்சேரிக்கு, மேல் ஓணாம்பட்டு கீழ் ஓணாம்பட்டு கழனிகளுக்கு பாயுறது. மஞ்சநீர் ஓரடிக் கால்வாயில் போகும் தண்ணீரில் காவட்டிக்காலில் மேல் விடுகாலில் ஒரு மெரி தண்ணி விட்டு மீதி ராத்திரி தண்ணி ஈராளச்சேரிக்கு பாயுறது. ஆக சிங்க மதகு 1.

4. வடக்கு மூலை மதகு 1க்கு பின்தொட்டியில் பிரிவு வடபுறம் சேரிக்கு பாயுற பாதிரிகால்வாய் 1 இதில் வரும் தண்ணீரில் அபிராம்சேரி ஊருக்கு வடக்கே ஒரு காணிக்கு பாயுற தண்ணி வருகிற கால்வாய் 1. தென்கரையோரம் கட்டளைக்கு பாயுற கால்வாய் i. இதன் கிழக்கு மதகுக்கு எதிரடியில் கட்டளைக் கரையோரங்காலுக்கும் பாதிரி காலுக்கும் யேறுகிறத்துக்காக மதகுக்கு ஒரு கெஜத்துக்குள்ளாக 5 கருங்கல் குண்டு போட்டு போற பெருங்கால்வாய் 1ல் கட்டளைக்கி பாதி தண்ணி பாயுற கால்வாய் 1. இதில் சேரி சபாவிறத்தி கழனிக்கு

பாயுற 1 அடி தண்ணிக்கால்வாய் 1. இதில் கட்டளை 2 காணிக்கு பாயுற வரை தண்ணீர் போற கால்வாய் 1 பெருங்கால்வாயில் மத்த பாதி தண்ணி கால்வாய் 1க்கு மதியம் 30 நாழிகையும் சேரிக்கி பாயுறது. இராத்திரி 30 நாழிகையும் அபிராமச்சேரிக்கி பாயுறது. கட்டளை கழனி கால்வாய்கள் கழிவு (drainage) தண்ணீரை கட்டளைக்கு வடக்கே சேரிப்பாட்டையண்டை காவேரிப்பாக்கம் அபிராமச்சேரி தோட்டிகள் மடைபோட்டுக் கொண்டு வந்து விடுகிற தண்ணீரில் அபிராமச்சேரி கழனிகள் கழிவு தண்ணீரும் சேர்ந்து காவேரிப்பாக்கம் எல்லை வாயிச்சந்து மடுவுங்கரை கழனிகளுக்கு மதியம் 30 நாழிகையும் பாய்வது. இரவு 30 நாழிகையும் அபிராமச்சேரி கழனிகளுக்கு பாயுறது. அறம்பு நாளில் சேரிவழியண்டை கட்டளைக்கி கட்டியுள்ள கொண்டத்தில் 1 அடி தண்ணி எடுத்துவிட்டு காவேரிப்பாக்கம் எல்லை வாயிச்சந்து செவிட்டு தாக்கு கழனிகளுக்கு மதியம் 30 நாழிகையும் பாயவிடுகிறது. அபிராமச்சேரி மூலை கழணிக்கு இரவு 30 நாழிகையும் பாய விடுகிறது. ஆக மூலை மதகு 1.

5. வடக்கு பள்ளமதகு 1க்கு சேரிப்பாச்சல் பூறா கால்வாய் 1.

6. வடக்கு மேட்டுமதகு 1க்கு சேரிப்பாச்சல் பூறா கால்வாய் 1.

7. வடக்கு மாகாணிப் பட்டு மதகு 1 துரைப்பெரும்பாக்கம் மதகு 1 ஆக மதகு 2க்கு கால்வாய் 1க்கு சின்ன கடவாசலுக்கு மேலண்டை சேரி, வண்ணாரப்பட்டி முருகன், செங்கலடி வேம்பன் வகையறா கழனிக்கு பாயுற முதுமான் கால்வாய் 1 இதில் மாகாணிப்பட்டு தட்டான் கோவில் கழனிக்குப் பாயுற ஆனந்தப்பிள்ளையார் கோவிலுக்கு மேல்புறம் வழியாக போற ஓரடி தண்ணிக் கால்வாய் 1 மடுவோரம் முருக்கண்டை சேரி மேல்படி முதுமான் கால்வாயில் செங்கலடியோரம் 1 மெரி தண்ணி மாகாணிப்பட்டு தட்டான் கோவிலடி கழனிக்கு பாயுற கால்வாய். 1 இதில் சேரி வெள்ளம் கழனிகளுக்குப் பாயுறது. ஏரி கலங்கல் கழிவு தண்ணியை உப்புறந்தாங்கல் கொண்டத்துக்குக் கீழ் வரும் கழிவு தண்ணி சின்னகடவாசல் கொண்டம் கட்டி அந்த கழிவு தண்ணியும் முன் எழுதிய மாகாணிப்பட்டு துரைப் பெரும்பாக்கம் 2 மதகு நீரும் சேர்ந்து வரும் கால்வாயில் சேரி முதுமான் கால்வாயிக்கி 1 அடி தண்ணி போக மீதியை மேல்படி கொண்டத்துக்கு அணையினாலே கிழக்கே போற கால்வாய் 1க்கு தண்ணீர் பங்கு 11க்கு கொண்டத்தோரம் புதுப்பட்டுக்கு போற கால்வாய் 1க்கு தண்ணி பங்கு 1. இதன் கிழக்கே போற பெருங்கால்வாயில் மாகாணிப்பட்டு புத்தேரி பள்ள கழனிக்கு போற கால்வாய் 1க்கு தண்ணி பங்கு (1). இந்த தண்ணீரில் சிரஸ்தவலாபுரம் பங்குக்கு 1 மெரி தண்ணி விட்டு புத்தேரி பள்ளக்கால்வாயில் கொண்டு வந்து சிரஸ்தவலாபுரத்துக்கு பாயுறது. மீதி மாகாணிப்பட்டுக்கு பாயுறது. இது போக மீதி தண்ணி போற கால்வாயில் சேரி வடக்கு மேடு நெல்லித்தோப்பு கழனிகளுக்கு 1 மெரி தண்ணி பாயுற கால்வாய் 1. இதன் கிழக்கு பெருகால்வாயில் போற தண்ணீரில் மாகாணிப்பட்டுக்கு பாயுற (படப்பை) கால்வாய் 1க்கு பங்கு 2 இதில் மேல்படி மாகாணிப்பட்டு மேல் கழனி அத்திப்பட்டி கழனிக்கு பாசனத்திற்கு ஏற்ப தண்ணி நெருங்கி துரைப்பெரும்பாக்கம் கால்வாயில் விடுகிறது. இதன் கிழக்கே ஆலப்பாக்கம் கால்வாய் 1க்கு பங்கு 2ல் சத்துவானத்தத்துக்கு 6ல் 1 பங்கு தண்ணி போக மீதி தண்ணி ஆலப்பாக்கத்துக்கு பாயுறது. இதன் கிழக்கு தழுவணாம் பட்டு கால்வாய் 1க்கு பங்கு 1. இதன் கிழக்கு துரைப்பெரும்பாக்கம் கால்வாய் 1ககு தண்ணி பங்கு 4. மகாணிப்பட்டு மேலக்கழணி அத்திப்பட்டி கழனிக்கு பாசனத்திற்கு ஏற்ப தண்ணீர் விட்டு மற்ற தண்ணீரில் சிரஸ்தவலாபுரத்துக்கு 6ல் 1 பங்கு தண்ணி கால்வாய் 1, மற்ற 5 பங்கில் முருஞ்சுபாக்கத்துக்கு 1 பங்கு தண்ணி பாயுறது. மீதி 4 பங்கு துரைப்பெரும்பாக்கத்துக்கு பாயறது. ஆக மகாணிக்கப்பட்டு துரைப்பெரும்பாக்கம் மதகு 2.

8. வடக்கு பறச்சேரி மேட்டு பொம்மை மதகு 1க்கு சேரிபாச்சல் பூறாகால்வாய் 1.

9. கலங்களுக்கு வடக்கு பண்ணியூர் மதகு 1க்கு கால்வாய் 1க்கு சேரி கொசவன் பட்டரை கழனி 5 காணிக்கு கால்வாய் 1. மீதி கிழக்கே போற தண்ணியில் கெங்காதரநல்லூர் பங்கு 1க்கு கால்வாய் 1. பண்ணியூர், கீழ் வீராணம் பங்கு 2க்கு கால்வாய் 1க்கு பண்ணியூர் ஏரி எதிர்வாயில் கீழ் வீராணம் பங்கு 1க்கு கால்வாய் 1. பண்ணியூர் பங்கு 1க்கு கால்வாய் 1. ஆக பண்ணியூர் மதகு 1. ஆக ஏரி மதகு 10.

10. இந்த கால்வாய்கள் அவரவர் மாமூல் தரத்துக்கு கால்வாய் வெட்டி பாய விடுகிறது. மாமூலுக்கு அதிகமாய் ஆழ வெட்டக்கூடாது.

11. அறம்பு நாளையில் இராமாபுரம் தர்கா, சித்தேரி எலந்தடி, இந்த கழனிகள் கழிவு நீரை ராமாபுரம் பறச்சேரியண்டை கொண்டாபுரம் வகையரு கிராமங்களுக்கு போற கால்வாயை கொண்டம் கட்டி காவேரிப்பாக்கம் கோட்டைபாகத்து அஞ்சூர் அன்றாசுக்கு (மிராசுக்கு) 1/2 காணிக்கு இறைத்துக் கொள்ளலாம். மீதி தண்ணி காவேரிப்பாக்கம் கழனிகளுக்கு பாய விடுகிறது.

12. ஏரி கோடி பண்ணியூர் மதகுக்கு தெற்கே பறச்சேரி மேட்டு மதகுக்கு வடக்கு கலங்கல் கல் 125 (local measure) இதில் ஏரிக்கு ஜலம் அதிகமாகி கலங்கல் எடுக்கிறபோது கூடுமானவரை பின்னூர் கலங்கல் கட்டின பின் மேல்படி கசிவு நீரும் போகும் மடுவில் கொண்டம் கட்டி பாய விட்டும், ஏற்றம் வைத்து இறைத்தும் பாய்ச்சுவது மாமூல் முறை.

Type 1 : உப்புறந்தாங்கல் கொண்டம் 1. இது உப்புறந்தாங்கல் பாச்சலுக்கு சொந்தம்.

Type 2 : சின்ன கலங்கல் 1. இதில் மாகாணிப்பட்டு, துறைப்பெரும்பாக்கம் மதகு நீரும் கலந்து பாயும் கிராம முறை ஏழாம் காலத்தில் தாக்கல். இது தவிர மேல்படி சின்னகடவாசல் தண்ணி தேக்கத்தில் உப்புறந்தாங்கலார் பட்டடை (several names jointed) கழனிகளுக்கு இறைத்துக் கொள்ளும் ஏற்றம் 2.

Type 3 : சின்ன கடவாசலுக்கு கீழே சேரி கொண்டம் 1. இந்த கொண்ட நீர் சேரிப்பாச்சலுக்கு சொந்தம். இது தவிர ஷ சேரி கொண்டத்துக்கும் சின்ன கடவாசல் கொண்டத்துக்கும் மத்தியில் மாகாணிப்பட்டு மக்கள் இறைத்துக் கொள்ளும் ஏற்றம் 4. சேரி கிராமத்தார் பாசனத்திற்கு ஏற்ப ஏற்றம் வைத்து இறைத்து கொள்ளலாம்.

Type 4 : சேரி கொண்டத்தில் கழிவு நீர் வர்ர கால்வாய் 1க்கு அபிராமச்சேரி மடு ஓரம் கழனிக்கு பாயும் சாமம் 7 காவேரிப்பாக்கம் மேலேரு மூணாங்கொண்டம் கழனிக்கு மத்தியானத்தில் உதயம் முதல் பாயுற சாமம் 1 ஆக கிராமம் 2க்கு சாமம் 8க்கு நாள் 1. இது சரகத்தில் காவேரிப்பாக்கம் மேலேரு மூணாங்கொண்டத்துக்கு இறைக்கும் ஏற்றம் 1.

Type 5 : சேரி கொண்டத்துக்கு கீழ் மூஞ்சிபாக்கம், துரைப்பெரும்பாக்கம், காவேரிப்பாக்கம் கொண்டம் 1. இதில் விடும் நீர் மூஞ்சிபாக்கத்துக்கு இரவு 30 நாழிகையும் பாயும். துரைப்பெரும்பாக்கம் கழனிகளுக்கு மதியம் 20 நாழிகை பாயும். காவேரிபாக்கம் ஏரி 3ம் கொண்டம் கழனிக்கு மதியம் உதயம் முதல் 10 நாழிகை பாயுறது. ஆக கிராமம் 2க்கு மதியம் நாழிகை 30 ஆக கிராமம் 3க்கு நாழிகை 60க்கு நாள் 1.

Type 6 : இதன் கீழ் ஈராளச்சேரி, காவேரிபாக்கம் கொண்டம் 1க்கு ஈராளச்சேரிக்குப் பாயும் நாழிகை 50 காவேரிபாக்கம் ஏரி 3ம் கொண்டம் கழனிக்கு

மதியத்தில் உதயம் முதல் பாயுற நாழிகை 10 ஆக கிராமம் 2க்கு நாழிகை 60க்கு நாள் 1. இது தவிர மேல்படி ஈராளச்சேரி வகையரு கொண்டத்துக்கும், மூஞ்சிபாக்கம் வகையரு கொண்டத்துக்கும் மத்தியில் துறைப்பெரும்பாக்கம் ஏற்றம் 4, அபிராமச்சேரி ஏற்றம் 1, காவேரிபாக்கம் கீழேரு மூணாங்கொண்டம் ஏற்றம் 1, மேலேரு 3ம் கொண்டம் ஏற்றம் 1 ஆக காவேரிப்பாக்கம் ஏற்றம் 2ஆக கிராமம் 3க்கு ஏற்றம் 7.

Type 7 : ஈராளச்சேரி வகையறா கொண்டத்துக்கு கீழே உத்திரம்பட்டு ஒகளப்பட்டு தற்மநிதி கொண்டமும் இதன் பின்னும் அநேக கிராமங்களுக்கு கொண்டங்கள் முதலானது இருக்கிறது.

13. காவேரிப்பாக்கம் ஏரிக்கு தண்ணீர் வருகிற வயணம் பாலாற்றிலே தண்ணீர் திருப்புகிற சரகத்து திருயெகம்பச்சேரி சரகத்தில் சாத்தம்பாக்கம் ஆத்துக்கால்வாய் பாலத்துக்கு மேற்கு முதல் ஆத்துக்கு தென்கரை ராயசி தோட்டம் சரகத்துவரை. அணைகட்டி திருயேகம்பச்சேரி சாத்தம்பாக்கம், பூண்டி, கடப்பேரி, சுமைதாங்கி ஏரி, பில்லேரி இந்த ஏழு கிராமத்து சரகத்து எல்லைகளில் வருகிற கால்வாய் 1. இந்த கால்வாயில் நடக்கிற மாமூல்:

Type 1: அணைகட்டுகிற முகத்துவாரத்தில் தென்கரையோரமாக போற திருயேகம்பச்சேரி ஆத்துகால்வாய் 1.

Type 2 : வடவண்டை கரையோரமாக போகும் சாத்தம்பாக்கம் ஆற்றுக் கால்வாய் 1.

Type 3 : இதை விட்டு கூன்மடைக்கு கிழக்கே பூண்டி கிராமத்துக்கு உபகிராமமான வெள்ளியந்தாங்கல் நிலம் ஆத்துக்கால்வாயிக்கு தென்புறம் இருக்கிற நிலத்துக்கு தண்ணீர் பாய வடபுறத்திலிருந்து தொட்டி கட்டி ஆத்துக்காலுக்கு குறுக்கே பூமிக்குள்ளே மதகு கட்டி பூண்டி ஏரி தண்ணியை கொண்டு போகிற மதகு கால்வாய் 1.

Type 4 : இதன் மேற்கு ஆற்றுகால்வாயில் அறம்பு காலத்தில் பூண்டியர் ஊற்று தோண்டி தோட்டகாலுக்கு இறைக்குற தொட்டி ஏற்றம் 1.

Type 5 : இதன் வடக்கே கடப்பேரி சரகத்தில் கால்வாய் தண்ணீர் வடிந்த பிறகு அதில் வரும் கழிவு தண்ணீரும் கடப்பேரி தண்ணீரும் விட்டு ஆற்றுகால்வாயிக்கு கீழ் கரையிலிருக்கிற மேற்படி கடப்பேரி நிலம் பாயுறத்துக்காக கட்டியிருக்கிற கல்கொண்டம் 1. இந்த கொண்டம் ஆத்துக்கால்வாய் தண்ணி ஏரிக்கு வரும்போது மேல்படி கல்கொண்டத்துக்கு மேலே போட்டு கட்டியிருக்கிற கதவுகளை எடுத்துவிடுகிறது. மீண்டும் வடிகின்ற தண்ணீரை கடப்பேரியார் கட்டிக் கொள்ளலாம்.

Type 6 : இதற்கு வடபுறம் மேற்படி கடப்பேரியார் மட்கொண்டம் 1. மேற்படி கொண்டம் தை மாதம் 8ம் தேதிக்கு மேல் கட்டிக் கொள்ளுகிறது. மீண்டும் தண்ணீர் வந்தால் எடுத்து விடவேண்டும். மேற்படி மட்கொண்டத்து மண், ஏரியில் வருடாவருடம் வந்து விழுந்து ஏரிக்கு சேதமானபடியால் மேல்படி கொண்டம் வருடம் 1க்கு 1முறை கட்டிக் கொள்ளுகிறது. மீண்டும் கட்டுவது வழக்கம் இல்லை.

Type 7 : இதற்கு வடக்கு சுமைதாங்கி சரகத்தில் 121/2க்கு வடக்கே பரச்சேரிக்கு அடுத்த வடக்கு மேல்படி ஆத்துக்கால்வாயில் கீழ்கரையில் இராமாபுரம் செங்கல் மதகு 1. இதில் ஆத்து கால்வாயில் வருகிற கழிவு தண்ணீர் இராமாபுரத்துக்கு 2 பங்கும் சுமைதாங்கிக்கு 1 பங்குமாய் ஒரே கால்வாயில் பாய்வது மாமூல். இவை, ஆற்றுக்கால்வாயில் தண்ணீர் வரும் பொழுது, வருகிற தண்ணீர் சேதமின்றி மதகு வகையராக்கள் பாய்ச்சலுக்கு அதிகப்பட்டது சேமித்து (பந்து பண்ணி) 1 வேளை, கால்வாயில் அதிக நீர் வந்தால் மேலெழுதிய மதகு வகையராக்கள் சுத்தமாக

பந்து பண்ணிப்போட்டு வடிவிலே மேலெழுதியபடி நடக்கிற மாமூல். சுமைதாங்கி ஏரி, பில்லேரி இந்த கிராமத்தார் ஆற்றுக் கால்வாயில் ஏற்றம் வைத்து இறைக்கும் மாமூல் இல்லை. ஆக ஆற்றுக் கால்வாய் 1.

14. மூலை மதகுக்கு வடமேற்கு ஏரி எதிர்வாய் ஒழுகூர், செங்காடு, அம்மூர், வாங்கூர், வள்ளுவம்பாக்கம் இந்த எல்லைகளில் மேற்படியூர் வகையரா கிராம ஏரிகள் நிரம்பி கோடிகள் விடுகிற நீரும் மழை பெய்தால் மழை வெள்ளம் முதலானதும் கழனி கழிவு தண்ணீரும் இவை வரும் கால்வாய் 1.

15. கடப்பேரி, சுமைதாங்கி, மருதேரி ஏரி ஆகியவற்றின் நீர் காவேரிப்பாக்கம் ஏரிக்கு வருகிறது.

16. ஏரி உள்வாயில் வள்ளுவம்பாக்கம் எல்லையிலிருந்து கச்சந்திக்கால் ஊத்துக்கால் வந்து மூலை மதகு மூலம் வந்து சேரி கிராமத்துக்கு அறம்பு நாளில் தோண்டி பாய விடுகிற சிகஸ்து கசக்கால்வாய் 1.

17. ஏரி வடிகாலிலே ஏரி நீரை எதிர்வாய் நிலங்களுக்கு இறைக்கும் கிராம மாமூல்.

Type 1 : வள்ளுவம்பாக்கம் சரகத்தில் மேல்படியூரார் இறைக்கும் ஏற்றம் நவாபு சாயடி நாளையில் பரீதாபி வருசம் வரை இறைக்கலாம். மடுவில் ஏற்றம் 5, கூனங்குட்டை ஏற்றம் 2ஆக ஏற்றம் 7க்கு மேற்படி வருடம் அதிக ஏற்றம் வச்சு தகராறுபட்டு நிஜாம் மோதின்கான் சாயபு சபாவில் பிராதாய சோளிங்கபுரம் நரசிம்ம சுவாமி மலையில் வள்ளுவம்பாக்கம், குடி குள்ளப்பமுதலி, 16 ஏற்றம் வருடாவருடம் இறைத்துக் கொள்வதென பிரமாணம் எடுத்து அதுமுதல் மேற்படி கச்சந்தி கால்வாயும் கசக்கால்வாயும் கூடுகிற சரகத்து வரைக்கும் அந்தக்கால்வாயில் மடுவில் தண்ணியிருக்கிறபோது இறைத்துக் கொள்ளும் ஏற்றம் 14 மடுவில் தண்ணீர் நின்றால் இறைக்ககூடாது. கூனங்குட்டையில் இறைத்துக் கொள்ளும் ஏற்றம் 2ம் வடிவில் ஏத்துக்குழிக்கு தண்ணி வருகிற மட்டும் இறைக்கிறது. தண்ணீரில்லாமல் போனால் இறைக்கக்கூடாது. ஆக வள்ளுவம்பாக்கம் தொட்டி ஏற்றம் 16.

Type 2 : சேரியை சுற்றி குளத்துக்கு வடபுறம் பொன்னப்பந்தாங்கலார் மேல்படியூர் நிலம் காவேரிப்பாக்கம் ஏரி உள்வாயில் இருக்கிற பட்டடைற்கு பாய்வதற்கு இறைக்கும் தொட்டி ஏற்றம் 2. ஏரி வடிவில் கால்வாய் மாமூலுக்கு அதிக பள்ளமாக எடுக்காமல் ஏரி தண்ணி ஏத்துக் குழிக்கு வருகிற மட்டுக்கும் இறைத்துக் கொள்ளலாம்.

Type 3 : இடையந்தாங்கலார் மேல்படியூர் குட்டைக் கரையோரம் நிலத்துக்கு தென்பக்கம் வைத்து இறைக்கும் தொட்டி ஏற்றம் 2. ஏரி வடிவில் கால்வாய் மாமூலுக்கு அதிக பள்ளமாயெடாமல் ஏரி தண்ணீர் ஏற்றுக்குழிக்கு வருகிற மட்டுக்கும் இறைத்துக் கொள்ளுகிறது.

Type 4 : சேரி கிராமத்தார் ஏரி உள்வாயில் சுத்துக்குளம் முதல் மாகாணிப்பட்டு மதகு எதிரடி வரைக்கும் இருக்கும் சுமார் 40 காணி நிலம் பட்டரை ஏரி வடிவுத் தண்ணீரில் சாகுபடிக்கு தக்கதாய் தொட்டி ஏற்றம் வைத்து இறைத்து சாகுபடி செய்கிறது வழக்கம்.

Type 5 : சுமைதாங்கியார் ஷ ஈச்சந்தாங்கல் ஓடையில் தொட்டி ஏற்றம் 1 சடைச்சி பட்டரை தொட்டி ஏற்றம் 1 ஆக தொட்டி ஏற்றம் 2ம் ஏற்றுக் குழிக்கு தண்ணி வருகிற வரைக்கும் மாமூலுக்கு அதிக பள்ளமாக போடாமல் இறைத்துக் கொள்ளுவது.

Type 6 : சேரி, பில்லேரி இந்த இரண்டு கிராமத்தாரும் காவேரிப்பாக்கம் ஏரி தண்ணீரை இறைத்துக் கொள்ளும் மாமூல் இல்லை. இந்த முறையில் ஏரியின்

வடிவு கொடுக்குங் காலத்தில் உள்வாய் நிலங்களுக்கு ஏற்றம் வைத்து இறைப்பது மாமூல்.

18. மேல்படி ஏரி நீர் பிடிக்கிற சரகத்து கிராம முறை, காவேரிப்பாக்கம் ஏரி, பில்லேரி, சுமைதாங்கி ஏரி, பொன்னப்பந்தாங்கல், இடையன் தாங்கல், பாகவெளி, முசிறி, வள்ளுவம்பாக்கம் இந்த கிராம நிலங்களில் ஏரிக்கு நீர் கட்டுவதற்கு ஏற்ப நிலங்கள் பிடிச்சுப் போறது. வடிவுக்காலத்தில் அந்தந்த கிராமத்தார் 17 கலப்படிக்கு ஏற்றம் வைத்து இறைத்தும் பின்பு கிணறுகள் எடுத்தும் நஞ்சை நீர் பயிர். தோட்டக்கால் நஞ்சை, புஞ்சை, சோளம், கம்பு வகையரா பயிர் செய்து கொள்ளுகிறது.

19. மேற்படி ஏரிக்கு பாலாற்றிலிருந்து தண்ணிகட்டி திருப்பும் ஆள் முறை 100. இவை. காவேரிப்பாக்கம் ஆள் 10. கொண்டாபுரம் ஆள் 12. அத்திப்பட்டு 4. விளாகம் 4. சிறுகரும்பூர் 4. இராளச்சேரி 4. பெருகரும்பூர் நண்டேரி 2. அபிராமச்சேரி 4. சேரி 15, மாகாணிப்பட்டு 6. தழுவணாம்பட்டு 2, துரைபெரும்பாக்கம் 6, சிரஸ்தவலாபுரம் 2. புதுப்பட்டு 4. கீழ்வீராணம் 1. பண்ணியூர் 1. கெங்காதாரநல்லூர் 1 ஆக கிராமம் 19க்கு ஆள் 100. இந்த முறையில் ஆற்றுக்காலுக்கு வேண்டிய ஆள் போய் தண்ணி கட்டி திருப்பி கலுங்கல் தளவரிசைக்கு, தண்ணி கட்டுகிற வரைக்கும் மேலெழுதிய கிராமத்துக்கு ஆட்கள் இருந்து வேலை செய்ய வேண்டும். கலுங்களுக்கு தண்ணி கட்டினால் வடமஹாணத்தாரில் கட்டளை, அபிராமச்சேரி, மாகாணிப்பட்டு, தழுவணாம்பட்டு, துரைபெரும்பாக்கம், சிரஸ்தவலாபுரம், கீழ் விராணம், பண்ணியூர், கெங்காதரநல்லூர், சேரி இந்த கிராமத்து ஆட்கள் கலங்கலண்டை வந்து அவரவர் சரகத்துக்கு அவரவர் கொத்தி, மெரிச்சு கொட்டகை போட்டுக் கொள்ளலாம். மற்ற கிராம ஆட்கள் ஆற்றிலே தண்ணி வரும் வரை கொண்டம் கட்டி கால்வாயிலே திருப்பிக் கொண்டிருக்கிறது. ஆற்றிலே தண்ணீர் வரும் காலத்தில் 25 ஆள் முதல் எத்தனை ஆள் ஆத்துக்கால்வாயிக்கு போன போதுக்கும், ஆலப்பாக்கத்தார் இரண்டு ஆள் இருந்து கொண்டு மேலே எழுதிய ஆட்களுக் முன் வைக்கலை கொழந்து பிடித்துக் கொண்டு காத்துக் கொண்டு அளவு பிடித்துக் கொண்டிருந்து இதில் மேல் வெள்ளம் வந்து சேர்கிற சமாச்சாரம் அப்போதைக்கப்போ அவ்விடத்தில் இருக்கும் ஆட்களுக்கும் சர்க்கார் மக்களுக்கும் (service guards) தெரிவிக்க வேண்டும். மேலும் அவர்கள் சாப்பிட வேண்டியிருந்தாலும், இரவு தூங்க வேண்டியிருந்தாலும் ஒருவர் மாறி ஒருவர் ஆஜராக வேண்டும். ஆற்றிலே நீர் வருகிறபோது மேலெழுதிய ஆட்களுடன் கூட ஒரு மிராசு செம்பிடவன் (head fisherman) ஆஜராக வேண்டும். மேலும் முக்கிய காரியங்களுக்கு தக்கதாய் அந்தந்த கிராமத்தாரைக் கொண்டு சர்க்கார் நடத்திக் கொள்வது.

20. ஏரிக்கு நீர் வருகிறபோது பெரியாத்துக்கால்வாய் சமாச்சாரம் ஒருவனும், 121/2 சமாச்சாரமும் (demorcated ayacut area), சிங்கக் கிணத்தளவு ஒருவனும், கச்சி நதிக்கால்வாய் சமாசாரம் ஒருவனும் இந்த முறைக்கு நீர் வந்ததும் வடியும் காலம் (depletion period) சொல்லிக் கொண்டு கலுங்கல் கொத்தி, மெரிச்சு இரவு பகல் ஒருவன் அவ்விடத்தில் ஆஜராயிருந்து அடிக்கடி பொம்மை மதகு (Paracheri Sluice) அளவு பிடித்து கலுங்கல் சாவடியண்டை இருக்கும் சர்க்கார் மனுஷியாளுக்கும். கிராமத்தாருக்கும் சொல்லிக் கொண்டு பின் மதகு மூடுவது. திறப்பதுமான காரியங்களில் மிராசு செம்படவன் கலந்து கொள்ள வேண்டும். சுமைதாங்கி, சடயேரி முழு கடையாகாமல் இப்போ இருக்கிற கரைக்கு அதிகமாக புது மண் போட்டு கட்டக்கூடாது (bund level should not be raised).

21. ஏரிக்கரையின் பேரில் குடிசை போட்டு காக்கிறது முறை. இதன்படி, கடவாசலருகில் நீர் கட்டியபின் இராமாபுரம் மதகருகில் அவ்வூர் தோட்டி,

தலையாரிகளும்; நரிமதகருகில் கொண்டாபுரம், அத்திப்பட்டு, விளாகம், சிருகரும்பூர் இந்த 4 கிராம தோட்டி, தலையாரிகளும்; கிங்கமதகருகில் காவேரிப்பாக்கம் தோட்டி, தலையாரிகளும்; கம்முகட்டிகளும்; கட்டளை சரகத்தில் கட்டளை தோட்டி, தலையாரிகளும்; சேரி கன்னிகா கோவிலருகில் மாகாணிப்பட்டு துரைப்பெரும்பாக்கம் தோட்டி, தலையாரிகளும்; மேட்டு மதகு மூலை திருப்பத்தருகில் சேரி தோட்டியும்; பண்ணியூர் மதகருகில் பண்ணியூர், கீழ் வீறாணம், கங்காதரநல்லூர் தோட்டி தலையாரிகளும் அவரவர் சரகத்தில் குடிசை போட்டுக் கொண்டு அங்கு ஆஜராகி இருக்க வேண்டும். அப்போது முரசு அடித்து அவரவர் சரகத்தில் நடக்கும் வித்தியாமானம் கலுங்கல் சாவடியில் இருக்கிற பேருக்கு தெரியப்படுத்த வேண்டும். கலங்கலருகில் சர்க்கார் சாவடி அருகிலே ஆஜராகி இருக்கும் மக்கள், மிராசு செம்பிடவன் 1. மாகாணிப்பட்டு, துரைப்பெரும்பாக்கம், சேரி, அபிராமச்சேரி தலையாரி மேல்படியூர் குடிமி மேல்படியூர் பெரியதனக்காரர், கணக்கப்பிள்ளை சர்க்கார் மனிதர் இவர்கள் ஆஜராக வேண்டும். அந்த இடத்தில் கரையின் கீழ் ஆலப்பாக்கத் தோட்டி குடிசை போட்டுக் கொண்டு எல்லா நேரத்திலும் கும்பிட்டி நெருப்பு (coal fire) வைத்துக்கொண்டு இருக்க வேண்டும். சர்க்கார் சாவடி அருகில் ஆஜராகும் தோட்டி தலையாரிகள் முதலான பறையர் இருப்பதற்காக கரையின் கீழே குடிசை போட்டுக் கொண்டு இருக்க வேண்டும். வேண்டிய தீவட்டி (Torch) (கொருங்கட்டை) சர்க்கார் தாலுக்காவிலுருந்து அழைத்துக் கொள்ள வேண்டும். சர்க்கார் மனிதர்கள் (Neerkatties) இரவிலும் தீவட்டி பிடித்துக் கொண்டு கரையில் அலைந்து கொண்டு குடிசைகளில் உள்ள தோட்டி தலையாரிகளை ஜாக்கிரதை படுத்திக்கொண்டு காரியங்களுக்கு ஏற்ப நடந்து கொள்ளுதல் வேண்டும்.

22. கலங்கல் கட்டுகிற மாமூல் தெற்கு முதல் சேரிகல் 30. கட்டளை 6. அபிராமச்சேரி 4. மாகாணிப்பட்டு 26. தழுவணாம்பட்டு கல் 9. துறைப்பெரும்பாக்கம் கல் 21. மூஞ்சிபாக்கம் 5. சிரஸ்தவலாபுரம் 9. பண்ணியூர் 5. கீழ்வீராணம் 5, கெங்காதரநல்லூர் 5, ஆக கிராமம் 11-க்கு கல் 125 இந்த படிக்கு அவரவர் கொத்தி மெரிச்சுக் கட்டுகிறது. கலுங்கல் எடுக்கிற போது வடக்கு முதல் எடுக்கிறது. கட்ட வேண்டிய போது 125 கல் கொத்தி மெரிச்சு நீர் மட்டத்துக்கு தடுத்து அதுக்கு மேல் அவரவர் அவரவரது பந்தோபஸ்த்தாய் கட்டி கொட்டகை போட்டுக் கொள்ளுகிறது. கலங்கல் முழுமையும் எடுத்துவிட்டால் அவரவர் அவரவர்களே தடுத்து பந்தோபஸ்து பண்ணிக் கொள்ள வேண்டும். கலங்கல் எடுத்துக்கட்டும் காலத்தில் கட்டுவதற்கு புளியமிலார் வேண்டிய அளவு மரங்களிலிருக்கிற கிராமத்திலிருந்து சர்க்காரால் அழைத்து போட்டு வைக்கப்படும்.

23. ஆற்றுக்கால்வாய் கட்டுவதற்கும். கலங்கல் கட்டுவதற்கும் ஒரு வேளை மடைகள் எடுத்துக்கொண்டால் அவைகளை கட்டுவதற்கும், ஏரிக்கரை வகையரு குடிசை முதலானது கட்டுவதற்கும் ஏரியின் கீழ் கிராமத்தில் அந்தந்த ஊர் வழக்கப்படிக்கு வைக்கல் கொடுத்துக் கொண்டு வருகிறது.

24. ஏரிக்கால்வாய் சொந்தமாக வெட்டுகிற மாமூல் முறை, நரி மதகு கால்வாய் பெருகரும்பூர் நண்டெரியார், காவேரிப்பாக்கம், கீழ்புரம், காஞ்சீபுரம் அத்துவரைக்கும் சொந்தமாக வெட்டிக் கொள்வது. இதற்கு மேல், அகரத்து மடைவரை காவேரிப்பாக்கத்தாரும், பெருகரும்பூர் நண்டெரியாரும் பேர் பாதியாய் வெட்டுகிறது. இதற்கு மேல் மதகு வரை கொண்டாபுரத்தாரும் அத்திப்பட்டு, விளாகம், சிறுகரும்பூராரும் பேர்பாதியாய் வெட்டுவது. சிங்கமதகு கால்வாய் பெருங்கால் காவேரிப்பாக்கம் பாப்பார் தக்கை மட்டுக்கும் இரளாச்சேரியார் சொந்தமாய் உப்புமேடு பங்கு மாமன் கால்வாய் வெட்டுகிறது. அதற்குமேல் காவேரிப்பாக்கத்தார் சொந்தமாய் மதகு வரை வெட்டுகிறது. கூஷ ஊர் கால்வாயில் பிள்ளை புதைக்கிற

மேட்டு வரைக்கும் இராளச்சேரியார் சொந்தமாய் வெட்டுகிறது. அதற்கு மேல் காவேரிப்பாக்கத்தார் சொந்தமாய் வெட்டுகிறது.

25. காவேரிப்பாக்கம் சிங்கமதகு கால்வாயிலே முறைபோடுகிற முறை. அறம்பு நாளிலே சுமார் 40 நாளைய தண்ணீர் இருக்கும் போது காவேரிப்பாக்கத்தார் முறை போட்டுக் கொள்ளும் பங்கு வயணம். உப்புமேடு பங்குமுறை 1. நடுபங்கு முறை 1. அரைப்பங்கு கடைக்கால் பங்குமுறை 1 பூந்தோட்டம் பங்குமுறை 1/2, கன்னியாகோவில் பங்குமுறை 1/2 ஆக பங்கு 5க்கு முறை 4 இந்த முறையில் பாயவிட்டுக் கொள்ளுகிறது. முறை போட்டால் இராளச்சேரிக்கி தண்ணீர் விடுவதில்லை.

26. மேற்படி ஏரிக்கு கரையில் வெட்டுமடை (cut open bund) வெட்ட வேண்டியதில்லை.

27. 16ம் காலத்தில் எழுதிய கச்சந்திக் கால்வாயில் வருகிற நீரில் வள்ளுவம்பாக்கம் சரகத்தில் பொன்னப்பன் தாங்கல் ஏரிக்கு தண்ணீர் போவதற்காக கட்டி இருக்கும் கொண்டம் 1. இதில் திரும்புற வெள்ளம் பொன்னப்பந்தாங்கல் ஏரிக்கு தண்ணி போறதை யாரும் தடங்கல் பண்ணக்கூடாது.

28. ஆத்துகால்வாய் கூன்மடையருகில் அத்திப்பட்டு, விளாகம். சிறுகரும்பூர், கொண்டாபுரம். இந்த 4 கிராமத்து தோட்டிகள் காக்கும் குடிசை 1. இதில் கூஷ கிராமத் தோட்டிகள் ஆஜராயிருந்து கால்வாய் நிலமைகளை சர்க்காருக்கு (கும்பினி) (company) தெரியப்படுத்துகிறது. இந்தப்படி எங்கள் இராசியில் எழுதிக் கொடுத்த மாமூல் நாமா.

தூசி - மாமண்டூர் ஏரி

1 தூசி - கம்பெனி சர்க்காருக்கு

(1) அனுமந்தபேட்டை கிராமத்திலிருக்கும் முக்கிய தண்ணீர் வரத்து கால்வாய் மூலம் (main water supply channel) தண்ணீர் கொடுக்கப்படவேண்டிய முறை பற்றி மனகே வேலாயுத முதலியார், இசாராதார் கணக்கு குருநாதபிள்ளை, மற்றும் அனுமந்தபேட்டை தாலுக்கா திருவத்தூர், துக்குடி, சித்தூர் மக்கள் எழுதிக் கொடுத்த மாமூல் நாமா, யாதெனில்:

(2) அனுமந்தபேட்டை மாமண்டூரை சேர்ந்த கிராமங்களுக்கு தண்ணீர் பாய பெரிய ஏரி 1ற்கு பாலாற்றிலிருந்தும், வெள்ளக்காலின் மூலமும் தண்ணீர் கிடைக்க வழி செய்யப்பட்டுள்ளது. தூசி கிராமத்திற்கு சொந்தமாக ஊருக்கு வடக்கே தென்வடலாக கரையிடப்பட்ட 5 மாதங்கள் தண்ணீர் கொடுக்கக்கூடிய சித்தேரி ஒன்று (1) உள்ளது. இதற்கு மழைக்காலத்தில் தூசி ஏரிக்கு வடக்கே தூசியின் எல்லை ஓரம் முதல், ஆங்கில கம்பெனி விளத்தோடு எல்லை ஓரம் வரை வெள்ளக்கால் (catchment) இருக்கிறது. மேலும் கொல்லை மேட்டு வெள்ளங்கள், காட்டு வெள்ளங்கள் எல்லாம் சேர்ந்து தண்ணீர் கொடுக்கிறது. மேலும் (தாலுக்கு மசுக்கூரிலே சேர்ந்து வாகை, அப்துல்லாபுரம் ஏரியில் அதிகமாக உள்ள வழிந்தோடும் (surplus) தண்ணீரும் வந்து சேருகிறது. அப்துல்லாபுரம் தண்ணி பிடிக்கும். அத்து (boundary) மேற்படி ஊர் சித்தேரி எதிர்வாயில் காசி ராமேஸ்வர பாட்டையில் அரைமட்டு தண்ணீர் பிடிக்கிறது.

(3) இசா (a share in joint property right) பெரிய ஏரியில் தூசி வகையறா கிராமங்களுக்கு மேல்படி ஏரிக்கரையின் மலைக்கு வடக்கே பலகை போடுகிற மேட்டு பெரிய மதகு ஒன்றும். மலைக்கு தெற்கே நாழிகை வாசல் (time bound)

மரம் போடுகின்ற பள்ளமதகு ஒன்றும் ஆக 2 உள்ளன. பள்ள மதகு மூடி மேட்டு மதகு திறந்து தண்ணீர் பாய்கின்ற பொழுது, மேட்டுமதகிற்குள் 3 அடிக்கு 11/2 அடி என்ற கணக்கில் தண்ணீர் எடுத்து பின் தொட்டியின் கிழக்கே தண்ணீர் பெறப்படுகிறது. அப்பொழுது தூசி நிலத்தில் மலையோரங்காலுக்கு ரெண்டு கூடை தண்ணீர் இரவும் பகலுமாக செல்லும். அதன் கிழக்கு பலகடைப்பு கொண்டத்துக்கு கீழே இலவமரத்து மடையில் தண்ணீர் விடுவதில்லை. தை மாதம் பிறந்தால் எறும்புக்கொல்லை மானாவாரி கல்லேரி மடுவுக்கு இரண்டடி அகலம் நான்கு விரைக்கடை உயர்த்தி தண்ணீர் விடுகிறது. அதற்கு கிழக்கே சம்போடை காலுக்கு ஒரு மெரி தண்ணீர் விடுகிறது. நத்தக் கொல்லை தக்கை கால்வாய்க்கு இரண்டு கூடை தண்ணீர் (a unit of flow measurement) பாய்கிறது. அதற்கு கிழக்கே போகும் தண்ணீரில் பள்ள மதகு கால்வாயில் நாலு கூடை தண்ணீர் எடுத்து கரை ஓரம் காலுக்கும் ஒட்டி பத்துகாலுக்கும் பாய்கிறது. அதற்கு கிழக்கே சட்டி குட்டையில் தூசி கால்வாயும், பல்லாவரத்தான் கால்வாயும் பிரிகிறது. அடையாளக்கல்லுக்கு வடபுரம் பல்லாபுரத்தான் காலுக்கு தென்புரம் தூசிக் காலின் வடபுரம் பல்லாவரத்தான் காலில் தண்ணீர் பாதியாய் போகிறது. அதன் கிழக்கே பல்லாவரத்தான் காலில் பாசனம். வாகை எல்யையில் வாகை ஏரிகீழ் பல்லாவரத்தான் காலுக்கு மேல்புரம் வாகை ஏரி. பங்காறு தாங்கள் இவைகள் உலர்ந்து போன பின் வாகையார் இறைத்துக் கொள்ளும் தொட்டி 3. அதன் கிழக்கே பேட்டை கிணத்தடி தொட்டி 3, ஆக தொட்டி 6ம், வாகை ஏரி பங்காறுத்தாங்கள் உலர்ந்து போனால் தொட்டி மாற்றி தொட்டி இரண்டு தொட்டி இறைத்துக்கொள்கிறது. அதன் கிழக்கே தூசி நிலத்தில் சென்னாங்குட்டைக்கு 2 தொட்டி. 11/4 காணிக்கு இறைத்து கொள்ளலாம்.

அதற்கு எதிர் தூசி நிலத்துக்கு காணி 1/2க்கு பல்லாபுரத்தான் காலில் தக்கை வைத்து (wooden obstruction) பாய விடுகிறது. அதன் கிழக்கே குயவன் புழக்கடை கழணிக்கு சித்தேரி உலர்ந்து போனால் 4 தொட்டி இறைத்துக் கொள்கிறது. அதன் கிழக்கே பிரம்ம வாஸ்து புழக்கடைகளுக்கும் இறைத்துக் கொள்ளலாம். அதன் கிழக்கே பள்ளி தெருகாலில் பாச்சல் கொண்ட கழணி ஆறு காணிக்கு பல்லாவரத்தான் கால்வாயை தேக்கி பாயவிட்டு எடுத்து விடுகிறது. தெருக்காலுக்கு பல்லாவரத்தான் காலில் அடையாளக் கல்லுக்கு மேலே பல்லாவரத்தானுக்கு தண்ணி போறது. தெருக்காலிலே (Road channel) 4 தொட்டி தண்ணி போறது. குழியன் கன்னி காலுக்கு 1 மேரி தண்ணி விடுகிறது. அதன் கிழக்கே கருமான் புழக்கடை கழனிகளுக்கு 2 தொட்டி தண்ணி இராபகலாய் போகும். புழக்கடை திருத்து கன்னி காலுக்கு 1 தொட்டி தண்ணி விடுகிறது. கன்னி மூலைகாணி 1க்கு பல்லாவரத்தான் காலில் கொண்டம் கட்டி பாயவிட்டு தண்ணி விலக்கி விடுகிறது. திருத்து கன்னி காலுக்கு 1 தொட்டி தண்ணி பாயவிட்டு விலக்கி விடுகிறது. அதன் கிழக்கு பெரியபுலம் கன்னி காலுக்கு ஒரு தொட்டி தண்ணி விட்டு விலக்கி விடுகிறது. அதன் பின் தூசி எல்லை தாண்டி குரங்கணில் முட்டம் எல்லையில் போகும் தண்ணீரில் அறம்பு காலத்துக்கு குரங்கணில் முட்டத்தார் நாளில் ஒன்றும் (1/4) பல்லாவரத்தார் பாதியும். கணிகிலுப்பையார் 1/4 பங்குமாக பாயவிட்டுக் கொள்ள வேண்டும்.

தூசி பெருங்கால் சட்டிக்குட்டைப் பிரிவு பல்லாவரத்தான் பிரிவு காலுக்கு கிழக்கு பாச்சல் முறை. துக்கழிகுட்டை கண்ணிக்காலுக்கு 1 மேரித் தண்ணி விடுகிறது. அதன் கிழக்கே வாகை எல்லையில் 10 கன்னிக் காலுக்கு 10 சப்பாற தண்ணி (Local Measurement) பாயவிட்டு தண்ணி விலக்கி போடுகிறது. அதன் கிழக்கு தூசி எல்லையில் வடக்கு பட்டரை கன்னிக்கால் நாலுக்கு 4 சப்பாறதண்ணி விடுகிறது. சாலை வழி கண்ணாறுக்கு 2 தொட்டி தண்ணி விடுகிறது. அதன் கிழக்கே குளக்கீழ் கண்ணாறு 2 தொட்டி தண்ணி விடுகிறது. அதன் கிழக்கு

குளக்கீழ் கண்ணாறு 3க்கு 3 தொட்டி தண்ணி விடுகிறது. அதன் கிழக்கே வாணியன் கழனி கண்ணாறுக்கு ஒரு தொட்டி தண்ணி விடுகிறது. அதன் கிழக்கே அரசடி கண்ணாறுக்கு 2 தொட்டி தண்ணி விடுகிறது. அதன் கிழக்கே சேவிறான் கண்ணாறு 2க்கு 2 தொட்டி தண்ணி விடுகிறது. அதன் கிழக்கே சந்திகரைக்கு பெருங்காலியேலிருந்து பாயுறது. அதன் கிழக்கே குறை தண்ணீர் குறங்கனில் முட்டத்தார் பெருங்காலுக்கு கிழக்கே பில்லேரிமேடு கண்ணிக்கால் 1க்கு 1 தொட்டி தண்ணி விடுகிறது. அதன் கிழக்கே குறங்கணில் முட்டத்தார் காவல்மாணியம் கண்ணாத்தில் தோட்டி மாணியம். தட்டான் மாணியம், ஓச்சளிமான்யம் பாய்ந்து குறை தண்ணீர் குறங்கனில் முட்டத்தாருக்கு பாயறது. பெருங்காலுக்கு தெற்கு காக்குறுட்டிமேடு கன்னிகாலுக்கு ஒரு தொட்டி தண்ணி விடுகிறது. அதன் கிழக்கே உதிரிபட்டி கண்ணாறுக்கு 1 தொட்டி தண்ணி விடுகிறது. அதன் கிழக்கே குறங்கணில் முட்டத்தாருக்கு பாயறது.

பள்ளமதகு பாச்சல்: பள்ளமதகு திறந்தால் பின் தொட்டிக்கு 2 மார் (two times) கீழ்புறம் மடுவிலே கீழ்குப்பத்தாருக்கு 11/2 அடி அகலமும் 6 விரக்கடை உயர்த்தியும் மடுவிலே விடுகிறது. அதன் கிழக்கு ஆச்சாங்கொல்லை நூக்கை கல்லேரி கன்னிக்காலுக்கு 2 தொட்டி தண்ணி இராபகலாய் விடுகிறது. அதன் வடபுறம் மானாவாரி கண்ணாறு 2க்கு 2 மலையோரம் பாச்சலுக்கு மேட்டு மதகு உலர்ந்து போனால் அந்த பயிருக்கு தொட்டி வைத்து இறைத்துக் கொள்ளலாம். அதன் கிழக்கே கோவிந்தன் கொண்டத்தில் எறும்பு கொல்லைக்கும் மடுவுக்கும் 2 தொட்டி தண்ணி விடுகிறது. இதன் கிழக்கே பாறைபட்டி கன்னிக்கால் 2க்கு 2 தொட்டி தண்ணி விடுகிறது.

மேட்குமதகு உலர்ந்து போனால் அப்போதைக்கு இறைத்துக் கொள்ளலாம். பெருங்கால் வடபுரம் மேட்குமதகு தண்ணி பெருங்காலோடே வந்து சேர்த்து மேலெழுதியபடிக்கு தூசி எல்லைவரை பாய்ந்து மீதி தண்ணீர் குறங்கணில் முட்டத்தார் 4ல் ஒன்றும் பல்லாவரத்தார் பாதியும் கணிகிலுப்பையார் 4ல் 1 பங்கும் பாய விடுகிறது. கீழ்கன்னிகால்கள் எல்லாம் மாமூல்படிக்கு இராபகலாய் தண்ணி பாய்ச்சிக்கொள்வது. பள்ளக்கால் மாமூல்படி தூசி பாஞ்சு மீதி தண்ணியை பல்லாவரத்தார் 2 பங்கும் குரங்கணில் முட்டத்தார் 1 பங்கும் கணிகிலுப்பையார் 1 பங்குமாக பாய விட்டுக் கொள்வது.

ஷயூர் சொந்த சித்தேரி 1க்கு மேட்டுமதகு 1. பள்ளமதகு 1. ஆகமதகு 2. இந்த 2 மதகும் திறந்து வர்ற தண்ணி ஷ ஏரி கீழ் நஞ்சை மேட்டு கழணிகளுக்கு தோட்டி கம்முகட்டியை வைத்து 1 மடை பாச்சலாகி சமமாக பாயவிட்டுக் கொண்டு வருவது. மேல்படி ஊர் ஏரி கீழ்கழணிகளுக்கு பெரிய ஏரி தண்ணி வந்து பாயுறது. இந்த சித்தேரி 2 மதகுலேயும் 15 நாள் தண்ணீர் இருக்கும் போது கழணிகளுக்கு தண்ணீர் வராது. மேல்படி சித்தேரியின் கீழ் இட்டபயிர் உலர்ந்து போகாமல் இசா பெரிய ஏரித்தண்ணீர் கூஷயூர் பள்ளிக்கூடத்தருகுள்ள தெருக் கால்வாயில் வாற தண்ணியிலே படப்பைசரக அத்து முதல் பிடாரி கோவிலுக்கு கீழ்புரம் வரை 2 அல்லது 3 தொட்டி வைத்து இறைத்துக் கொள்ளுவது.

4. மேல்படி ஊரிலே வாழைத் தோட்டம், கரும்புத் தோட்டம் இல்லை. வெற்றிலைத் தோட்டத்துக்கு பெரிய ஏரி தண்ணி நஞ்சை கழணி ஒப்பந்தம் பாய்கிறது. கூஷ ஊர் சொந்த சித்தேரி தண்ணி அசல் கிராமங்களுக்கு போய் பாயுற மாமூல் இல்லை. இந்த முறையில் எங்கள் இராசியில் நாங்கள் எழுதிக் கொடுத்த தண்ணி மாமூல் நாமா.

வெங்கடகிருஷ்ணய்யங்கார்

கணக்கு குருநாத பிள்ளை.

2. குறங்கணில் முட்டம்: கம்பெனி சர்க்காருக்கு

1. செங்கல்வராய குருக்கள். கிருஷ்ணய்யர். வெங்கய்யா குடிகள். கணக்கு வெங்கடாசலப்பிள்ளை. குறங்கணில் முட்டம் தாலுக்கா திருவத்தூர் துக்குடி சித்தூர் மக்கள் எழுதிக் கொடுத்த தண்ணி மாமூல் நாமா யாதெனில் கூஷி ஊருக்கு தண்ணீர் இருக்கிற தாய்க்கால் தபசில் மேற்படி ஊருக்கு சொந்த ஏரி தாங்கல் பூஜ்ஜியம்.

2. மாமண்டூர் தூசி இசா பெரிய ஏரி தண்ணி விடுகிறது. தூசி மாமூல் நாமாவை எழுதியிருக்கிற முறைக்கு வாகை தூசிக்கு தண்ணீர் பாய்ந்து தூசிக்கு கிழக்கே பல்லாவரத்துக்கு வர்ரக்கால்வாயில் மேல் படியூர் வடமோடு நஞ்சை கழனிகளுக்கு பாயுற கன்னிகால்களாலே நஞ்சை கழனிகளுக்கு தண்ணீர் பாயுது. இதன் கிழக்கே பொலிகாரன் கோயில்சரகத்து கழணிகளிலே கொண்டமாய் தண்ணீர் ஏறாத கழனிக்கு பல்லாவரம் கால்வாயைத் தேக்கி தண்ணீர் பாயவிட்டு கொண்டம் எடுத்து விடுகிறது. தண்ணீர் அறம்பு நாளில் பல்லாவரம் காலில் 1/4 பங்கு தண்ணி பாயவிடுகிறது. தூசி பெருங்கால் பாச்சல் தூசி தோட்டி மானியத்துக்கு கிழக்கே மேல்படி யூருக்கு பாயுற தாய்க்கால் ஊருக்கே வடக்கே வழிக்கரை கால் 1. கீழ்க்காணித் தாங்கல் கால் 1. இதன் தெற்கே உதிரிபட்டி மோளகாறனத்துக்கால் 1. ஆக கால் 3. இந்தகாலுக்கு கூஷியூர் நஞ்சைக்கழணிகளுக்கு தண்ணீர் பாயுறது. தூசி நடுக்கழணி வொட்டு (illegible) பள்ளகாலிலே தண்ணி ஒதுக்கி வந்த நஞ்சை கழனிகளுக்கு பாயுற தண்ணி அறம்பு நாளையில் ஒட்டு பத்து காலில் வாற தண்ணி பல்லாவரத்துக்கு 2 பங்கு போக மேற்படியூருக்கு 1 பங்கு தண்ணீர் பாயும் தண்ணி அறம்பு நாளில் மேல்படியூர் எல்லையிலே பல்லாவரம் மரு கொண்டத்தில் கிழக்கத்தியாருக்கு மட்டில் குறைந்து, வராமல் நின்று போனால் அதன் பின் மேற்படி ஊருக்கு மடுவிலே 2 துலாம் வைத்து இறைத்துக் கொள்வது.

3. மேற்படி ஊருக்கு கரும்பு தோட்டம் வாழைத்தோட்டம் இல்லை. வெற்றிலை தோட்டத்துக்கு நஞ்சை கழணிகள் ஒப்பந்தம் தண்ணீர் பாயுது.

3. அரிகரபாக்கம்: கம்பெனி சர்க்காருக்கு

1. மார்க்க சகாயர். அரிகரபாக்கம் கதிம்குடி வெங்கடேச கவுண்டன். முத்து கவுண்டன். கணக்கு திருமலைப்பிள்ளை. அரிகரபாக்கம் தாலுகா திருவத்தூர் துக்குடி சித்தூர் மக்களாகிய நாங்கள் எழுதிக்கொடுத்த தண்ணீர் மாமுல் நாமா யாதெனில் ஷெயூருக்கு தண்ணியிருக்கிற தாய்க்கால் தபசில் ஷெயூருக்கு மேற்கு தெற்கு வடக்கு கரைபோட்டு இருக்கிற 5 மாதம் பாயுற ஏரி 1.

2. மேற்படி ஏரிக்கு தண்ணி வருவது. மழைகாலத்தில் ஏரி எதிர் வாயில் தாலுக்கா மதுக்கூரிலே நெல்லி கும்பினி சாகிரிலே (Land belongs to company) சேர்ந்த திருவடி இராமாபுரம் 2 கிராமத்து ஏரிகோடிகள் எடுத்து வரும் தண்ணீரும் கடகால் தண்ணீரும், வந்து நிரம்புகிறது. கூஷி ஏரி தண்ணி பிடிக்குற அத்து (Limit). கூஷி ஏரி எதிர்வாயில் காஞ்சீபுரம் பாட்டை (road) வழியில் முகந்தாப்போல் பிடிக்கும்.

3. மேற்படியூர் நஞ்சை கழனிகளுக்கு தண்ணி பாயுற முறை. மேற்படி ஏரிக்கு பள்ளமதகு 1. இதன் வடக்கு ஆனைகல் வைத்துள்ள மேட்டு மதகு 1. ஆக மதகு 2. ஏரியில் தண்ணீர் முழுவதுமாக நிரம்பி இருக்கும் பொழுது கூஷி ஏரியின் கீழ் பள்ள கழணிகள் எல்லாம் தாலுக்கா மசுக்கூரிலே சேர்ந்த மாமண்டூர் ஏரி தண்ணீர் முழுதும் பிடிக்கும். இன்னும் மேட்டு பட்டறை இறைப்பு கழனிகளுக்கு மேட்டு மதகு திறந்து வருகிற தண்ணீர் சிறிது கழனிகளுக்கு தொட்டி வைத்து இறைத்துக் கொள்வது. மாமண்டூர் ஏரி தண்ணீர் கூஷியூர் எல்லையில் பிடித்து இருக்கிறபடியினாலே அதை அருகிலுள்ள மேட்டு கழனிகளுக்கு தொட்டி வைத்து

இறைத்துக் கொள்ளலாம். இன்னும் மாமண்டூர் ஏரித்தண்ணீர் வடிவில் (depletion) கூஷியூர் ஏரி பள்ளமதகு திறந்து பள்ளக் கழனிகளுக்கு தோட்டி வறண் சமமாக பாய விடுகிறது. சிறிது கழனிகளுக்கு மாமண்டூர் ஏரி தண்ணீர் பிடித்து இருக்கிறதினாலே தொட்டி வைத்து இறைத்துக் கொள்வது. இன்னும் மேற்படியூருக்கு கிழக்கே மாமண்டூர் ஏரி எதிர்வாயில் கூஷி ஊர் எல்லை காரை குட்டைக்கு கிழக்கே எல்லையிருக்கிற படியினாலே எல்லை சரகத்தில் தண்ணியிருக்கிற மட்டுக்கும் ஒரு துலை (Shallow Pond) தொட்டியும், இருதுலை தொட்டியும் வைத்து இறைத்துக் கொள்வது.

4. மேற்படியூர் ஏரித்தண்ணீர் வேறெந்த கிராமத்திற்கும் பாய்கிற உரிமை இல்லை. கூஷி ஏரித்தாங்களுக்கு வரும் புதுப்பட்டு கடைகால் வெள்ளமும் கொல்ல மேட்டு வெள்ளமும் சேர்ந்து வருகிறபடியினாலே அந்த நீரை யாரொருவரும் தடுக்கக்கூடாது.

எங்களூர் ஏரி நரம்பினால் உள்வாயில் மாமூலமாக தண்ணீர் பிடிக்குற ஊருக்கு வடமேற்கு ஏரி முழுவதும் ரொம்பினால் உள்வாயில் எங்களூர் எல்லை கசிவு பட்டரை முத்து நாயக்கன். வெங்கட்ராம் நாயக்கன் பத்துகட்டு நஞ்சை (occupancy holding), மூன்று தளை கீழ்புறம் வறப்பு வரை தண்ணீர் பிடிக்கும்.

ஊரின் கிழக்கு வால்தாங்கல் முழுவதுமாய் நிரம்பினால் உள்வாயில் எங்களூர் எல்லை கிளர் செடி வரைக்கும் தண்ணி கட்டுகிறது. எங்களூர் எல்லை முழுவதுமாக ஏரி நீர் பிடிக்குறது மாமூல் இல்லை.

4. மாமண்டூர்

1816ம் ஆண்டு நவம்பர் 16ம் தேதி வெங்கடாச்சாரியார். அருள்பு வெங்கடசேசய்யங்கார். வெங்கய்யர். சின்ன அப்பனய்யங்கார். எம்பெருமாளய்யங்கார். குருவி வெங்கிடசாரி. கணக்கு ராமசாமி பிள்ளை மற்றும் ராமகிருஷ்ண பிள்ளை ஆகியோர் ராசியில் எழுதிக் கொடுத்த மாமண்டூர் ஏரியில் மாமண்டூருக்கு உட்பட்ட பகுதிக்கு தண்ணீர் விடப்படுவது பற்றிய மாமூல் நாமா.

மாமண்டூர் மற்றும் தூசிக்கு மேற்கு தெற்கே இருக்கின்ற பெரிய ஏரிக்கு கீழ்கண்ட முறையில் தண்ணீர் கிடைக்கின்றது. 1) வங்களத்தூர், முத்தவேடு, பிச்சவாடி ஊருக்கருகிலுள்ள பாலாற்றில் வெள்ளம் வருகின்ற பொழுது குறுக்கே அணையிட்டு தண்ணீரை திருப்பி கால்வாயின் மூலம் பெறப்படும் தண்ணீரும் 2) வெங்களத்தூர் ஏரி கலங்கல் தண்ணீரும். கழனிகளின் கழிசல் தண்ணீரும். திருப்பனங்காடு ஏரி கலங்கல் தண்ணீரும். உமையாள்புரம் கழனி கழிசல் தண்ணீரும். ஒழுகம்பாக்கத்தில் வந்து விழுந்து ஏரியில் சேர்கிறது 3) மேலும் வெள்ளக்குளம் கழனியில் வருகிற கழிவு தண்ணீரும். திரும்ப முறை வெள்ளமும், மாவந்தல் கொல்லை வெள்ளமும் சேர்ந்து ஏரிக்கு வருகிறது. 4) மேலும் பாலாறு பெருக்கெடுக்கும் பொழுது பெருமளவு தண்ணீர் இராஜாக் கால்வாய் மூலம் பெறுகிறது.

பாலாற்றில் வெள்ளம் வராதபொழுது மேற்படி ஏரிக்கு (இராஜாக்காலில்) கொண்டம் கட்டப்படும். இதன்படி, திருப்பணங்காடு கொண்டம் (1). வள்ளகுளம் கொண்டம் (1). ஆக இரண்டு கொண்டம் கட்டப்படும். இதன் மூலம் ஏரி தண்ணீர் பெரும். அரசங்குப்பம் மதகு (2) செட்டி தாங்கல் மதகு (1) கும்பினி சர்க்கார் ஒழுக்கம்பாக்க மதகு (1) வெள்ளகுளம் மதகு (1) ஆக 5 மதகுகளும் மேலும் ஏரிக்கு எதிர்வாயில் உள்ள சித்தாத்தூர் ஓடையின் ஹரிஹரபாக்கம் ஏரி. சித்தாத்தூர் சோழபுரம் பெருமலிமேடு செல்லப்பெருமழிமேடு. அழிஞ்சில்பட்டு ஆகிய இந்த கிராமங்களின் ஏரி கலங்கல் தண்ணீரும் மற்றும் இவ்வூர்களின் கழனி கழிசல்

தண்ணீரும் சித்தாத்தூர் ஓடையில் விழுந்து பெரிய ஏரியை அடைகிறது. இது மாமூல் முறை. இம்முறையில் தண்ணீர் வந்து ஏரி நிரம்பினால் 18 மாதம் தண்ணீர் பாயும்.

மாமண்டூர் ஏரியின் எதிர்வாயில் பிடிக்கிற அத்து, மேற்கு சித்தாத்தூர் பாட்டை அத்துக்கு பிடிப்பு, தெற்கு அழிஞ்சல்பட்டு, இளந்தோப்பு அத்து, செல்ல பெருமிலிமேடு ஊர் அத்து, பெருமிலிமேடு தாங்கல் கீழ் அத்து, சோழபுரம் வடவண்டை ஆலமரத்து அத்து, கணக்கம்பாக்கம் நமூண்டி ஊர் அத்து, ஹரிஹரப்பாக்கம் எல்லை அத்து, அய்யங்குளம் கண்ணப்பன்தாங்கல் வடக்கு அத்து இந்த சரகத்தில் தண்ணீர் பிடிக்கிறது. (மேலே குறிப்பிட்டுள்ள பகுதிகள் வரை ஏரியின் அளவு பரந்து உள்ளது)

மாமண்டூர் ஏரிக்கு வருகின்ற கால்வாயில் மேலே கூறப்பட்ட இரண்டு கொண்டமும் ஆடிமாதத்தில் திறக்கப்படுகிறது. அதில் தைமாதம் வரை அணை கட்டக்கூடாது. தை மாதத்திற்கு மேல் கொண்டம் கட்டிக் கொள்ளலாம். சித்திரையில் மழை பெய்து கால்வாயில் தண்ணீர் வந்தாலும். வெங்கலத்துர் கலங்கல் மற்றும் கழிசல் தண்ணீர் வந்தாலும் கொண்டம் கட்டக்கூடாது எடுத்துவிட வேண்டும். தண்ணீர் வரத்து நின்று போனால் கொண்டம் கட்டிக்கொள்ளலாம் இது மாமூல் முறை. மாமண்டூர் ஏரிக்கு மதகு முறையில் மேற்படி ஏரியில் தெற்கு கோடியில் பலகை போடுகின்ற மேட்டு பெரிய மதகு ஒன்று, நாழிகை வாசல் மரம் போடுகின்ற மதகு 1. ஆக மதகு 2. மூலையில் வடகோடி கலங்கலுக்கு பலகை போடுகின்ற பெரிய மதகு 1. இதில் தெற்கு மலைக்கு தென்பகுதியில் நாழிகை வாசல் மரம் போடுகிற மதகு 1. ஆக தூசி மதகு 2. பெரிய ஏரி 1க்கு மதகு 4.

ஏரி முழுவதுமாக நிரம்பி இருக்கும்போது மாமண்டூர் எல்லையில் தென்பகுதியில் பலகை போடுகிற மதகிலே தண்ணீர் திறந்து பின் பகுதியில் பின் தொட்டியில் வாயிற்படிக்கு மேலே பிள்ளையார் தும்பிக்கை தொட்டுவற அளவு தண்ணீர் திறந்து விடுவது. அப்பொழுது சோதியம்பாக்கம், அஞ்சுறான் கால்வாயில் கொண்டம் கட்டி பள்ளக்கால் பாசனத்திற்கு அந்த கொண்டத்தில, அகல அளவில் ஆறேகாலடி, 21/2 அடி உயரம், இரண்டு சாண்விட்டு குறை தண்ணீர் மேல்படி கால்வாயிலே கொண்டு போய் மாமண்டூர் எல்லையிலே காவல் மானியத்திற்கு பாய்கின்ற நாலு காலடி குட்டையில் உள்ள தொட்டி 3க்கு நீர் பாயுது. 2) அஞ்சுரான் கால்வாயிற்கு சோதியம்பாக்கம் கால்வாய் மூலம் பாயிற படியினாலே, அஞ்சுரான் கால்வாயில் நாலு பங்கு தண்ணீரும் சோதியம்பாக்கம் கால்வாயிக்கு 11/2 பங்கு தண்ணீரும் மாறாமையாகும்.

சோதியம்பாக்கம் கால்வாயில் போகிற தண்ணீரில் மேல்படி தாலுக்காவில் சேர்ந்த நரசமங்கலம் எல்லையில் தொட்டி வச்சி இறைக்கிற வயனம். வாணியம் பிறவிடை தொட்டி 3. ஊருக்கு மேலண்டை தொட்டி 2. அதன் வடபுற தொட்டி 1. களத்து மேட்டுக்கு கீழ்புறம் ஆலமரத்தண்டை தொட்டி 1. அதற்கு தெற்கே தென்மைரத்தண்டை தொட்டி 1 அம்மன் கோயிலுக்கு கிழக்கே தொட்டி 2. இதற்கு தெற்கே தோட்டி மான்யத் தொட்டி 1. ஒட்டத்தாங்கல் உள்வாயில் தொட்டி 1. வெங்கட்றாயன் சத்திரத்து தோப்புக்கு மேல்புற தொட்டி 1 ஆக தொட்டி 13க்கு ஏககாலத்தில் ஒரே முட்டாய் 13 தொட்டியும் இறைக்கக்கூடாது. நாள் 1க்கு நாழிகை 60க்கு ஒன்று விட்டு ஒன்று இறைக்கும் முறையில் தினமும் விடாமல் இறைத்துக் கொள்ளலாம். இப்பால் தாலுக்கு மசுகூர் பூண்டி எல்லையில் மேற்படியூர் கழனிகளுக்கு (நிலங்களுக்கு) கால்வாயின் மேல் பனந்தக்கை வைத்து தண்டுக்கரை மூலமாய் ஒரு கூடை தண்ணி போறது. இதன் மேற்கே மேல்படியூர் எல்லையில் ஈச்சன்குட்டை கழனியன்டை பனந்தக்கை வைத்து ஒரு கூடை தண்ணி போறது. இப்பால் தாலுக்கா மஸ்கூர் வாழவந்தல் எல்லையில் மேற்படியூர் கழனிக்கு

மூலமடையன்டை அஞ்சூரான் கால்வாயிக்கு மேலே பனந்தக்கை வச்சு கால் மேலே கால் வச்சு ஒரு மேரி அகலமும் ஏழு விரக்கடை உயர்த்தியும் தண்ணி விடுகிறது. அப்பால் தாலுக்கா மசுக்கூர் சோதியம்பாக்கம் எல்லையில் தண்ணி போயி கால்வாயண்டை சோதியம்பாக்கம் கழனிகளுக்கு பாயறது. மேல்படியூருக்கு 3 அடி அகலமும் ஆறு விரக்கடை உயர்த்தியும் தண்ணி பாயறது. தாலுக்கா மஸ்கூர் பகவந்தபுரத்துக்கு இரண்டு அடி அகலமும் ஆறு விரக்கடை உயர்த்தியும் தண்ணி போகும். இது சோதியம்பாக்கம் கால்வாயின் முறையாகும்.

அஞ்சூரான் கால்வாயில் 4 பங்கு தண்ணியில் நரசமங்கலம் எல்லையில் மேல்படியூருக்கு பாய்ச்சல் வயனம். வண்ணான் துறை மடை 1. அதன் தெற்கு மடை 1. அதன் கிழக்கு மடை 1. அதன் வடக்கு மடை 1. ஆக மடை 4. பிடாரி கோவிலண்டை மடை 1. அதன் கீழண்டை மடை 1. அதற்கு கிழக்கு மடை 1 அதன் கிழக்கு வழியண்டை மடை 1. அதன் கீழண்டை மடை 1. பக்கிரி மான்யத்து மடை 1. இதன் தெற்கு மடை 1. ஆக மடை 11. இந்த மடை 11ம் பாசன வேலையில் எடுத்து பாய விடலாம். மற்ற வேலைகளில் மூடி விட வேண்டும். கோமுட்டி கழனியன்டை அஞ்சுரான் கால்வாய் குறுக்கே கட்டி மேல்படி ஊருக்கு 2 காணிக்கு பாயவிட்டு எடுத்து விடவேண்டும். அந்த இடத்தில் தனிக்கால்வாய் 1 வெங்கட்ராயன் தோட்டம் மேலண்டைக் கருமா மான்யத்து மடை மேல்படியூருக்கும் மாமண்டுருக்கும் பாயற மதகில் உள்ளது. அகல அடி 1 உயரத்தில் விரக்கடை 6 அளவு தண்ணி விடுகிறது. மேல்படி நரசமங்கல எல்லையில் சின்ன மன்னாறடி கணக்கு மானியம் கன்னிக்கால் 1. இந்த முறையில் தண்ணீர் நரசமங்கலம் கழனிகளுக்கு பாய்கிறது.

இதன் கிழக்கே மாமண்டுர். பூண்டி, வாழவந்தல் ஆகிய 3 கிராமத்திற்கு தண்ணீர் பாய்கிற கருமான் மான்யம் 1க்கு மட்டு அகல அடி 3 உயரத்தில் விரக்கடை 9 இந்த முறையில் தண்ணீர் மேல்படி 3 கிராமத்திற்கும் போகிற வாய்க்காலிலே மாமண்டுர் பூண்டிக்கு போகிற வாய்க்கால் 1க்கு மட்டு அகல அடி 2 உயரம் விரக்கடை 6 இந்த முறையில் கால்வாய் தண்ணீர் கொண்டு போய் மாமண்டுர் கால்வாயிலே போற தண்ணியோடு மேல்படியூர் கழனிகளுக்குப் பாய்கிறது. பூண்டி கால்வாயிலே போற தண்ணீர் 2 கிராமத்திற்கும் சமமாகப் போகும். பூண்டிக்கு காணி 3/8. மாமண்டுர் வையாளி கால்வாயை கட்டி பாயவிட்டு எடுத்துவிட வேண்டும். மேல் எழுதின மாமண்டுர் பூண்டி, வாழவந்தல் ஆகிய மூன்று கிராமத்துக்கும் போற கால்வாயிலே பூண்டி எல்லையில் நஞ்சை காணி 1க்கு கொண்டம் கட்டி பாயவிட்டு பின் எடுத்து விட வேண்டும். மேல்படி கால்வாயில் நஞ்சை காணி 1/2க்கு அஞ்சுரான் கால்வாயும் மேல்படி கால்வாயிலே குறுக்கே கட்டி பாயவிட்டு எடுத்துவிடுகிறது. மேலெலுதின 3 கிராமத்துக்கும் செல்லும் கால்வாயிலே கொண்டத்துக்கு கிழக்கே பூண்டிக்கு பாயும் கன்னிக்கால்வாய் 1. இதன் கிழக்கே கால்வாயை குறுக்கே கட்டி பாயுற காணி 3/8 பாயவிட்டு எடுத்துவிட வேண்டும். இதன் கிழக்கே கன்னிகால்வாய் 1. அதிலே கொண்டாம் பாயுற காணி 1/2-க்கு மேல்படி கால்வாயை குறுக்கே கட்டி மட்டுக்காலும் குறுக்கே கட்டி பாயவிட்டு எடுத்து விடுகிறது. இதன் கிழக்கே தண்டுக்கரை கன்னிக்கால்வாய் 1. இதன் கிழக்கே கன்னிக்கால் 1. அதற்கும் கிழக்கே சேரியண்டை கன்னிக்கால்வாய் 1. ஆக பூண்டிக்கு பாயுற கன்னிகால்வாய் 5க்கு தேவையானபோது பாசனத்திற்கு ஏற்ப தண்ணீர் பாய்ச்சிய பின் மற்ற வேளைகளில் மூடி விடவேண்டும்.

அதன் கிழக்கே மாமண்டுர், சோதியம்பாக்கம் கழனிக்கும் வாழவந்தலுக்கும் பாயுற பிரிவு வயனம். மாமண்டுர் சித்தாலபாக்கம் கழனிக்கு போற தண்ணி கால்வாயிலே போகும். வாழவந்தல் கால்வாயிலே மாமண்டுர் கிரிபுதாசு மான்ய புலத்துக்கு பாய்கிற கன்னிக்கால்வாய் 1. இதன் கிழக்கே வாழவந்தலுக்கு போற

கால்வாயிலே தண்ணீர் போகும். இந்தக் கால்வாயில் மேலே எழுதின கிராமம் 3க்கும் பாயுற முறை. மேலே எழுதின அஞ்சுறான் கால்வாயிலே வாழவந்தல் கிராமம் 1க்குப் பாயும் கால்வாயில் தாலுக்கா மசுக்கூர் வாழவந்தல் எல்லையில் பூந்தோட்டத்துக்கு தென்பகுதியில் மேட்டுக்காரன் ஓடைக்குப் பாய்கிற கால்வாய் 1க்கு மட்டு அகலம் விரக்கடை 6. உயரம் விரக்கடை 4. இந்த முறையில் தண்ணீர் விடப்படுகிறது. இதற்கு தெற்கே மேல்படி எல்லையில் கால்வாயில் அண்டை வச்சு பாயவிட்டு பின் எடுத்துவிட வேண்டும். அதன் கிழக்கே பூந்தோட்டத்துக்கு கிழக்கே மூலமடை கால்வாய் 1க்கு ஆள் காலடி அகலமடி 1 உயரம் விரக்கடை 6 அளவில் தண்ணீர் விடப்படும். வாழவந்தல் எல்லையில் மேல்படியூருக்கு பாய்கிற கால்வாய் 2. இதன் கிழக்கே போகிற கால்வாயில் கும்பினி சர்க்கார் கல்பாக்கம் எல்லையில் மேல்படி கிராமத்துக்கு பாய்கிற கால்வாய் 1க்கு மட்டு அகலஅடி 1 உயரம் விரக்கடை 6 அளவில் தண்ணீர் பாயும். இதன் கிழக்கே கால்வாய் 1க்கு அகலம் 6 விரக்கடை, உயரம் 6 விரக்கடை அளவில் தண்ணீர் விடப்படுகிறது. ஆக கல்பாக்கத்துக்கு பாய்கிற கால்வாய் 2 இதன் கிழக்கே கும்பினி சர்க்கார் மண்ணூர் எல்லையில் மேல்படியூருக்கு பாய்கிற கால்வாய் 1க்கு மட்டு ஆள் கால்மேலே கால்வைச்சு ஒரு மெரி அகலமும். உயரம் விரக்கடை 6 அளவிலும் தண்ணீர் விடப்படுகிறது. இதன் கிழக்கே தாலுக்கா மசுக்கூர் நாயக்கன் பாளையத்துக்கு மேற்படி எல்லையில் பாய்கிற கால்வாய் 1க்கு மட்டு அகல அடி 1, உயரம் விரக்கடை 6. இந்த வகையில் தண்ணீர் விடப்படுகிறது. இதன் கிழக்கே தாலுக்கா மசுக்கூர் கிரிசாபுரத்துக்கு மேற்படியூர் எல்லையில் கன்னிக்கால்வாய் 1.

இதன் கிழக்கே கால்வாய் 1க்கு மட்டு வயனம். அகலம் ஆள் கால்மேலே கால் வைத்து ஒரு மெரி அகலமும், உயரம் விரக்கடை 6 அளவிலும் தண்ணீர் பாயும். ஆக கால்வாய் 2க்கு ஒன்று விட்டு ஒன்று பாய விடுகிறது. இதன் கிழக்கே கால்வாய் 1க்கு அகலம் ஆள் காலடி 1 உயரம் விரக்கடை 6 இந்த அளவில் தண்ணீர் பாயறது. ஆக கிரிசாபுரத்து எல்லையில் மேல்படி ஊருக்கு பாயற கால்வாய் 3க்கு பாய்கின்ற கால்வாய் 2. இதன் கிழக்கே மேல்படியூர் எல்லையில் தாலுக்கு மஸ்கூர் மேனல்லூருக்கும் கும்பினி சர்கார் பூனைத்தாங்கலுக்கும் இக்கால்வாயிலே மரம் போட்டு மேனல்லூருக்கு 1 அடி அகலம், உயரம் விரக்கடை 6 அளவிற்கு தண்ணீர் விடப்படுகிறது. பூனைத்தாங்களுக்கு ஒரு அடி அகலமும், உயரம் விரக்கடை 6 அளவில் 2 கிராமத்திற்கும் தண்ணீர் போகிறது. மேனல்லூர் பிரிவு கால்வாயிலே கிரிசாபுரத்துக்கு பாயுற காணி 3/4க்கு கிரிசாபுரம் ஒரு அடி கால்வாயை தண்ணீரைக் கட்டி மேனல்லூர் பூனைத்தாங்கல் கால்வாயிலே கொண்டு வந்து அதில் 3ல் ஒருபங்கு தண்ணீர் விட்டு மீதி 2 பங்கு தண்ணீரை மேனல்லூர் கால்வாயிலே கொண்டு வந்து மேல்படியூருக்கு பாதி தண்ணீர் விட்டு மீதி தண்ணீர் கிரிசாபுரம் கால்வாயிலே கொண்டு வந்து கிரிசாபுரம் காணி 11/2க்கும் பாயவிட்டு அந்த கன்னி கால்வாயை அடைத்து மேல்படியூர் ஒரு அடி மடைக்கால்வாயை எடுத்து விடுகிறது.

ஏரி முழுதுமாய் தண்ணீர் இருக்கும் பொழுது மேல் எழுதின முறையில் தண்ணீர் பாய வேண்டும். தண்ணீர் பற்றாக்குறை நாளையில் மேல்படி கால்வாயிலே தண்ணீர் பாய்கிற மாமூல் பின் வருமாறு. நரசமங்கலத்துக்கு பாய்கிற மடை 11க்கு தண்ணீருக்கு ஏற்ப தண்ணிவிட்டு மீதியை கால்வாயிலே கொண்டுபோய் மாமண்டூர், பூண்டி, வாழவந்தலுக்குப் பாய்கிற கிராமத்துக்குப் போற கருமான் மான்ய மடைக்கு தண்ணீருக்கு ஏற்ப மாறாமல் தண்ணிவிட்டு மீதி தண்ணியை கால்வாயிலே கொண்டுபோய் வாழவந்தல் கால்வாய்களுக்கும் தண்ணீருக்கேற்ப மாறாமல் தண்ணீவிட்டு மற்ற தண்ணீரை கால்வாயிலே கொண்டுபோய் வாழவந்தல் எல்லையில் மூலமடை காணி 1/4க்கு மேல்படி கால்வாய் மூலம் கட்டி பாயவிட்டு எடுத்துவிடுகிறது. இதன் கிழக்கே போற கால்வாய் கும்பினி சர்க்கார் கல்பாக்கம்

எல்லையில் கால்வாயை குறுக்கே கட்டி முழுவதுமாக 40 நாழிகை (16 மணி) பாயவிட்டு பின் எடுத்துவிட வேண்டும். அதன் கீழ் கால்வாயில் கொண்டு போற தண்ணீரை கும்பினி சர்க்கார் மன்னூர் எல்லையில் கால்வாயை குறுக்கே கட்டி, கல்பாக்கம் கால்வாயையும் குறுக்கே கட்டி இருபது நாழிகை (8 மணி) பாயவிட்டு கால்வாயை எடுத்துவிட்டு மண்ணூரான் கால்வாயையும் குறுக்கே கட்டி தண்ணீர் ஒதுக்கி கொண்டு போய் நாயக்கன் பாளையத்தான் எல்லையில் கால்வாயை குறுக்கே கட்டி முழுமை தண்ணியும் மேல்படி தாலுக்கா மசுக்கூர் நாயக்கம் பாளையத்துக்கு முப்பது நாழிகை பாயவிட்டு கால்வாயை எடுத்துவிட்டு மேற்படியூர் கால்வாயையும் கட்டி தண்ணி ஒதுக்கிக் கொண்டு போய் கிரிசாபுரம் எல்லையில் கால்வாயை குறுக்கே கட்டி முழு தண்ணீரையும் மேற்படியூருக்கு 60 நாழிகை (1 நாள்) பாயவிட்டு கால்வாயை எடுத்துவிட்டு மேல்படியூர் கால்வாயையும் கட்டி தண்ணீரை ஒதுக்கி கொண்டுபோய் தாலுக்கா மசுக்கூர் மேனல்லூர் எல்லையில் கால்வாயை குறுக்கே கட்டி மேற்படியூர் கால்வாயில் முழு தண்ணீரும் 30 நாழிகை பாயவிட்டு கால்வாயை எடுத்துவிட்டு மேற்படி ஊர்க்கால்வாயை குறுக்கே கட்டி தண்ணீரை ஒதுக்கி கொண்டு போய் பூனைத்தாங்கல் எல்லையில் 30 நாழிகை பாய விடுகிறது. தண்ணீர் அறம்பு (பற்றாக்குறை) நாளையில் இந்த அஞ்சூரான் கால்வாயிலே இந்த கிராமங்களுக்கு இந்த முறை பிரகாரம் இந்த கிராமத்தார்கள் அருகிலிருந்து தண்ணீர் பாய்ச்ச வேண்டும் இது மாமூல் முறை.

பள்ளக்காலுக்கு தண்ணீர் பற்றாக்குறையானால் கிழக்கு பகுதியில் பிடுங்கி பள்ளத்திலே விட வேண்டும். தண்ணீர் அறம்பு நாளையில் மேட்டுக்காலுக்கு தண்ணீர் ஏறாமல் போனால் பின்தொட்டியில் கீழ் கண்ணை பிடுங்கி நரசமங்கலம், மாமண்டூர், பூண்டி இந்த மூணு கிராமத்திற்கும் பாயரத்துக்காக பள்ளக்கால்வாயை கொண்டம் கட்டி தண்ணீரை திருப்பி மாமண்டூர், வாழவந்தல், நாயக்கன்பாளையம் வகையராக்களுக்கு தண்ணி பள்ளக்காலுக்கு விட்டு மீதி தண்ணீரை கால்வாய் புறத்தில் மாமண்டூர் எல்லையில் காவல் மானியத்துக்கு பாயுற மடையை பாய விடுகிற போது பாயவிட்டு மற்ற தண்ணீரை இறைக்கிற தொட்டி 2. இதன் கிழக்கே நரசமங்கலம் எல்லையில் மேற்படி ஊர் காவல் மானிய தோப்பை மடை 1 பள்ளக் கோனேரி மடை 1. அதன் கிழக்கு மடை 1. அதற்கும் கிழக்கு மடை 1. அதற்கும் கிழக்கே கன்னிக் கால்வாய் 1. அதன் கிழக்கே கன்னிக் கால்வாய் 1. அதன் கிழக்கே கன்னிக் கால்வாய் 1. அதன் கிழக்கே கன்னிக் கால்வாய் 1. அதன் கிழக்கே காவல் மானிய கன்னிக் கால்வாய் 1. அதன் கிழக்கே கூத்தாடி மானிய கன்னிக் கால்வாய் 1 ஆக நரசமங்கலம் எல்லையில் கூடி ஊர் கன்னிக்கால்வாய் 6. மடை 4. இந்த மடை நாலும் கன்னிக் கால்வாய் ஆறும் பாய விடுகிற போது பாயவிட்டு மற்ற வேளையில் கட்டி விட வேண்டும். மேல்படி நரசமங்கலம் எல்லையில் தொட்டி வச்சி எறைக்கறது முறை. பிடாரி கோவிலண்டை தொட்டி 1 இதை தேவைக்கேற்ப இறைக்கிறது. இதன் கிழக்கே பக்கிரி மானியத்தண்டை தொட்டி வச்சி மேலெழுதிய தொட்டி இறைக்கிறதை நிறுத்தி இரண்டாம் தொட்டியை இராபகல் 60 நாழிகையும் எறைச்சிக் கொள்கிறது. இதன் கிழக்கே மன்னாரடியில் ஒரு தொட்டி வச்சு மேல் இறைக்கிற 2 தொட்டியையும் நிறுத்தி இந்த தொட்டியை இறைத்துக் கொள்ள வேண்டும். ஆக தொட்டி 3கு. ஒன்று விட்டு ஒன்று இறைக்கிற தொட்டி 2, இடைவிடாமல் இறைத்து கொள்கிறது 1. இதன் கிழக்கே நரசமங்கலம் மாமண்டூருக்குப் பாயற கருமான் மானியம் கால்வாய் 1க்கு கீழேகல் வைத்து ஒரு அடி அகலமும். நாலு விரக்கடை உயரமும் உள்ள அளவில் தண்ணீர் விடுகிறது. இதன் கிழக்கே பூண்டி கால்வாயில் 3ல் ஒரு பங்கு தண்ணீரை விட்டு மாமண்டூர் கால்வாயிலே 2 பங்கு தண்ணீர் விடுகிறது. மேற்படி கால்வாயில் அறம்பு சமயத்தில் நரசமங்கலம் கன்னிக்கால்வாய் 6, மடை 4, மாமண்டூர் மடை 5 இவைகளுக்கு தண்ணீருக்கு ஏற்ப விட்டு குறை தண்ணீரையும், பூண்டிக் கால்வாயையும் கட்டி

நரசமங்கலத்துக்கு பாய்கிற நாள் 1. மேற்படி ஊருக்கு பாய்கின்ற கால்வாயையும் கட்டி மாமண்டூருக்கு பாய்கிற நாள் 2. இந்த 2 கிராமத்துக்கும் (நரசமங்கலம்+மாமண்டூர்) இந்த முறையில் தண்ணீர் பாய்வது மாமூல். இது பெரிய வாய்க்கால் பாசன முறை (வயனம்=method).

நரசமங்கலம், பூண்டி, மாமண்டூர் வாழவந்தல் இந்த 4 கிராமத்துக்குப் பாய்கிற மாமண்டூர் ஊர் ஓரமாய் வருகிற கால்வாய் பாசன வயணம். மேற்படி மேலெழுதின மேட்டு மதகு கால்வாயிலே கொண்டங்கட்டி பள்ளமதகுக்கு தண்ணி விடுகிற அகலம் அடி 21/2, உயரம் சாண் 2 (18 inches) இந்த முறையில் தண்ணீர் விடப்படும். அந்த தண்ணீரை பள்ள மதகின் பின் தொட்டியில் 2 பங்கு தண்ணீர் மாறாமல் விட்டு மேலெழுதின கால்வாயிலே ஒரு பங்கு தண்ணீர் விடுகிறது. இது பாய்ச்சல் முறை.

மாமண்டூர் எல்லையில் மல்லந்தாங்கல் மடை 1. அதன் கீழ் மடை 1. அதன் கீழ் மடை 1. அதன் கீழ கன்னிகால்வாய் 1 ஆக மடை 2. கன்னிகால்வாய். 1. இதன் கிழக்கே நரசமங்கலம் எல்லையில் கால்வாயை குறுக்கே கட்டி. மேல்படி ஊருக்கு பாயுற கொண்டம் மூலம் காணி 2 1/2 பாய விட்டு எடுத்து விடவேண்டும். இதன் கிழக்கே மாமண்டூர் எல்லையில் பாயற கணக்கு மானியம் காணி 1/4. வார பத்து காணி 1/4 ஆக, காணி 1/2 கால்வாயை குறுக்கே கட்டி. கொண்டம் பாயற காணி 1/2 காணி பாயவிட்டு எடுத்து விடுகிறது.

கன்னிக் கால்வாய் 1, இதன் கிழக்கே சித்தமான்யம் மடை 1. இதன் கிழக்கே நாத்தங்கால் மடை 1. இதன் கிழக்கே கால்வாயை குறுக்கே கட்டி கொண்டம் பாயற புத்து மேடு காணி 1 ஆகிய இவை பாய விட்டு எடுத்து விடவேண்டும். இதன் கிழக்கே புத்து மேடு கன்னிக்கால்வாய் 1 இதன் கிழக்கே அஷ்ட்ட கூறு கன்னிக்கால்வாய் 1 இதன் கிழக்கே கால்வாயை தேக்கி கொண்டம் பாயற பெருமா மான்ய புலம் காணி 3/8 பாயவிட்டு எடுத்து விடுகிறது. இதன் கிழக்கே கன்னிக்கால்வாய் 1, இதன் கிழக்கே கால்வாயை தேக்கி ஊர் மக்கள் பாயவிட்டு எடுத்து விடுகிறது. இதன் கிழக்கே பள்ளக்களுக்கு கன்னிக்கால்வாய் 1 இதன் கிழக்கே காவாங்கரை கொண்ட காணி 2ல் காணி 1க்கு கால்வாயை குறுக்கே கட்டி பாயவிட்டு எடுத்து விடுகிறது. இதன் கிழக்கே போகும் கால்வாய் 1க்கு மட்டு வயனம் ஒரு அடி அகலம் உயரம் 8 விரக்கடை அளவு தண்ணீர் விடுகிறது. இதன் கிழக்கே ஊர் கன்னிக்கால்வாய் 1. இதன் கிழக்கே ஓரடிக் கன்னிக்கால்வாய் 1. இதன் கிழக்கே கொசவன் மடை கால்வாய் 1க்கு மட்டு அகலம் அடி 2 உயரம் 8 விரக்கடை அளவு தண்ணீர் விடுகிறது. இதன் கிழக்கே பாகல்மேடு கன்னிக்கால்வாய் 1 இந்தக் கால்வாயிலே கிழக்கே போற பெரிய கால்வாயை தேக்கி மேலெழுதின கன்னிக்காலிலே பாய்கிற கொண்டம் காணி 11/4 பாயவிட்டு எடுத்து விடுகிறது. (கூஷ கன்னிக்கால்வாயிலே தண்ணி போறதுண்டு) இதன் கிழக்கே தண்டுக்கரை கன்னிக்கால் 1. மேலெழுதின கன்னிக்கால்வாயிலே கிழக்கே போற பெரிய கால்வாயை தேக்கி கொண்டம் கட்டி பாய்கிற பகல மேட்டு காணி 1/2 பாயவிட்டு எடுத்து விடுகிறது. (மேலெழுதின கன்னிக் காலிலே தண்ணி போறதுண்டு). இதன் கிழக்கே வெட்டியா மூட்டி மாண்யம் காணி 1க்கு பெரிய கால்வாயை தேக்கி பாயவிட்டு எடுத்து விடுகிறது. இதன் கிழக்கே வெட்டியாமூட்டி கன்னிக்கால்வாய் 1 மேல்படி கால்வாயிலே கொண்டம் கட்டுகிறது. முறை. கிழக்கே போகிற பெரிய கால்வாயை தேக்கி கொண்டம் கட்டி காணி 1 பூண்டிக்கு பாயும் காணி 3/4 ஆக காணி 13/4ம் பாயவிட்டு எடுத்துவிட வேண்டும். மாமண்டூர் எல்லையில் பெரிய கால்வாய் 3, கன்னிக்கால்வாய் 11, கால்வாய் 14 மடை 6 இந்த முறையில் தண்ணீர் வந்து மேற்படி மாமண்டூர் கழனிகளுக்கு விடுகின்ற சமயத்தில் அவரவர் பாய்ச்சிக் கொள்ள வேண்டும்.

மாமண்டூர் மேட்டுக்கால் வையாளி கொல்லையிலும் பையடி கொல்லைக்குகீழ் இந்த படி பயிர் செய்திருந்து ஒரு தண்ணி 2 தண்ணி வந்தால் மேற்படி பெரிய கால்வாயில் தொட்டி வைத்து இறைப்பது முறை. கொளக்கீழ்த் தொட்டி 1. நாவிதன் மானியத்தண்டை தொட்டி 1. கொளக்கீழ் தொட்டி 1. இலுப்பை பட்டி தொட்டி 1. கொசவன் மடை வையாளி கொல்லை தொட்டி 1. ஆக தொட்டி 5. இந்த 5 தொட்டியும் இறைத்துக் கொள்ளலாம். சேடை விட்டு இறைக்கக்கூடாது. இதை விட்டு கிழக்கே பூண்டி எல்லையில் மேற்படி ஊருக்கு கிழக்கே போகிற பெரிய கால்வாயை தேக்கி பாய்ச்சும் கணக்கு மான்யம் காணி 1. வார பத்து காணி. 1. ஆக இவை 2 காணியும் பாய்ச்சிவிட்டு எடுத்துவிட வேண்டும். இதன் கிழக்கே மாமண்டூர் எல்லையில் பிறப்பேறி கன்னிக்கால்வாய் 1. இதன் கிழக்கே பெரிய கால்வாயை தேக்கி கொண்டம் பாயற காணி 31/2 பாயவிட்டு எடுத்து விட வேண்டும். இதன் கொண்ட வாயிலிலே கன்னிக்கால்வாய் 1. இதன் கிழக்கே பெரிய கால்வாயை தேக்கி கொண்டம் பாயுற காணி 1-ம் பாயவிட்டு எடுத்துவிட வேண்டும். இதன் கிழக்கே வாழவந்தலுக்கு பாயும் கால்வாய் 1க்கு ஒன்றறை பங்கு தண்ணீரும் மாமண்டூர் தூசி மேட்டுக் கால்வாய் 1க்கு 1 பங்கு தண்ணீரும் போய்ப் பாயுறது மாமூல். தூசி மேட்டு கால்வாயிலே கொண்டம் பாயுற காணி 1 1/2க்கு மேலே எழுதிய வாழவந்தல் கால்வாயை கட்டி பாயவிட்டு எடுத்துடனும். இந்த கால்வாயிலே மேலெழுதின காணி 191/4க்கு கொண்டம் கட்டும் பொழுது மேலெழுதின தோட்டக் கால்களையும் (direct channels), கன்னிக் கால்களையும் (field channels), மடைகளையும் கட்டி பாயவிட்டு அந்தந்த கால்வாயிலே தண்ணீர் எடுத்து போடுகிற மாமூல் படி கால்வாயிலே மேட்டு மதகு தண்ணி வர்றது நின்று போனால் மதகிலே நாழிகை வாசல் மரம் தூக்கி கொண்டம் கட்டி கடை வாசல் கழு மேலே மட்டுகல் போட்டு மாமண்டூர் வடகல்பாக்கம், மனதூரசுபா மஸ்கூர் நாயக்கம்பாளையம், கிரிஜாபுரம் ஆகிய கிராம பாசனத்திற்கு உள்ளடக்கியது. ஆகவும் மதகிலே தண்ணி புறப்படுவதற்கு முன்னதாக மாறாமல் தண்ணி விட்டு மீதியை மாமண்டூர், நரசமங்கலம், பூண்டி, வாழவந்தல் ஆகிய 4 கிராமத்திற்கு மாமண்டூர் ஊர் ஓரமாய் வர்ற பெரிய கால்வாயிலே கொண்டு வந்து கன்னிக்கால்களுக்கும் மட்டு போட்டு, இருக்கிற கால்வாய்களுக்கும், மடைகளுக்கும் கால்வாயில் வர்ற தண்ணீருக்குத் தக்கினதாய் தண்ணிவிடனும். தண்ணீர் பற்றாக்குறைவின் பொழுது மேல்படி கால்வாயில் தொட்டி வைத்து நீர் எடுத்து பயிர் செய்யறது.

நரசமங்கலம் எல்லையில் மேற்படி ஊருக்கு தொட்டிவைப்பது முறை. ரெட்ட தொட்டி 1க்கு தொட்டி 2. காவல் மானியம் தொட்டி 1. ஆக தொட்டி 3க்கு இடைவிடாமல் இறைக்குற தொட்டி 2. இது இறைக்கும் பொழுது தொட்டிமேல் தொட்டி வைத்து இறைக்கக்கூடாது. நீர் பயிர் (wet land) செய்யும் காணி 2க்கு சேடை இட்டு நீர் பயிர் செய்கிறது. பூண்டி எல்லையில் மேற்படியூருக்கு இறைக்க தொட்டி 1 இந்த தொட்டி 1ம் இராபகல் 60 நாழிகையும் இறைத்துக் கொள்ளலாம். தொட்டி மேல் தொட்டி வைத்து இறைக்கக்கூடாது. நீர் பயிர் செய்யும் காணி 1.

கொண்டம் கட்டுகிறவர்கள் நரசமங்கலம், பூண்டி, மாமண்டூர் இந்த 3 கிராமத்திற்கும் தண்ணீர் பாய ஆறு நாளைக்கு ஒரு தரம் கொண்டம் கட்டிக் கொள்கிறது. 3 காணிக்கு பாய்வது. மேலும் கால்வாய் தண்ணீர் கொண்டு போய் மேலெழுதினபடி வாழவந்தலுக்கு 11/2 பங்கு தண்ணீர் போக 1 பங்கு தண்ணீர் மாமண்டூர் தூசி மேட்டுக்காலுக்கு கொண்டுவந்து கூஷியூர் கழனிகளுக்கு பாய்கிறது. மேலும் அறம்பு (பற்றாக்குறை - Scarcity) வந்தால் வேழல் வெட்டுக் கால்வாய் வெட்டி மேற்படியூர் ஓரமாய் வர்ற பெரிய கால்வாயிலே கொண்டு வந்து மேல்படியூர் நஞ்சைக் கழணிக்கு பாய்கிறது. பள்ளக்கால் வயணம். மேலெழுதின மேட்டு மதகு கால்வாயிலே கொண்டம் கட்டி பள்ளக்கால் பாய்ச்சலுக்கு விடுகிற அகலம் அடி

21/2, உயரம் சாண் 2 இந்த முறையில் விடுகிற தண்ணீரில் ஊர் ஓரமாய் வர்ற பெரிய கால்வாயிக்கு 1 பங்கு தண்ணீர் போக மற்ற 2 பங்கு தண்ணீரை பள்ளமதகு பின் தொட்டியிலே வந்து விழுகிறது. பின் தொட்டியிலிருந்து புறப்படும் இத்தண்ணீர் மாமண்டூர் முக்கிய மேட்டுக் கால்வாய் 1க்கு பாசனத்திற்கு ஏற்ப தண்ணீர் விடனும். மேற்படியூர் நாலா மேட்டுக் கால்வாய் 1க்கு பாசனத்திற்கு ஏற்ப தண்ணீர் விடனும். ஆக மாமண்டூர் கால்வாய் 3. இந்த கால்வாயில் வருகிற தண்ணீர் மேல்படியூர் கழனிகளுக்கு பாய்கிறது. மீதித் தண்ணீர் பள்ளக்காலுக்கு வருகிறது. இது முறை.

மாமண்டூருக்கு மேல், மேற்படியூர் காணி மாணியத்து கால்வாயிக்கு பாசனத்திற்கு ஏற்ப தண்ணீர் விடப்படுகிறது. கூஷ காணி மான்யத்து கால்வாயில் கொண்டம் பாய்கிற காணி 41/2க்கு பள்ளக்கால்வாயை குறுக்கே கட்டி பாயவிட்டு எடுத்து விடுகிறது. காணி மான்யத்து கால்வாயிக்கும் தண்ணீர் போறது. இதன் கிழக்கே மாமண்டூர் கோன வட்டம் கால்வாய்க்கு பாசனத்திற்கு ஏற்ப தண்ணீர் விடப்படுகிறது. பள்ளகால்வாயும், கோன வட்டம் கால்வாயும் குறுக்கே கொண்டம் கட்டி பாயறது. மேலும் காணி 3/4 பாயவிட்டு எடுத்து விடப்படுகிறது. இதன் கிழக்கே பள்ளக்கால்வாயிக்கு போற தண்ணிக்கு மட்டு வயணம் அகலம் அடி 3, உயரம் விரக்கடை 12 இந்த அளவிற்கு விட்டு மீதி தண்ணீர் மாமண்டூர் வடக்கு நிலங்கலுக்கு போகும். மேல்படியூர் கழனிகளுக்கு பாயுறது. மேலெழுதின பள்ளக்கால் மட்டு தண்ணியிலே கால்வாயிலே குறுக்கே கொண்டம் கட்டி பாயுறது 9. மேற்படி காணி 11/4 பாயவிட்டு எடுத்துடனும். இதன் கிழக்கே கொடலக்கரை கால்வாய் அருகே பள்ளக்காலுக்கு மட்டு வயணம், அகலம் அடி 21/2, உயரம் 8 விரக்கடை இந்த அளவில் பள்ளக்காலுக்கு தண்ணீர் விட்டு மீதி தண்ணீர் மாமண்டூர் கொடலக்கரை கால்வாயிலே போய் மேற்படியூர் கழணிகளுக்கு பாய்கிறது. மேற்படி கொடலக்கரை கால்வாயில் கொண்டம் கட்டுகிறவர்கள் மேற்படி மட்டு போட்ட பள்ளக்கால்வாயை தேக்கி கொண்டம் பாயுற செங்கழனீர் பள்ளம் குடலக்கரை காணி 81/2 பாயவிட்டு எடுத்து விடுகிறது. மேற்படி குடலக்கரை கால்வாயிலே தண்ணி போறது.

மேலெழுதின பள்ளக்காலிலே மட்டு தண்ணீர் போறதில் கிழக்கே மாமண்டூர் கரம்பு காணி கால்வாய் 1க்கு மட்டும் மரம் போட்டு ஒரு அடி அகலமும், நாலு விரக்கடை உயரமும் தண்ணீர் விட்டு குறை தண்ணீரை கம்பெனி சர்க்கார் வடகல்பாக்கம், மண்ணூர் தாலுக்கா மசுக்கூர், நாயக்கன்பாளையம், கிரிஜாபுரம் இந்த 4 கிராமத்திற்கும் போகிற கால்வாய் 1க்கு மட்டு மேலெழுதின ஒரே அளவு போட்டு அகலம் அடி 3, உயரம் விரக்கடை 4 இந்த முறையில் கிராமத்திற்கு தண்ணி விடுகிறது. மேலெழுதின கரம்பு காணி கால்வாய் 1 இதிலே போற தண்ணியிலே மாமண்டூர் கழனிகளுக்கு விடுகாலிலே (வடிவுகாலிலே) பாயறது போக, பாயாத கழனிகளுக்கு மேற்படி கரம்பு காணி கால்வாயைக் கட்டி பாயவிட்டு எடுத்துவிடணும். மேலெழுதின கரம்பு 1 கால்வாயிலே ஏறாத கழனிகளுக்கு மேற்படி கால்வாயும் 4 கிராமத்துக்கு ஆன கால்வாயையும் குறுக்கே கட்டி கொண்டம் பாயற கரம்பு காணி 3 பாய விட்டு இரண்டு கால்வாயையும் எடுத்துவிடனும் கரம்பு காணி கால்வாயிலே போகிற தண்ணீரை மாமண்டூர் கழனிகளுக்குப் பாயவிட்டு மற்ற தண்ணீரை கம்பெனி சர்க்கார் கல்பாக்கம் கழனிகளுக்கு போய் பாய விடுகிறது.

பதினாலு கிராமத்துக்கும் போற கால்வாயில் வடகல்பாக்கம் எல்லையில் மேற்படி ஊருக்கும் மண்ணூருக்கும் பாய்கிற முறந்தூக்கி கால்வாய் 1க்கு மட்டு வயணம் 1 அடி அகலமும் 4 விரக்கடை உயரமும் தண்ணீர் விட்டு மேல்படி கால்வாயிலே போற தண்ணீரை கல்பாக்கம் கழனிகளுக்கு பகல் 30 நாழிகையும், இரவு 10 நாழிகையும், மண்ணூர் கழனிகளுக்கு இராத்திரி 20 நாழிகையுமாக பாய விடுகிறது. இதன் கிழக்கே போகிற கால்வாயிலே மேற்படி வடகல்பாக்கம் எல்லையில் மேற்படியூர் கன்னிக் கால்வாய் 5க்கும் பாசனத்திற்கு ஏற்ப தண்ணீர் விடுகிறது.

இதன் கிழக்கே போற கால்வாயும், முறந்தூக்கி கால்வாயும் மேல்படியூர் கன்னிக்கால்வாயும் குறுக்கே கட்டி வடகல்பாக்கம் கொண்டம் பாயுற காணி 3 பாய விட்டு மேலெழுதின கால்வாய்களை எடுத்து விட்டு மேல்படி கால்வாயெல்லாம் மேலெழுதினபடிக்கு தண்ணீர் பாய்கிறது. இதன் கிழக்கே மண்ணூர் எல்லையில் மேல்படியூருக்கு பாயும் கன்னிக்கால் 2ற்கு பாய்ச்சலுக்கு தக்கினதாய் தண்ணீர் விடவேண்டும். கிழக்கே போற கால்வாயும் மேற்படியூர் கன்னிக் கால்வாயும் வடகல்பாக்கம் முறந்தூக்கி கால்வாயையும் தேக்கி மேல்படி மண்ணூருக்கு கொண்டம் பாயுற காணி 1 பாயவிட்டு எடுத்து விடுகிறது. இதன் கிழக்கே நாயக்கன்பாளையம் எல்லையில் கூஷியூருக்கு பாய்ச்சலுக்கு கால்வாய் 1க்கு மட்டு போடுகிறது முறை. அகலம் அடி 1 உயரம் 6 விரக்கடை அளவு தண்ணீர் விடுகிறது. மீதி தண்ணீரை நாயக்கன்பாளையம், கிரிஜாபுரத்துக்கு ஒரு காலிலே தண்ணீர் போயி அதிலே நாயக்கன்பாளையம் கொண்டம் பாயறது முறை. கிழக்கே போற கால்வாயை குறுக்கே கட்டி மேற்படியூர் மட்டு காலும் கட்டி மண்ணூர் கன்னிக்காலும் கட்டி கொண்டம் போட்டு மேலெழுதின முறையில் காலுகளிலே (வாய்க்கால்களிலே) தண்ணீர் எடுத்து விடுகிறது. கிழக்கே மேற்படியூருக்கு கன்னிக்கால்வாய் 1. இதன் கிழக்கே மேற்படியூருக்கு கன்னிக்கால்வாய் 1 ஆக நாயக்கன் பாளையம் கன்னிக்கால்வாய் 3 இதற்கு பாசனத்திற்கு ஏற்ப தண்ணீர் விடவேண்டும். இதன் கிழக்கே கிரிஜாபுரத்துக்கு பாசனத்திற்கு ஏற்ப தண்ணீர் பாயறது. இந்த கால்வாயிலே பள்ளமதகு திறந்து அறம்பு வந்தால் மாமண்டூர் கொண்டம் கட்டுகிறவர்கள் 6 நாளைக்கு 1 தரம் முக்கிய மேட்டுக்கால்வாயிலே கொண்டம் கட்டுகிற காணி 11/2க்கு பள்ளக் கால்வாயை கட்டி பாயவிட்டு எடுத்துவிடுகிறது.

நாலாம் மேட்டுக்கால்வாயிலே கொண்டம்ட கட்டுகிறது முறை. நாலாம் மேட்டு பூந்தோட்டம் காணி 11க்கு பள்ளக்காலிலே அண்டை வைத்து 6 நாளைக்கு ஒரு தரம் 2 காணி வீழ்க்காடு பாயவிட்டு எடுத்துவிடுகிறது. மற்ற கொண்டமெல்லாம் மேற்படி கால்வாயிலே போகிற தண்ணீரை 6 நாளைக்கு ஒரு முறை பாய்ச்சிக் கொள்ளலாம். காணி மான்யத்து கால்வாய் கொண்டங்களுக்கு 6 நாளைக்கு 1 தரம் பாய்ச்சிக் கொள்ளலாம். ஒன்பது வெட்டி கொண்டத்துக்கும் 6 நாளைக்கு ஒரு தரம் பாய்ச்சிக் கொள்ளும் கொணாவட்டம் கால்வாயில் கொண்டம் காணி 21/2க்கு 6 நாளைக்கு 1 தரம் பள்ளக்காலை அண்டைவச்சு பாயவிட்டுக் கொள்ளலாம். மேலும் பெருங்காலிலே கொண்டம் கட்டுகிறவர்கள் காணி 41/2க்கு நாளைக்கு 1 தரம் பள்ளக்காலை அண்டை வைத்து பாயவிட்டு எடுத்துடணும். மீதி கொண்டக்காரர் மேற்படி கால்வாயில் பாய்கிற தண்ணீரை 6 நாளைக்கு 1 தரம் பாயவிட்டுக் கொள்ளலாம். கொடலக்கரை கால்வாயிலே கொண்டம் கட்டி காணி 8 1/2க்கு பள்ளக்கால்வாயை குறுக்கிட்டு 6 நாளைக்கு 1 தரம் மிராசு காணி 11/2 விழுக்காடு பாயவிட்டு கொள்ளுகிறது. கரம்பு காணி கால்வாயிலே கொண்டம் கட்டுகிறவர்கள் காணி 31/2க்கு 6 நாளைக்கு 1 தரம் நாயக்கன்பாளையம் வகையரா கிராமங்கள் கால்வாயின் குறுக்கே கட்டி மிராசு ஒரு கொண்டம் கட்டிக் கொள்கிறது. மேற்படி கால்வாயிலே மீதி கொண்டங்களுக்கு போற தண்ணீரை பாயவிட்டுக் கொள்கிறது. நாயக்கன்பாளையத்துக்கு போகிற கால்வாயினை கட்டுவதில்லை. மீதி தண்ணீர் கல்பாக்கத்துக்கு போறது. இது மாமண்டூர் கழனிகளுக்கு தண்ணீர் பாயற மாமூல்.

மேட்டு மதகு தண்ணீர் நின்று போய் பள்ளமதகு திறந்தால் கம்பெனி சாக்கார் வடகல்பாக்கம், மண்ணூர் தாலுக்கா மசுக்கூர், நாயக்கன் பாளையம், கிரிஜாபுரம் ஆக இந்த 4 கிராமத்துக்கு கால்வாய் உண்டானாபடியினாலே ஒரு மடை பாய்ச்சலாகி கல்பாக்கத்துக்கு இரவு 30 நாழிகையும். பகல் 3 நாழிகையும் ஆக 33 நாழிகை பாயவிட்டுக் கொள்ளுகிறது. மண்ணூருக்கு முழுமை தண்ணியும் பகலிலே 17 நாழிகையாக பாய்ச்சிக் கொள்ளலாம். நாயக்கன் பாளையம் பகல் 10 நாழிகையும் ராத்திரி 30 நாழிகையும் மறுநாள் பகல் 10 நாழிகையும் ஆக நாழிகை 100 முழுமை

தண்ணீரையும் பாயவிட்டுக் கொள்ளுகிறது. கிரிஜாபுரத்துக்கு பகல் 20 நாழிகையும் முழு தண்ணியும் பாயவிட்டுக் கொள்ளுகிறது. நாழிகை 120க்கு நாள் 2 இந்த முறை வட்டமாய் பாய்ச்சிக் கொள்ளலாம். மேற்படியூர்களில் வாழைத்தோட்டம், வெற்றிலத்தோட்டம், கரும்புத்தோட்டம் இல்லை.

Main Points Stressed in the Water Mamulnamas

1. The ways of water reached to the tank - sources of supply.
2. Limit and the extent of catchment areas.
3. The type of sluice and its situation - used for irrigation.

(பலகை போடும் பெரிய மதகு, நாழிகை வாசல், மரம் போடுகிற பள்ளமதகு etc)

4. Method of water allocation from the sluice - How to & How much water to be released to the channels.
5. Allocation of water during scarcity period.

நீர் பகிர்வு முறையானது மதகில் புறப்பட்டு தொடர்ந்து எந்தெந்த பகுதிகளுக்கு, எவ்வளவு, எப்படி செல்லவேண்டும் என்பது விரிவாகவும், தெளிவாகவும் கூறப்பட்டுள்ளது. நஞ்சை கழனிகளுக்கான நீர்ப்பாசன முறை கூறப்பட்டுள்ளது.

Mamandoor: தேவையானபோது பாசனத்திற்கு ஏற்ப தண்ணீர் பாய்ச்சும் முறை கூறப்படுகிறது. பற்றாக்குறை தண்ணீர் காலத்திலான பகிர்வு விரிவாக உள்ளது.

Kaveripakkam: கால்வாய் வெட்டும் முறை கூறப்படுகிறது. கழிவு நீர் பயன்படுத்தும் முறையும் கூறப்படுகிறது. மேலும் பாலாற்று நீர் பெறும் முறைக்கு ஆட்கள் பகிர்வும் உள்ளது. ஆட்கள் செய்ய வேண்டிய முறையான வேலைகள் கூறப்பட்டுள்ளது. கலங்கல் கட்டுகிற மாமூலும் விவரிக்கப்பட்டுள்ளது.

ஏரிக்கால்வாய் சொந்தமாக வெட்டும் முறை கூறப்பட்டுள்ளது. அறம்பு காலத்தில் முறை போட்டு தண்ணீர் பாய்ச்சுவதும் கூறப்பட்டுள்ளது.

நாழிகை கணக்கு	சாமம் கணக்கு
24 நிமிடம் 1 நாழிகை	1 சாமம் = 3 மணி
1 மணி 21/2 நாழிகை	8 சாமம் = 24 மணி. 1 நாள்.
24 மணி 60 நாழிகை	4 சாமம் = 1/2 நாள்.
1 சாண் = 9 inches	இரவு 4 சாமம் = பகல் 4 சாமம்.

3 types of irrigation

(1) Normal storage (full tank level) period.

(2) Scarcity supply period.

(3) more than scarcity supply period.

Glossary - அருஞ்சொற்கள்

வெள்ளக்கால்	catchment
வழிந்தோடும் தண்ணீர்	surplus water
அத்து	boundary
இசா	a share in joint property right
நாழிகை வாசல்	time bound
இரண்டு கூடை தண்ணீர்	a unit of flow measurement
தக்கை வைத்து	wooden obstruction
தெருக்காலிலே	Road channel
சப்பாற தண்ணி	Local Measurement
2 மார்	two times
வொட்டு	illegible
நெல்லி கும்பினி சாகிரிலே	Land belongs to company
அத்து	Limit
பாட்டை	road
வடிவில்	depletion
ஒரு துலை	Shallow Pond
பத்துகட்டு நஞ்சை	occupancy holding
வயனம்	method
தோட்டக் கால்களையும்	direct channels
கன்னிக் கால்களையும்	field channels
நீர் பயிர்	wet land
பற்றாக்குறை	Scarcity
கசிவுக்கால்	spring channel
தபசில்	List
கொண்டம்	earthen weir
துண்டித்து	intercepting the flow
கழிவு	drainage
கலங்கல் கல் 125	local measure
பட்டடை	several names jointed
சர்க்கார் மக்களுக்கும்	service guards
மிராசு செம்பிடவன்	head fisherman
121/2 சமாச்சாரமும்	demorcated ayacut area
வடியும் காலம்	depletion period
பொம்மை மதகு	Paracheri Sluice
புது மண் போட்டு கட்டக்கூடாது	bund level should not be raised
கும்பிட்டி நெருப்பு	coal fire
தீவட்டி (கொருங்கட்டை)	Torch
சர்க்கார் மனிதர்கள்	Neerkatties
வெட்டுமடை	cut open bund
கும்பினி	company

TRANSLATED ENGLISH VERSION OF *MAMULNAMAS* FOR KAVERIPAKKAM AND DUSI-MAMANDUR TANKS

Mamulnamas is a record on the distribution of water which the ryots have accepted for many years. It was written during the early 19th century (1815), and approved and printed by the British government in the year 1907. It refers to shares of water supply to be allocated in different channels across villages from each tank. In order to understand its full importance, a rough translation of the *Mamulnamas* of Kaveripakkam and Dusi-Mamandur tanks is given below.

Mamulnamas for the Kaveripakkam tank

To the Company's government (1815).

The *Mamulnamas* executed by the leading *mirasidars*, viz., Manakay Vengu Ayyangar, lessee Mettukkudi Krishna Ayyangar, Vinayaga Pillai, Karnam Nataraja Pillai, Talaiyari Nallamuttan, Karuppu Muttan, Raghavan, Padavattan, Viran, Chinna Kandan, Totti, Tottiyan, Kandan and others of Kaveripakkam village, of the said taluk, is as follows:

Ilukum No. 8, received from Huzur regarding *Mamulnamas* on the 7th January 1815 and the *takid* (written document) and the *kistinama* (proclamation) received on the 27th July were read over to the Mirasidars by Ramappa, the Tahsildar of the said taluk, who then required the Mirasidars to execute a *Mamulnamas* in order that they might, without any dispute, conduct themselves, hereafter, according to the custom obtaining with regard to irrigation with the water of the tank, pond (tangal) and channel, and affix their signatures thereto and also obtain thereto the signatures of the neighboring villagers. Thereupon, all the Mirasidars have jointly ascertained the customary rights of wet, dry and garden lands of the villages, inspected in person, irrigation made from the tank, pond and spring-channel and drawn up the details below according to the custom obtaining.

Description of sluices and the outlet channels of the tank

1. ***Gaffor sluice***

First column: Village proper, 1, hamlet, nil, total number of villages 1. Particulars of irrigation made from the ten sluices of the one tank forming the source of irrigation of the wet classification of the above village: on the western side of the southern bund of the tank for Ramapuram-sluice 1, channel 1, entirely used for the irrigation of Ramapuram.

Note: According to the Up-dated Settlement Registers the total *ayacut* of Gaffor sluice is 67 acres in 1983 serving only a single village, namely Ramapuram (Kadaperi).

2. ***Fox sluice***

Second column: Particulars of the "turn" shares of water of the one big channel for the Fox sluice to the east hereof irrigating Kaveripakkam are (as follow): The number of shares of water for one channel of the Veeravarkoil *madai* (a channel for irrigation or a small sluice of a channel), is three. Out of the three shares of water due for one channel of the Abathu *madai* to the east hereof, one share pertains to Elandai channel irrigating Kaveripakkam. Out of two shares due for one channel going eastward, water that can be baled out by one picottah during 30 *naligai* (one naligai = 24 minutes) of the day-time is allowed for the garden lands of Kaveripakkam fort. Out of the remaining water, water that can pass through one human foot-space (in a dam) flows in a channel to Kondapuram at the *Kudiraikolam Maniyam* along the fort; and water that can pass through one human foot-space (in a dam) flows in a branch channel to irrigate Kondapuram at the Kosavan channel. Of the remaining water one share irrigates Kaveripakkam throughout the day-time, and one share Nanderi-Perukarumbur throughout the night. Out of this share irrigating during the night, water that can pass through one human foot-space (in a dam) is taken in a channel to Vennappatti field of Kaveripakkam at Pakkirithakka (a fakir's residence). From this water of the Nanderi-Perukarumbur channel the people of Kaveripakkam divide off, at day break the water going to Eralacheri and lead off separately the Perukarumbur channel. From the big channel going south of the Agaram *madai*, three shares of water pertain to the one branch channel irrigating Kondapuram north of the field belonging to the share of Periyatambi Mudaliar, at the border of Ramapuram. One channel with water one foot deep irrigates the Othipattu fields of Vilaham near Kondapuram-Sitheri to the south hereof. (One) share of water is taken for the one branch channel irrigating Sirukarumbur alongside of the garden lands within the limits of Vilaham to the south hereof. One share of water is taken for the one branch channel irrigating Athipattu near the garden lands of Vilaham to the south hereof. One share of water is taken for the one channel irrigating Vilaham on the eastern side. Thus the (total number of) shares of water for one sluice is 12.

Note: According to the Up-dated Settlement Registers the total *ayacut* of Fox sluice in 1983 is 1679 acres spread over in five villages, namely Kaveripakkam, Kondapuram, Eralacheri Athipattu and Sirukarumbur.

3. *Lion sluice*

Third column: For the Lion sluice 1 at the border of the eastern bund to the north hereof, a supply dam (*thoti*) is raised and Kaveripakkam is irrigated. For the one sluice to the north of the above thoti, there is a channel irrigating the Muranthukki and other fields. For the one sluice to the south of the above thoti, there is a channel irrigating the Kondakkal medu and other fields. (There are also) the above mentioned thoti and a sluice opening at the head. Branches for one big channel irrigating Kaveripakkam are (as follows): Kannika koil channel along the northern bund 1; to the east hereof, field channel with water one foot deep 1; to the east hereof, Manappantharai channel 1; to the east hereof, Devadanam channel 1; a channel one foot deep running northward in Manjanirthakkai to the east hereof 1; to the east hereof, (Nadukkalvai) central channel 1; to the south hereof, for Maman channel 1. Among the above mentioned channels irrigating Kaveripakkam, particulars of irrigation of Eralacheri are (as follows): Out of the water going in the Maman channel, one share is for Kaveripakkam Uppumedu for 27 *naligai* from the *Sunset* (24 minutes after the *Sunset*) to three *naligai* to day break (24 minutes before the Sun-rise), and after allowing for the other fields water that can pass through one human foot-space (in a dam), the remaining water irrigates Eralacheri. This is diverted at day break and irrigates Kaveripakkam. Out of the water flowing in the central channel, the water flowing in the Pilippattu channel as belonging to the share of Uppumedu irrigates the fields of Eralacheri, Mel (west) Onampattu and Kil (east) Onampattu for one night every three days. Out of the water flowing in the one foot Manjanir channel, water that can pass through one human foot-space (in a dam) is allowed for the channel branching west from the Kavetti channel, and the remaining water irrigating at night irrigates Eralacheri.

Note: According to the Up-dated Settlement Registers the total *ayacut* of Lion sluice is 1109 acres in 1983 spread over in Kaveripakkam and Eralacheri villages.

4. ***Moolai sluice***

Fourth column: For the one northern corner sluice, the Bathiri channel branches out and flowing northward irrigates Cheri lands. From the water flowing through this one channel irrigating a *Kani* (1.33 acres) land to the north of Abiramacheri (a hamlet of Cheri) and one channel irrigating Kattalai along the southern bund. For one sluice intended for water to pass into the channel running along the Kattalai bund and the Bathiri channel, opposite to the one sluice to the east hereof, there is a big channel laid with five granite boulders within a distance of one yard on the opposite side. Out of this, a channel supplies one-half of water to Kattalai. From the big channel, a *Oradi* channel with water one foot deep irrigates the Sabavirathi field of Cheri. From this, a channel flows with water sufficient for irrigating two *Kani* (2.66 acres) land of Kattalai. The remaining half of the water of the big channel flows irrigates Cheri for the 30 *Naligai* (12 hours) of the day time and the remaining 30 *Naligai* of the night for Abiramacheri. The water draining from the channels irrigating Kattalai lands is brought by the *Thotties* of Kaveripakkam and Abiramacheri by means of a *madai* at the Cheri road north of Kattalai, with which joins water from Abiramacheri during the 30 *Naligai* of the night. In times of water scarcity, one foot of water of the supply dam constructed for Kattalai near Cheri is diverted through a *Kondam*, into the lands located at Kaveripakkam border of *Vayichandu Sevittuthakku* for 30 *Naligai* of the day-time and the night water of 30 *Naligai* goes to Abiramacheri corner lands.

Note: According to the Up-dated Settlement Registers the total *ayacut* of Moolai sluice is 783 acres in 1983 irrigating three villages, viz., Kattalai, Cheri and Kaveripakkam.

5. ***Cheri low level sluice***

Fifth column: For the one northern low-level sluice, there is a whole channel irrigating Cheri.

6. ***Cheri high level sluice***

Sixth column: For the one northern high-level sluice, there is a whole channel irrigating Cheri.

Note: According to the Up-dated Settlement Registers the total *ayacut* of these two sluices is 447 acres in 1983.

7 + 8. *Maganipattu and Thuraiperumpakkam sluices*

Seventh column: North Maganipattu sluice 1, Thuraiperumpakkam sluice 1, in all, sluices 2. From these flows one Muduman channel, which feeds a channel with water one foot deep irrigating Vannarapatti Murukkan, Sengaladi, Vemban and other fields of Cheri to the west of Chinnakadavasal (small surplus weir), and flowing alongside of the pathway west of Anantha Pillaiyar temple. From the said Muduman channel in Cheri branches out a channel near the Murukkan (field) by the side of the deep pool, having a flow of water that can pass through one human foot-space (in a dam) alongside of Sengaladi and irrigating the Thattan Koviladi field in Maganipattu. The surplus water from the calingula (a surplus weir) of the tank, the water draining from the base of the Upparanthangal supply dam, and the water draining from the Chinnakadavasal (small surplus weir) a supply dam to be constructed, together with the waters issuing from the above mentioned two sluices of Maganipattu and Thuraiperumpakkam form a channel. Out of this channel, water one foot deep flows into the Muduman channel of Cheri. The remaining water feeds the channel flowing eastward on account of the said supply dam. The water thereof is divided into eleven shares or pangus (in the following manner).

One share of water is taken for one channel going to Puthupattu alongside of the supply dam. Two shares of water are taken for one channel which branches out of the big channel running to the east of the above and irrigates the Putheripallam fields of Maganipattu. Out of the share of water due to Sirasthavalapuram from the above (share of) water, water that can flow through one human foot space (in a dam), being allowed for and carried to the Putheripallam channel, irrigates Sirasthavalapuram. The remaining water irrigates Maganipattu. From the channel along which flows the remaining water left thereafter, water that can pass through one human foot space (in a dam) is allowed to flow out in a channel to irrigate Vadakkumedu and Nellitope fields in Cheri. Out of the water flowing in the big channel running to the east of the above, two shares of water are taken for one channel irrigating Maganipattu. Out of this, as much water as may be necessary for the irrigation of the western fields of the said Maganipattu and the fields of Athiyadi is dammed and diverted into Thuraiperumpakkam channel. Two shares of water are taken for the Alapakkam channel running to the east of the above. Out of this water, one-sixth share is taken for Sathuvanatham. The remaining water irrigates Alapakkam. One share of water is taken for the Thaluvanampattu channel running to the east of this. Four shares of water are taken for the Thuraiperumpakkam channel running to the east of the above. Out of this, as much water as may be necessary for the irrigation of the western fields of Maganipattu and the fields of Athiyadi is allowed to be taken, and out of the remaining water, one-six share is taken for Sirasthavalapuram by means of one channel. Out of the remaining five-sixths, one-sixth share of water irrigates Moonjipakkam. The remaining water irrigates Thuraiperumpakkam.

Note: According to the Up-dated Settlement Registers the total *ayacut* of the Maganipattu and Thuraiperumpakkam sluices is 1492 acres in 1983 which spread over in five villages, namely Cheri, Maganipattu, Alapakkam, Puthupattu and Thuraiperumpakkam.

9. *Paracheri mettu bommai sluice*

Eighth column: From the one high-level Bommai sluice of the north Paracheri, one full channel irrigates Cheri.

Note: According to the Up-dated Settlement Registers the total *ayacut* of Paracheri sluice is 41 acres in 1983.

10. *Panniyur sluice*

Ninth column: From the one channel from Panniyur sluice 1, north of the calingula (surplus weir), flows one channel irrigating five *Kanies* (6.65 acres) of the Kosavan Pattadai fields of Cheri. Out of the remaining water flowing eastward, there is one channel for the one share of Gangatharanellur. From the one channel for the two shares of Panniyur and Kilveeranam flows one channel for the one share of Kilviranam in the foreshore of the Panniyur tank, and one channel for the one share of Panniyur. These are the particulars of the one Panniyur sluice and of the ten tank sluices.

Note: According to the Up-dated Settlement Registers the total *ayacut* of Panniyur sluice is 164 acres in 1983, which is distributed in two villages.

Tenth column: These channels are used for irrigation, the respective parties digging a channel according to the *Mamul* right; but they should not dig deeper than *Mamul*.

Eleventh column: In times of water scarcity, the water draining from the *Ethuka*, Sitheri and *Elandai* fields of Ramapuram is utilised for irrigating the Arusur pine-apple half *kani* of the garden lands of Kaveripakkam Fort, by putting up a supply-dam at the Ramapuram Paracheri in the channel going to Kondapuram and other villages; and the remaining water is allowed to irrigate the Kaveripakkam fields.

Details on the Construction of Surplus Weir (Kondam) in KPT

Twelfth column: To the south of the Panniyur sluice and north of the Paracheri high-level sluice at the extremity of the tank is calingula 125 kal (length of the weir). When the tank receives excessive water supply and this calingula is opened and the Pinnur calingula (the tanks situated below this one are also getting filled) is closed as far as possible, the water thus oozing out is flowing in a pool; and this water is used for irrigation by means of a supply-dam being raised and by being baled out with picottah according to *mamul* in the manner following:

First mode: Upparanthangal supply dam (*Kondam*) 1. This supply dam pertains to the Upparanthangal irrigation.

Second mode: Small calingula 1. Particulars of villages irrigated by this (water) with which joins the water of the Maganipattu and Thuraiperumpakkam sluices are given in the seventh column. Besides this, the people of Upparanthangal bale out water for fields under cultivation with two picottah (*etrams* to lift water) from the stagnant water of the said Chinnakadavasal (small surplus weir).

Third mode: To the east of the small surplus weir is one supply-dam kondam of Cheri. The water of this *kondam* pertains to the irrigation of Cheri. Besides this, water is baled out by four *etrams* by the Maganipattu people between the *Kondam* of Cheri and that of the small surplus weir for (the irrigation of) Maganipattu. The people of the Cheri village put up *etrams* adequate for irrigation and bale out water.

Fourth mode: For the one channel through which flows the water drained from the Cheri *Kondam*, irrigation for the fields of Abiramacheri along the pool is for seven Jamam (21 hours) and irrigation for the Meleru *Third Kondam* fields of Kaveripakkam is for one *Jamam* in the noon from the Sunrise. Thus the total (period of irrigation) for two villages is eight *Jamam* making one day. Within these limits is one *etram* balling out (water) for the *Meleru Third Kondam* of Kaveripakkam.

Fifth mode: To the east of the Cheri *Kondam* is the one *Kondam* of Moonjipakkam, Thuraiperumpakkam and Kaveripakkam. The water flowing in this irrigates Moonjipakkam all the 30 *naligai* (12 hours) of the night. It irrigates Thuraiperumpakkam fields for 20 *naligai* (eight hours) in the day time and the remaining four hours supply for Kaveripakkam *Kondam* lands in the day time from Sunrise. Thus the total (period of irrigation) for two villages during day-time is 30 *naligai* and the total for three villages is 60 *naligai* making one day.

Sixth mode: To the east of this, for one *Kondam* of Eralacheri-Kaveripakkam, irrigation for Eralacheri is for 20 hours and irrigation for the *third kondam* fields of Kaveripakkam during day-time from the Sunrise is for 4 hours. Thus the total (period of irrigation) for two villages is one day. Besides this, between the *Kondam* of the aforesaid Eralacheri and other (villages), and that of Moonjipakkam and other (villages), there are four *etrams* belonging to Thuraiperumpakkam, one etram belonging to Abiramacheri, one to Kaveripakkam Keleru third Kondam and one to Meleru third Kondam; thus there are two etrams for Kaveripakkam and seven etrams for three villages.

Seventh mode: To the east of the Eralacheri vagaiyara *Kondams* there is a *Kondam* for Uthirampattu, Ogalapattu and Dharmanithi, and there are also *Kondams*, etc., for many other villages.

Sources of Supply to KPT

Thirteenth column: Particulars of water-supply for Kaveripakkam tank are (as follow): In that part of the region for diverting water in the Palar which lies within the limits of Tiruyekambacheri, a dam is constructed extending from the western side of the bridge over the river channel in Sattambakkam to Rayasithottam limits on the southern bank of the river, and a channel runs through the boundary limits of the seven villages, viz., Tiruyekambacheri, Sattambakkam, Poondi, Kadaperi, Sumaithangi, Cheri and pilleri. *Mamul* irrigation in force with respect to this channel is as (follow):

First mode: Tiruyekambacheri river channel 1; which goes along the southern bank at the head of the channel where the dam construction commences.

Second mode: Sattambakkam river channel 1; which goes along the northern bank.

Third mode: Further off, to the east of the Koonmadai is the one sluice channel which carries off the water of the Poondi tank through a sluice built underground across the river channel by constructing a reservoir from the northern side for the irrigation of the land south of the river channel belonging to Velliyantangal village which is a hamlet of Poondi village.

Fourth mode: To the west hereof, a *kondam* is used for irrigating garden lands by a spring dug by the Poondi people in the river channel in times of water scarcity.

Fifth mode: To the north hereof is one masonry *kondam* constructed for irrigating the Kadaperi fields on the eastern bank of the river channel, to which is taken the surplus water flowing in the channel after it is drained off within the said Kadaperi limits, as also the Kadaperi waters. When the water of the river channel flows to the tank, the turfs laid on the said masonry *kondam* are removed. When the supply of water drains away, the said Kadaperi people put them on again.

Sixth mode: To the north of this is one earthwork *kondam* of the said Kadaperi people. The said supply dam used to be put up after the eighth of Thai (20th January). Again when water is available, it is removed. As the mud of the said earthwork *kondam* is washed into the tank and cause damage to it year after year, the said *kondam* used to be constructed only once a year.

Seventh mode: To the north of this is the one brick built sluice of Ramapuram north of *Pannandarai* (twelve and a half) in the Sumaithangi limits and immediately north of the paracheri, on the eastern bank of the said river channel. Through this the water draining in the river channel flows in one channel and is utilized customarily in the proportion of two shares to Ramapuram and one share to Sumaithangi. In order that the water of the river channel might not be wasted when the channel conveys water-supply, the water that is in excess of the (quantity required for) irrigation by the sluices, etc., is secured, and if at any time excessive water-supply is conveyed in the channel, the aforesaid sluices, etc., are completely safeguarded, and when the supply of water drains away, the *mamul* described above is followed. It is not customary for the people of the villages of Sumaithangi tank and Pilleri to bale out water in the river-channel with *etrams*. These are the (particulars of irrigation from) the one river-channel.

Fourteenth column: To the north-west of the corner-sluice is the one channel into which flow the water escaping at the extremities of the tanks of the following and other villages at the foreshore of the tank, viz., Olugur, Chengadu, Ammur, Vangur and Valluvampakkam, within the limits of these villages, the hill floods, etc., when there are rains, and the water draining from the fields.

Fifteenth column: The water escaping at the extremities of the tanks of Kadaperi, Sumaithangi and Marutheri empties itself into the Kaveripakkam tank.

Sixteenth column: In the bed of the tank, from the border of Valluvampakkam the Kacha river channel-the spring channel- flows through the corner sluice into a channel which, not being in good repair, is dug for the irrigation of Cheri village in times of water scarcity.

Seventeenth column: When the tank is drying up, the village *mamul* in baling out the water of the tank for the lands in its foreshore is (as follow):

> **First mode:** Originally there were seven *etrams* used for baling water within Valluvampakkam limits, in which *Maduvu Etram* five and *Konankuttai Etram* two, by the people of the said place from the days of the *Nawab Sahip* to the Tamil year *Parithabi* 1782. During the said year more *etrams* than the total number of seven were put up (and water baled out); disputes arose and a suit was filed in the Darbar of the Nizam Muhi-ud-din Khan Sahip; Kullappa Mudali, a ryot of Valluvampakkam, swore on the *Solingapuram Narashima Swami Hill Temple* that sixteen *etrams* used to bale out water every year; and since then (the year 1782 onwards) 14 *etrams* bale out water within the region where the said *Kachchanadi* channel and the spring channel meet, when there is water in the deep pool (*Maduvu*) of that channel. When the (flow of) water in the *Maduvu* stops, water should not be baled out. The two *etrams* at the *Koonan Kuttai* bale out water, during the time when water is diminishing, as long as water reaches the Picottah *pit* (Ethukkuli). When there is no water, it should not be baled out. These are the (particulars of the) 16 reservoir picottah (*thoti etram*) of Valluvampakkam.
>
> **Second mode:** On the northern side of the Suttukulam (pond) of Cheri there are two *thoti etrams*, with which the people of Ponnappanthangal bale out water for the cultivable lands of the said village which lie in the bed of the Kaveripakkam tank. When the tank is getting dry, water is baled out as long as it reaches the Picottah *well* (*Ethukuli*) by a channel dug not deeper than *mamul*.

Third mode: Two *thoti etrams* are employed by the people of Edayanthangal to bale out water on the southern side of the land lying along the pond in the said place. When the tank is getting dry, water is baled out as long as it reaches the *Picottah well* by a channel dug not deeper than *mamul.*

Fourth mode: The people of Cheri used to put up *thoti etram* and bale out, from the water draining from the tank, water adequate for the irrigation of about 40 *Kani* of the cultivable (*Pattadai*) land which lie in the bed of the tank and extend from Suttukulam to the place opposite to the Maganipattu sluice, and cultivate the same.

Fifth mode: The people of Sumaithangi used to put up in the Eachanthangal Odai (a large watercourse) of the said place one *thoti etram* and one Sadaichi pattadai *thoti etram*, altogether two *thoti etrams*, and bale out water as long as water reaches the *Picottah well*, without digging deeper than *mamul.*

Sixth mode: It is customary for the two villages of Cheri and Pilleri to bale out water of KPT. When the tank is getting dry in this manner, it is customary to bale out water for the fields in the bed of the tank.

Free catchment areas of KPT

Eighteenth column: Particulars of the villages contributing to the free catchment area of KPT are (as follows): In the lands of the villages of Kaveripakkam Eri, Pilleri, Sumaithangi, Ponnappanthangal, Edayanthangal, Bhagaveli, Musiri, and Valluvampakkam an extent sufficient for the filling up of the tank is under water. When the water drains away (from this area), the respective villages put up *etrams* and bale out water as mentioned in column 17, and also sink wells, cultivate wet crops, garden crops and punjai crops like cholam, cumbu, kelvaragu, etc.

Procedures adopted to intercept water from the Palar to KPT

Nineteenth and Twentieth columns: Particulars of men employed for diverting the water of the Palar river for the supply of the aforesaid tank - for 100 men (to be supplied) the proportion is (as follows):

Water flowing in Palar was diverted to the KPT by a *Korambu* constructed, repaired or reconstructed as necessary by 100 *ayacutdars* from 19 ayacut beneficiary villages. Number of *ayacutdars* participated from each village is given as follows: Kaveripakkam (20); Kondapuram (12);

Athipattu (4); Villaham (4); Sirukarumbur (4); Eralacheri (4); Perukarumbur Nanderi (2); Kattalai (8); Abiramacheri (4); Cheri (15) Maganipattu (6) Thaluvanampattu (2); Thuraiperumpakkam (6); Sirasthavalapuram (2); Puthupattu (2); Kilveeranam (1); Panniyur (1); Gangatharanellur (1); and Alapakkam (2). During the flood season, whatever may be the number of men going to the river channel to divert water, two men from Alapakkam village were to remain at the tank and make Paddy *straw fire* to vigil at night on the tank bund to alert the people including the *Talaiyari* (village revenue functionary to assist the Village Administrative Officer), the *Karnam* (village accountant), and the *Maniyam* (village headman) and inform them about the flow measurements from the river channel. They took turns at eating and sleeping. It was compulsory that one of these two persons was always present to watch the water flow.

Whenever there was flow from the river, a *sluice guard* (head fisherman) was also to be present at the tank. His main function was to protect the sluices from damage and to remove the silt deposited in and around it. During the supply period if there was any problem in the sluice, only the *sluice guard* was authorised to undertake the needed repairs.

If supply continued even after it touched the surplus weir at a storage depth of 28 feet, the construction of surplus embankment was to be taken up to a height of 3 feet. Ten of the villages (Kattalai, Abiramacheri, Maganipattu, Thaluvanampattu, Thuraiperumpakkam, Sirasthavalapuram, Kilveeranam, Panniyur, Cheri and Gengatharanallur) had to construct the surplus weir according to the procedure detailed in the *Mamulnamas*. All these villages are at the tail end of the distribution network and therefore stand to benefit more from increased storage in the tank. The remaining persons, allotted for each village as stated above were to keep vigil at the diversion site to make sure that the *Kondam* (a temporary earthen embankment to raise the water level to irrigate elevated lands) built across the river is not breached as long as water flow continued in the river.

During the period when water flowed into the tank, one person was to inform the level of supply available to the inlet channel from the Palar. Another person was to inform the level of supply from two points: the *Singha Kinathalavu* and Kachchi *Nathi* channel. Thus three persons were required to inform the level of flow from their respective places to the officials stationed at Kaveripakkam village. These processes might be continued until water ceases to flow in the inlet channel. These persons also took note of the water level in KPT at *Bommai sluice* (Paracheri sluice). During the process of opening and closing of sluices the *sluice guard* might also be present. During the flood season the bunds of *Sumaithangi* and *Jadayeri* tanks (which are located adjacent to the KPT) should not be raised beyond its original capacity.

Policing the Tank

Twenty-first column: Particulars about watching from huts on the tank-bund are (as follow): After water is collected at *Kadavasal* (surplus weir), the *Thotties, Thalaiyaries* of Ramapuram should erect and live in huts near the Gaffor sluice. The *thoties* and *Thalaiyaries* of the four villages, viz., Kondapuram, Athipattu, Vilaham and Sirukarumbur should erect and live in huts near the Fox sluice. The *Thotties*, *Thalaiyaries* and *kammukutties* of Kaveripakkam should erect and live in huts near the Lion sluice. In Kattalai region the Kattalai service persons; near the Cheri Kanniga temple region the Maganipattu and Thuraiperumpakkam persons; in the Cheri mettu madagu region the Cheri service persons; in Panniyur sluice the Panniyur, Kilveeranam and Gangatharanellur service people should erect and live in huts.

When the *Thotties, Thalaiyaries* and *kammukutties* are present in (their) huts they should inform the happenings of the water flow level and other matters to villagers by beating drums and keeping the people in the calingulla *chavadi* (the place where the chief of a village holds his meetings) informed of what is taking place in their respective limits. In the premises of the Government *chavadi* near the surplus course region the persons to be present are: *miras* fisherman 1, the village headman, the accountant and the Thalaiyaries of the following four villages namely, Maganipattu, Thuraiperumpakkam, Cheri and Abiramacheri and the village servants of these villages. In the head of the surplus course region the Alapakkam *thoties* might erect a hut and stay there always having a *coal fire* (torch). In order to stay the *thotti, thalaiyaries* in the government quarters around the tank bund huts should be erected below the bund. Sufficient materials needed for making the *torch* be brought from the villages located all around the tank bund. The Government servants should carry the *torches* with them, patrol at night on the bund warning the *thotti, thalaiyaries* present in the huts and the government quarters and get things done according to exigencies.

Twenty-second column: The *mamul* for closing up the calingulla is given in Table 1.

In the said proportion construction work in the surplus weir should be done commencing from the south. The stones removed as per proportion,

Table 1 Personnel Allotment for Closing of Calingulla When the KPT is in Full Storage

Sl. No.	*Name of the village*	*Allotment of length (Kal=Stone Measurement)*
1.	Cheri	30
2.	Kattalai	6
3.	Abiramacheri	4
4.	Maganipattu	26
5.	Thaluvanampattu	9
6.	Thuraiperumpakkam	21
7.	Moonjipakkam	5
8.	Sirasthavalapuram	9
9.	Panniyur	5
10.	Kilveeranam	5
11.	Gangatharanellur	5
	TOTAL	125

viz., the 125 stones used by being trampled underfoot well when the *calingulla* was closed, should after the water is safeguarded to its existing level, be secured by the respective parties. In case if the weir is breached, it must be reconstructed by themselves according to the *mamul* as said above.

Twenty-third column: For the purpose of putting up a dam in the river channels, for closing the surplus weir and any *madai* which may occur and for constructing huts, etc., at the tank bund the materials necessary for the work, like the Tamarind sticks, gunny bags, etc., should be brought by the villagers under irrigation by the tank according to the practice of each village.

Maintenance of channels in the Fox and Lion sluices

Twenty-fourth column: The procedures followed to maintain the sluice channel below the tank is given only for a couple of sluices. The FOX sluice channel is dug by the people themselves of Perukarumbur-Nanderi to the east of Kaveripakkam up to the limits of Perukarumbur. Beyond that as far as Agaram madai, the people of Kaveripakkam Perukarumbur Nanderiyar should dig the channel each one-half. Further on up to the sluice the people of Kondapuram and the people of Athipattu, Vilaham and Sirukarumbur should dig each one-half. The big channel (perunkalvai) branching off from the Lion sluice channel up to the *Pappar thakkai* of Kaveripakkam, and the portion of *Maman kalvai* pertaining to Uppumedu up to the burial-ground of children in Kaveripakkam should be dug by the people of Eralacheri themselves. Beyond that, the people of Kaveripakkam should themselves dig the channel.

Twenty-fifth column: Particulars of turns in the Kaveripakkam Lion sluice channel are (as follow): Particulars of the use of water by turns by the people of Kaveripakkam in times of water scarcity when there is a water supply for about forty days are - Uppumedu share turn 1, intermediate share turn 1, half-share Kattalai kal share turn 1, flower garden share turn 1, Kannikoil share turn half; thus for five shares there are four turns. Irrigation is accordingly made. If turns are fixed, water is not supplied to Eralacheri.

Twenty-sixth column: On the bund of KPT, a *madai* need not be opened.

Twenty-seventh column: From the Kachchanathi channel mentioned in the fourteenth column, there is a channel opened for supplying water to the Ponnappanthangal tank within the limits of Valluvampakkam. Nobody should check the flow from the water diverted from this to Ponnappan Thangal tank.

Twenty-eighth column: Near the Koonmadai of the river channel, there is a hut for the *thotis* of Athipattu, Vilaham, Sirukarumbur and Kondapuram to (stay in and) keep watch. The *thotis* of the aforesaid villages should remain in this hut and give information to the Government about the channel.

To this effect is the Mamulnamas executed by the aforesaid *mirasidars* and the other people with free will.

Mamulnamas for the Dusi-Mamandur Tank

The *Mamulnamas* of Dusi-Mamandur Tank (DMT) was written by eight prominent landlords (Venkatachchariar; Arulbu Venkatasesaiyankar; Venkayyar; Chinna Appanaiyyankar; Kuruvi Venkatachchari; Karnam Ramasamy Pillai and Ramakrishna Pillai) on November 16, 1816. The document speaks about the supply of water to the tank and the allocation of water to different villages served by it.

Sources of Supply to DMT

For Dusi-Mamandur tank (DMT), we do not get any detail from the *Mamulnamas*, regarding the number of persons from each village required to participate in the diversion of flow, either from the Rajakkal channel or from the diversion spot at Palar Anicut. However, the general method to be adopted for getting supply to the tank is specified clearly.

According to the specification, when there was no supply from the Palar, two *Kondams* were to be built, one at Tiruppanankadu and the other at Vallakulam to divert the drainage and rain water flowing in the inlet (Thennampattu) channel to DMT. Further, along with the following five *madagus* (sluices) (two in Arasankuppam, one each in Settithangal, Olugampakkam, Vellakulam), the tank surpluses of Hariharapakkam, Cholapuram, Perumbulimedu, Sellaperumbulimedu, Sithathur and Alingilpattu villages and the field drainages also reach DMT through the foreshore area of Chithathur Odai. In the above said manner, if water reaches DMT at FTL, it should be feasible to irrigate the *ayacut* for 18 months.

The *Kondams* at Tiruppanankadu and Vallakulam were to be removed during the Tamil month *Aadi* (July), and the channel should not be obstructed by constructing *Kondam* till the Tamil month Thai (January). The *Kondam* could be built only after Thai. During the month of *Chithirai* (April), when rain water and the surpluses and drainages of Venkalathur village are available, the *Kondam* should be removed. Only if the water stopped itself as a result of poor rainfall or drainages the *Kondam* might be built.

Sluices of DMT: The southern end of the tank have two sluices. One is a plank inserted high level sluice and the other a low level *naligai vasal* (a term used to measure the flow of water which passes through the sluice) plank inserted low level sluice. The northern end of the tank also have two sluices. One is a plank inserted high level sluice and adjacent to it is the other, a plank inserted *naligai vasal* low level sluice. Hence a total of four sluices, of which the first two are called (elevated) DUSI sluices and the other two are called MAMANDUR (deep) sluices.

Method of Water Distribution during Normal Supply Period

When DMT is full, the *high level sluice* of Mamandur should be opened first, and the water shall be released up to the mark touching the *Pillaiyar Thumbikkai* (a carved stone structure available at the sight of the sluice). At that time *Kondam* should be built across the *Sothiyampakkam* channel and the *Anjuran* channel to irrigate the lands located at the lower region. For this purpose a *kondam* is built with 2.5 feet height and 6.25 feet breadth, in which 1.5 feet of water is diverted to the low lying lands and the remaining water is let into *Naalu Karadi Kuttai* (a place located at the Mamandur region of Kavalmaniyam lands), in which the water passes to 3 *thoti*, but supply available for irrigation is only 2 *thoti*. The share of water between *Anjuran* channel and *Sothiyampakkam* channel is 4:1.

Water Distribution through the Sothiyampakkam Channel

Water passing through the Sothiyampakkam channel is used by **Narasamangalam** village limit by through a thoti. There are in all 13 such *thoties*. **Three** are Vaniyam Piravidai *thoties*. **Two** are in the western side of the village, **one** further north, **one** further east (Alamarathu *thoti*); **one** further south near *thennamaram*, **two** further east of Amman temple, **one** further south (Maniyam *thoti*), **one** in Ottanthangal Ulvai, and **one** in the upper portion of *Venkatrayan Chattra Thoppu*. These 13 *thoties* should not be operated at a time, but shall be operated one after one. By this method, all the *thoties* can be operated for **a day of 60 *Naligai*** (measuring the time period of 24 minutes per *naligai*). Below this, *Oru Koodai* (a local flow measurement) water is diverted to Poondi lands and further west near the *Eachen Kuttai* lands *Oru Koodai* water is diverted by placing Palmyra woods across the channels. Further below in the **Valavandal** limit to irrigate the lands of that village and above the *Anjuran* channel water is diverted at a local measure of 1 *meri* (9 inches) breadth and 7 inches height by placing Palmyra woods. Below it, water goes to **Sothiyampakkam** limits to irrigate the lands of that village. The measurable limit of supply to this village is 3 feet breadth and 6 inches height. To the village of **Bagavanthapuram** water passes at a measure of 2 feet breadth and 6 inches height. **One share water of Sothiyampakkam channel is utilised in the aforesaid manner.**

Distribution of Water through Anjuran Channel

The total share of water in this channel is **four Pangus**, in which there is a *mamul right* to irrigate the lands in **Narasamangalam** through 11 *madais*. These are: Vannanthurai madai **one**, further south madai **one**, further east madai **one**, further north madai **one**, Pidari koil madai **one**, further lower portion madai **one**, further east madai **one**, further east madai **one**, further east madai **one**, Pakkirimanya madai **one** and further south madai **one**. These 11 madais should be opened during irrigation time. At other times they must be closed. In Narasamangalam village near *Komutti* lands 2 *Kani* is irrigated by blocking the Anjuran channel. After irrigation it should be removed. In that place in order to irrigate the lands of Narasamangalam and **Mamandur** a channel passes through the lands of Venkatrayan *thottam* upper portion of Karuma Maniyam at a measure of one foot breadth and 6 inches height.

In Narasamangalam limit a channel is available to irrigate the *Chinna Mandraradi Kanakku Maniyam* lands. To irrigate *Karuma maniyam* lands located in the three villages of Mamandur, Poondi and **Valavandal** a *mattu* was constructed at a measure of 3 feet breadth and 9 inches height. From this *mattu* at a measure of 2 feet breadth and 6 inches height of water is diverted separately in a channel to irrigate the lands of Mamandur and Poondi. The water flowing in the Poondi channel irrigates Poondi and Valavandal lands equally. The share of water to Poondi is 3/8 *Kani*, which should be irrigated by obstructing and diverting the Mamandur Vaiyali channel. After irrigation it should be removed. The water flowing to the above three villages, namely, *Mamandur, Poondi* and *Valavandal* is blocked near Poondi border to irrigate 1 Kani land of that village and after irrigation it should be removed. From this channel 1/2 *Kani nanjai* land might also be irrigated. In the channel, which irrigates the above said three villages, one *Kannikalvai* (small field channel) passes to the east of the *kondam* to irrigate Poondi lands. Further east, the channel is diverted to irrigate 3/8 Kani and then removed. Further east one *Kannikalvai*, in which 1/2 *Kani* land is irrigated by erecting a *kondam* and after irrigation it should be removed. Further east one *Kannikalvai*, further east near paracheri one *Kannikalvai*. Hence the total *Kannikalvai* for Poondi is 5, in which water to be diverted whenever it is required and in other times it must be closed.

Further east water flows to irrigate the lands of Mamandur, **Sothiyampakkam**, and Valavandal. Water passing to irrigate the lands in *Sithalapakkam* might go in a channel directly. The *Kiribudasu Maniyam* lands in Mamandur is irrigated by a kannikalvai, which takes-off from **Valavandal Channel**. Further east water flows to Valavandal channel to irrigate the above said 3 villages. The Anjuran channel irrigates Valavandal village. In this channel a *mattu* was built at a measure of 6 inches breadth and 4 inches height to irrigate the lands located at the border of Valavandal in the south of flower garden of *southern mettukaran odai*. Further south of Valavandal border, the channel is blocked by *andai* (a partial diversion) method to irrigate the lands and after irrigation it should be removed. Further east of the flower garden lands water is diverted to *moolamadai* channel at a measure of one foot breadth and 6 inches height. Within the borders of Valavandal there are two channels to irrigate the lands of that village.

Further, there is one east going channel to supply water to **Vadakalpakkam** lands in which a *mattu* was constructed at a measure of 1 foot breadth and 6 inches height. Further east, there is one channel with a measure of 6 inches breadth and 6 inches height to supply water. Hence the Vadakalpakkam channels are two.

Towards the east, to supply water to **Mannur** lands, a *mattu* was constructed in a channel with 9 inches breadth and 6 inches height from which water is diverted. Further east, to supply water to **Kilnayakanpalayam** a channel is diverted, for which a *mattu* was erected at a measure of 1 foot breadth and 6 inches height. Towards east there is a *kannikalvai* to supply water to **Girijapuram**. In the east going channel a *mattu* was raised with 9 inches breadth, and 6 inches height to divert water further down. Hence there are two channels for Girijapuram. These might be operated alternately. Further east, a channel supplies water with one foot breadth and 6 inches height. Hence at the border of Girijapuram village there are three channels in which only two supply water.

Further east of Girijapuram limit, to supply water equally to **Menallur** and **Punaithangal** villages a diversion structure was built by plank at a measure of 1 foot breadth and 6 inches height. In Menallur branch channel to irrigate 3/4 *Kani* in Girijapuram the *Oradi* channel is blocked and it should be merged with Menallur, and Punaithangal channel, from which 1/3 water to be diverted. The remaining 2/3 water is let along with the Menallur channel and diverted half the supply to Menallur and the remaining half to Girijapuram. After irrigation the *Kannikalvai* should be closed and the *Oradi* channel of Girijapuram should be opened.

This is the method of water distribution when DMT reaches its full capacity

During **Scarcity period** the water distribution procedures in *Anjuran* channel are as follows: The village **Narasamangalam** is having 11 *madais*, for which water is diverted according to the availability, i.e., if more supply comes more water is let into the channel and vice-versa. The remaining water goes to Karuma *maniyam madai* which supplies water to the villages of **Mamandur, Poondi and Valavandal**. Further remaining water is let into the Valavandal channel according to the availability of water, and the remaining water goes to Valavandal border. There the *Moola madai kani* 1/4 should be irrigated and then it is be allowed to pass further down.

Further east, at the border of **Vadakalpakkam** the channel is obstructed fully and the supplies are utilised for 40 *naligai* and then opened out. Further down the channel passes to **Mannur** village, where it is obstructed and the Vadakalpakkam channel is also obstructed to irrigate **20 naligai** (8 hours) then opened. The Mannur channel is closed and all waters are diverted to **Kilnayakanpalayam** border. At that point the channel should be fully obstructed to irrigate the lands for 30 *naligai* (12 hours) in Kilnayakanpalayam and then the channel should be opened.

Water further passes to **Girijapuram** border. There it is stopped to irrigate the lands of that village for 60 *naligai* and then let out to **Menallur** there the water is obstructed fully to irrigate that village lands for 30 *naligai* then the water further flows to **Punaithangal** border. There it irrigates that village lands for 30 *naligai*.

This is the *mamul* method and procedure adopted for water distribution in Anjuran channel during scarcity period.

During scarcity period, when the low lying lands face inadequate supply, the eastern side of the cistern should be opened. During scarcity when the supply does not reach the elevated lands, the **back cistern** (*thoti*) of the **Low Level Sluice** should be opened to irrigate the lands of **Narasamangalam, Mamandur and Poondi**. For this purpose, in the low lying channel, a *kondam* is built to irrigate the fields of the aforesaid villages and then the flow is diverted to the low lying lands of Mamandur, Valavandal and Kilnayakanpalayam. The remaining water irrigates the *Kaval maniyam madai* lands in Mamandur village. Further remaining water is lifted by 2 **channel cisterns** (*thotis*).

Further east in Narasamangalam border there are 4 *madais* and 6 *Kannikalvai* available to irrigate that village. These are: Maniyam Thoppai *madai*; Palla Koneri *madai*; further east *madai*; and a further east *madai*. Towards the east of these *madais* there is **one** *kannikalvai*, further east **one** *kannikalvai*, further east one *kannikalvai*, further east **one** *kannikalvai*, further east **one** Kaval maniyam *kannikalvai*, and further east **one** Koothadi maniyam *kannikalvai*. These 6 *kannikalvai* and 4 *madais* are operated according to the needs of irrigation, i.e., whenever it is required. In other times it should be closed.

Within the Narasamangalam region there is a customary practice to lift water by installing channel cisterns (thoti etram). By this arrangement, there is one cistern near *Pidari koil*. It should be operated according to the needs of irrigation. Further east near *Pakkiri maniyam* lands there is one cistern. It can be operated continuously throughout day and night. However, when the cistern is in operation the above said first cistern (pidarikoil) should not be operated. Further east, in *Mannaradi* there is one cistern. It could be operated according to the needs of irrigation. When the last said cistern is in operation the first two cisterns must be stopped. Hence there are three *channel cisterns* in which 2 can be operated alternately and the remaining one can be operated throughout day and night.

Further east, one Karuman maniyam channel, which irrigates the lands of Narasamangalam and Mamandur. For which there is a measure to divert water at 1 foot breadth and 4 inches height in the channel. Towards east in the Poondi channel, 1/3 of water is let out and the remaining water is released to Mamandur channel.

During scarcity period, in Anjuran channel, the 6 kannikalvai and 4 madais of Narasamangalam and 5 madais of Mamandur should be given proper supply and the remaining water along with the Poondi channel water should be obstructed to irrigate the Poondi lands for a day, and Mamandur lands for two days. This is the *mamul* procedure to irrigate the lands of Poondi and Mamandur by the big channel irrigation method (*Perunkalvai Pasana Murai*).

To irrigate the lands of Narasamangalam, Poondi, Mamandur and Valavandal a channel passes through the border of Mamandur. In the Mamandur Upper sluice (elevated) channel there is a *kondam* built to supply water to the lower sluice channel. For which water is diverted at a measure of 2 and 1/2 feet breadth and 1 and 1/2 feet height. This water is let into the lower sluice back cistern. In which two shares are let into the lower sluice channel and one share might be given to the upper sluice channel.

In Mamandur region there are 3 *madais* and 1 *kannikalvai*. The *madais* are: Nallanthangal *madai*, further below one *madai*, and further below one *madai*. Further below one *kannikalvai*. Further east in Narasamangalam region the channel is blocked to irrigate 2 and 1/2 *Kani* lands in that village by erecting a *Kondam* and after irrigation it should be removed. Further east in Mamandur region 1/4 *kani Kanakku maniyam* and 1/4 *kani Veerapathu* can also be irrigated by erecting a *kondam*. Then it should be removed.

At the border of Narasamangalam there is a *mamul* to construct thoti to irrigate the lands of that village. There are three such thotis available in which two are under operation. During irrigation time according to *mamul* water should not be lifted by using one thoti above which another thoti. Two acres of nanjai lands are given *sedai* (irrigating harvested dry paddy field) to raise crops by that land. At Poondi border one thoti is operated throughout day and night to irrigate that village. But water should not be lifted by using one thoti above which another *thoti*. One *kani* nanjai land at Poondi can also be irrigated. The villagers of Narasamangalam, Poondi and Mamandur may construct *kondam* once in 6 days to irrigate 3 *kani* lands. Further water passes to Valavandal in the above said procedure and 1 and 1/2 share of water is given to that village and one share of water is diverted to Mamandur high level lands for irrigation. During scarcity period Mamandur farmers dig the channel and it should be joined with the big channel which passes through the border of Mamandur village. From this channel the wet lands of Mamandur are irrigated.

Operational procedures for low level sluice: After depletion of water supply in the upper sluices, the low level sluices are opened out. During that time Vadakalpakkam, Mannur, Kilnayakanpalayam and Girijapuram villages also get the supply by *Oru Madai Paichal* method. In which Vadakalpakkam gets supply for 30 *naligai* (12 hours) during night and three *naligai* in day time. Mannur village gets the entire supply in the day time of 17 *naligai* (6.8 hours). Kilnayakanpalayam gets a full supply of 10 *naligai* in day time, 30 *naligai* in the night and 10 *naligai* the next day. Girijapuram gets 20 *naligai* in day time of full supply. Hence a total of 120 *naligai* or two days (48 hours) are required to complete all the aforesaid lands for one full irrigation. In the above said villages no **banana**, **sugarcane**, and **betel leaf** cultivation.

Appendix 7

Appendix 7.1

Method of Measuring Cropping Intensity

Cropping intensity (CI) refers to the actual area cropped in the three seasons - counted four months each - in a year. CI is obtained by dividing the GCA (in terms of season hectares) by NSA.

CI illustration: A farmer own 2 hectares of land holdings and is cultivating paddy crop in all the 3 seasons. He grows 1 hectare of long duration (5 months) paddy in the first season, 1 hectare of medium duration (4 months) paddy in the second season, 1 hectare of short duration (3 months) paddy in the third season, and leaves the remaining land uncultivated in each season. In season hectares method, we assume 4 months duration for each season. CI is thus calculated in the following manner:

Method	NSA	Season I	Season II	Season III	GCA	CI = GCA/NSA X 100
1	2	3	4	5	6	7
Season-Hectares	2	1	1	1	3	CI = (3/2) X 100 = 150%

The same procedure is also applied for annual crops such as sugarcane and banana. For example, if a farmer raised 2 hectares of banana or sugarcane in the first season, the same hectare is counted for the second and third seasons also. Hence the GCA is 2+2+2 = 6 (season) hectares out of his total land holdings. The CI for this cultivated land is 300%.

Appendix 7.2

Method of Calculation of Gross Value Output

The output of different crops on each sample holding harvested in the latest year 2015-16 and in the preceding supply year, 2005-06 and also the survey year 2016-17 is valued at the actual wholesale prices prevailing in Kancheepuram (the nearest taluk centre) during 2016 and 2006. Although the survey year of the latest two year varies, for analytical purposes we used only 2015-16 wholesale prices for both the tanks. This procedure is followed to get the value of output of all crops, viz., paddy, groundnut, kelvaragu, vegetable, and gingelly, except banana and sugarcane.

For banana and sugarcane the output value is arrived at in the following manner:

Banana: For this crop we directly got the actual selling price of banana, instead of asking 'bunches of banana' available from the land. Suppose the crop was not yet harvested the approximate expected value of the crop in the latest reference year and the actual price fetched for banana including banana leaves and other by-products in the preceding supply year were asked. Hence, banana value was directly available in monetary terms.

Sugarcane: A few farmers in the KPT villages sold the cane for the Vellur *Ammundi* Sugar Factory, which fixed a rate on the basis of government directive every year and all the farmers reported getting the exact price without any difficulty. In the latest survey year (2015-16) one tonne sugarcane was valued at Rs.2850. In the preceding year (2005-06) it was Rs.850 per tonne. The yield of sugarcane was asked in terms of tonnes per hectare in each sample holding. On the basis of the above measure we have arrived at the Gross Value of Output (GVO) for sugarcane. In DMT, banana and sugarcane were not at all raised by any farmer.

Although we have got the value of banana and sugarcane crops in the above said method, we have come across wide variations in the number of seasons when the crops available on the land, especially, during the latest two tank supply years. For instance, at the time of interview we got 4 types of responses of the crop availability on the sample holdings, viz., (1) increase in area in the second or in the third season than in the first; (2) decrease in area in the second or in the third season than in the first; (3) harvesting of banana and sugarcane in the first season itself (as it was planted previously); and (4) harvesting of banana and sugarcane in the second season itself (as it was planted previously).

As a result of these varied responses we faced some problem in measuring the value of output for these two crops. So, to arrive at the nearest value the following procedure was used:

In the sample holdings respondents reported that the total number of months required for getting one banana crop (only Erode Munthan variety was cultivated in all the villages) varied from 14 to 18 months. Hence we averaged it for 16 months as a normal period. However, in some cases respondents reported the exact number of months, which is certainly below the average, i.e., 13 months only. In such cases we have also used that to arrive at the value of output. For sugarcane the period of crop did not exceed more than a year and the minimum number of months required to harvest was reported as eight by a couple of farmers for the transplanting variety of sugarcane.

On the whole, the procedure followed to get the output value is as follows:

(1) Both for banana and sugarcane, yield was converted into money value.

(2) To calculate the "month" yield, money value was divided by the number of months when the crop was available on the land.

(3) To arrive at the "seasonal" value, month yield was multiplied into four (i.e., four months per season).

(4) When the crop was raised in the second or third season we have calculated the value of the crop by using the "month yield" procedure.

(5) If the crop was available on the land throughout the year the value was averaged for the three seasons.

(6) In some cases area under banana varied in each season, where the value was arrived at by using the following procedure:

  (a) First to get the value of output per season;

  (b) Seasonal value is divided by the total hectare under the crop;

  (c) From the value of per hectare output, the value of seasonal output is derived by multiplying the per hectare value into actual extent of land cultivated in each season.

Overall, the method used to arrive at the value of per hectare output is, after getting the money value of each crop it is aggregated across the sample holdings and divided by the reported gross sown area in the year to get the value of production per unit of gross cropped area (GCA).

GCA is in terms of "season hectares" of cropping.

Bibliography

Agricultural Research Centre. 1962. *Dusi Village: A Re-Survey*. Unpublished Report. Madras: University of Madras.

Agro-Economic Research Centre. 1960. *Dusi: A Resurvey*. Mimeo-graphed Report. Madras: University of Madras.

Athreya, V.B. *et al.* 1990. *Barriers Broken: Production Relations and Agrarian Change in Tamilnadu*. New Delhi: Sage Publications.

Chambers, Robert. 1977. Men and Water: The Organisation and Operation of Irrigation. In B.H. Farmer. (ed.). *Green Revolution? Technology and Change in Rice-Growing Areas of Tamil Nadu and Sri Lanka*. pp. 340-61. London: The Macmillan Press Ltd.

____. 1980. Basic Concepts in the Organisation of Irrigation. In E.W. Coward. (ed.). *Irrigation and Agricultural Development in Asia*. pp. 28-50. New York: Cornell University Press.

____. 1988. *Managing Canal Irrigation: Practical Analysis From South India*. Cambridge: Cambridge University Press.

Coward, Jr. E.W. 1980. *Irrigation Development: Institutional and Organisational Issues* In E. W. Coward. (ed.). Irrigation and Agricultural Development in Asia: Perspectives from the Social Sciences. pp. 15-27. New York: Cornell University Press.

____. 1986. State and Locality in Asian Irrigation Development: The Property Factor. In K.E. Nobe and R.K. Sampath. (eds.). *Irrigation Management in Developing Countries: Current Issues and Approaches*. pp. 491-508. Boulder: Westview Press.

____. 1987. Action Experiments in Irrigation Development: Iden-tifying Policy and Program Implications. In W.R. Jordon. (ed.). *Water and Water Policy in World Food Supplies*. pp.361- 66. Texas: A & M University Press.

Kumar. Dharma. 1965. *Land and Caste in South India: Agricultural Labour in the Madras Presidency During the Nineteenth Century.* Cambridge: Cambridge University Press.

David Mosse. 1999. Colonial and Contemporary Ideologies of 'Community Management': The Case of Tank Irrigation Development in South India. Modern Asian Studies, Vol. 33, No. 2 (May, 1999), pp. 303-338

Dikshit, G.S. *et al.* 1994. *Tank Irrigation in Karnataka: A Historical Survey*. Bangalore: Gandhi Sahitya Sangha.

Dhawan, B.D. 1988. *Irrigation in India's Agricultural Develop-ment: Productivity, Stability, Equity*. New Delhi: Sage Publica-tions.

Downing, Theodore E and Gibson, McGuire. (eds.). 1974. *Irrigation Impact on Society*. Tucson: University of Arisona Press.

Elumalai. G. 1982. *Farmers' Organisations for Efficient Water Use in Irrigated Agriculture: A Case Study in Tamil Nadu*. Banga-lore: Indian Institute of Management.

Esman, M.J. and N. Uphoff. 1984. *Local Organisations: Intermedi-aries in Rural Development*. New York: Cornell University Press.

Fernea, Robert. 1970. *Shaykh and Effendi: Changing Patterns of Authority among the El Shabana of South Iraq*. Cambridge: Harvard University Press.

Freeman, D.M. and M.K. Lowdermilk. 1981. Sociological Analysis of Irrigation Water Management: A Perspective and Approach to Assist Decision Making. In C.S. Russell and N.K. Nicholson. (eds.). *Public Choice and Rural Development: Resources for the Future*. Pp.153-73. Boulder: Westview Press.

Freeman, D. M. *et al.* 1989. *Local Organisation for Social Devel-opment: Concepts and Cases of Irrigation Organisation*. Boulder: Westview Press.

Geertz, C. 1967. *Organisation of the Balinese Subak*. In E. W. Coward. (ed.). Irrigation and Agricultural Development in Asia: Perspectives from the Social Sciences. pp. 70-90. New York: Cornell University Press.

Gray, R.F. 1963. *The Sanjo of Tansanyika: An Anthropological Study of an Irrigation-based Society*. London: Oxford University Press.

Guhan, S and K. Bharathan. 1984. *Dusi: A Resurvey*. Working Paper No. 52. Madras: Madras Institute of Development Studies.

Hart, Henry. 1961. *Administrative Aspects of River Valley Develop-ment*. Bombay: Asia Publishing House.

____. 1978. Anarchy, Paternalism or Collective Responsibility under the Canals? *Economic and Political Weekly*. 13(51-52): A 125-34.

Hunt, Eva and Hunt, Robert, C. 1976. Canal Irrigation and Local Social Organisation. *Current Anthropology*. 17(3): 125-34.

IIMI (International Irrigation Management Institute). 1994. *International Conference on Irrigation Management Transfer*. (Three Volumes). Draft Conference Papers. Wuhan. China.

Janakarajan, S. 1993. In Search of Tanks: Some Hidden Facts. *Economic and Political Weekly*. 28(26): A 53-60.

Jayaraman, T.K. 1981. Farmers Organisation in Surface Irrigation Projects: Two Empirical Case Studies from Gujarat State. *Economic and Political Weekly*. 16(39): 89-98.

Jegadeesan, M. and Fujita Koichi. 2011. Deterioration of the Informal Tank Institution in Tamil Nadu: Caste-based Rural Society and Rapid Economic Development in India. Southeast Asian Studies, Vol. 49. No. 1. June. Pp. 93-123.

Kelly, W. W. 1980. *Water Control in an Agrarian State: Irrigation Organisation in a Japanese River Basin*. Ph.D. thesis. London: University Microfilms International.

____. 1982a. Irrigation Management in Japan: A Critical Review of Japanese Soil Science Research. *East Asian Papers No. 30*. New York: Cornell University.

____. 1982b. Water Control in Tokugama, Japan: Irrigation Organi-sation in a Japanese River Basin, 1600-1870. *East Asian Papers No. 31*. New York: Cornell University.

Korten, F.F. 1982. Building National Capacity to Develop Water Users' Associations: Experience from the Philippines. *World Bank Staff Working Paper No. 528*. Washington, D.C.

Krishnaswami, S.Y. 1947. *Rural Problems in Madras*. Madras: Gov-ernment Press.

Leach, E.R. 1961. *Pul Eliya: A Village in Ceylon*. Cambridge: Cambridge University Press.

Levine, Gilbert. 1980. The Relationship of Design, Operation and Management. In E.W. Coward. (ed.). *Irrigation and Agricultural Development in Asia*. New York: Cornel Uni. Press.

Lusk, M.W. and P.J. Riley. 1986. Irrigation Development and Public Choice Theory. *The Rural Sociologist*. 6(4): 280-89.

Maass, A. and R. Anderson. 1978. ... *and the Desert Shall Re-joice: Conflict, Growth, and Justice in Arid Environment*. Cam-bridge: MIT Press.

Madras Institute of Development Studies. 1986. *Tank Irrigation in Tamilnadu: Some Macro and Micro Perspectives*. Mimeo. Madras.

____. 1988. *Tamilnadu Economy: Performance and Issues*. New Delhi: Oxford and IBH Publishers.

Martin, E. and R. Yoder. 1983. Water Allocation and Resource Mobilisation for Irrigation: A Comparison of Two Systems in Nepal. *Nepal Studies Association*. Madison: University of Wiscon-sin.

McNeill, W. 1963. *The Rise of the West: A History of the Human Community*. Chicago: University of Chicago Press.

Millon, Rene. 1962. Variations in Social Responses to the Prac-tice of Irrigated Agriculture. In Richard B. Woodbury. (ed.). *Civilisation in Desert Lands*. USA: Utah State University.

Mukundan, T.M. 1988. *The Eri System of South India*. Madras: PPST Bulletin No. 16.

Nadkarni, M.V. 1979. *Impact of Irrigation: Canal, Well and Tank Irrigation in Karnataka*. Bombay: Himalaya Publishing House.

Narayana, D. *et al*. 1982. An Approach to Study of Irrigation: Case of Kanyakumari District. *Economic and Political Weekly*. 17(39) A 85-102.

Narayanamoorthy, A and Palli. 2017. "Can 'Kudimaramath' refill tanks in TN?". The Hindu Business Line. August 24. P. 9.

Narayanamoorthy, A and R. Suresh. 2016. Does Urbanisation Affect Tank Irrigation Development in Tamil Nadu? A Macro-Level Analysis. MIDS Journal. *Review of development & Change:* Vol. XXI. No.2. July-December. Pp. 11-34.

Nickum, J.E. 1980. Local Water Management in the People's Republic of China. In E.W. Coward. (ed.). *Irrigation and Agricultural Devel-opment in Asia*. New York: Cornell University Press.

Olson, Mancur. 1975. *The Logic of Collective Action: Public Goods and the Theory of Games*. Cambridge: Harvard University Press.

Ostrom, Elinor. 1992. *Crafting Institutions For Self-Governing Irrigation Systems*. San Francisco: ICS Press.

Owens, E. and R. Shaw. 1972. *Development Reconsidered: Bridging the Gap Between Government and People*. Oxford: Oxford University Press.

Palanisamy, K. 1983. *The Tanks of South India: A Potential for Future Expansion in Irrigation*. Minnesota: University of Minneso-ta.

Pant, N. and Verma R.K. 1983. *Farmers' Organisation and Irriga-tion Management*. New Delhi: Ashish Publishing House.

Patil, R.K. 1981. *Group Management of Irrigation Water Distribu-tion*. Bombay: National Institute for Bank Management.

Rajagopal, A. 1991. *Water Management in Agriculture with Special Reference to Irrigation Institutions: A Comparative Study of Canal and Tank Commands in Cauvery Basin, Tamilnadu*. Ph.D. the-sis. New Delhi: Jawaharlal Nehru University.

Rajendran, S. 2018. Tamil Nadu Revives Kudimaramathu: Ancient Wisdom of Water Management. E&PW, Vol LIII. No. 6. 10, February. Pp. 18-20.

Raju, Sarada. 1941. *Economic Conditions in the Madras Presidency. 1800-1850*. Madras: University of Madras.

Ramanathan, S. n.d. *Controversy around Kudimaramat in Madras Presidensy*. Unpublished.

Ramanathan, S. and Ghose. 1994. Irrigation Management Turnover: A User's Perspective: The Case of the Indira Gandhi Canal, Rajas-than. India. In *IIMI and Wuhan University of Hydraulic and Elec-trical Engineering*. Wuhan. China. Volume I: 173-84.

Ratna Reddy. V. and Bhagirathi Behera. 2009. The economic and ecological impacts of tank restoration in South India. European Journal of Development Research · February.

Reddy, D. Narasimha. *et al.* 1988. *The Decline of Tank Irrigation: The Continuation of the Colonial Legacy in the South*. Mimeo. Hyderabad: University of Hyderabad.

____. *et al.* 1990. Decline of Traditional Water Harvesting Systems: Tanks in the Drought Prone Areas of Andhra Pradesh. Seminar on *Traditional Water Harvesting Systems in India*. October 9-11. New Delhi. Centre for Science and Environment.

Reidinger, R.B. 1974. Institutional Rationing of Canal Water in Northern India: Conflict between Traditional Patterns and Modern Needs. *Economic Development and Cultural Change*. 23(1): 79-104.

Sakthivadivel. R. P. Gomathinayagam and Tushaar Shah. 2004. Economic and Political Weekly. July 31. Pp. 3521-3526.

Schumacher, E.F. 1973. *Small is Beautiful: A Study of Economics as if People Mattered*. London: Blond and Briggs.

Seenivasan. R. and J. Kanagavalli. 2014. Dying Tanks in Urban Areas: What can be done with them? MIDS Journal. *Review of Devpt & Change:* Vol. XIX. No.1. Jan-June. Pp. 109-122.

Sengupta, Nirmal. 1980. The Indigenous Irrigation Organisation in South Bihar. *The Indian Economic and Social History Review*. 17(2): 157-89.

____. 1982. *Tank Irrigation in Gangetic Bihar*. (Mimeo). Patna: A.N. Sinha Institute of Social Studies.

____. 1984. *A Political Economic Study of an Indigenous Irriga-tion System* Ph.D. thesis. Calcutta: Indian Statistical Institute.

____. 1985. Irrigation: Traditional vs Modern. *E&PW*. 20(45-47): 1919-38.

____. 1991. *Managing Common Property: Irrigation in India and the Philippines*. New Delhi: Sage Publications.

____. 1993. *User-Friendly Irrigation Designs*. New Delhi: Sage Publications.

Shankari, Uma and Esha Shah. 1993. *Water Management Traditions in India*. Madras: PPST Foundation.

Singh, P.B. 1988. *Betrayal of Rural India*. New Delhi: B.R. Publishing Corporation.

Sivasubramaniyan, K. 1985. *Economics of Water Management under Canal Irrigation*. M. Phil. Dissertation in Economics. Tiruchirappalli: Bharathidasan University.

———. 1995. *Irrigation Institutions in Two Large Multi-Village Tanks of Tamil Nadu: Structure, Functioning and Impact*, PH.D. Thesis (unpublished), MIDS, Madras: University of Madras.

———. 1997. "Irrigation Management under Different Conditions: A Study of Large Tank Systems in Tamil Nadu". Nepal Water Conservation Foundation Journal. *Water Nepal:* Vol.5. No.2. July-December.

———. 1998. "Maintenance of Irrigation Networks under Major Tanks in Tamil Nadu". MIDS Journal. *Review of development & Change:* Vol.III. No.2. July-December.

———. 1998. "Irrigation under the Palar Anicut System: A Case Study from Tamil Nadu". Gokhale Institute of Politics and Economics Journal. *Artha Vijnana:* Vol. XL. No.4. December.

———. 2000. Tank Irrigation under the Periyar Vaigai Basin: A Historical Perspective". Institute for Social and Economic Change Journal.

Journal of Social and Economic Development: Vol. III. No.2. July-December.

———. 2000. "Impact of Irrigation on Cropping Intensity, Cropping Pattern and Productivity under Tank Commands". Gokhale Institute of Politics and Economics Journal. *Artha Vijnana*. Vol. XLII. No.4. December.

———. 2001. "Tank Irrigation in South India: An Overview". In A.Vaidyanathan (ed.) *Tanks of South India*. Centre for Science and Environment. New Delhi. (Co-authored with A. Vaidyanathan).

———. 2002. "Alapakkam - A Resurvey". Working Paper: No.173. *MIDS*. August 2002.

———. 2003. (With Vaidyanathan. A. and S. Mariasusai). "Conditions and Characteristics of Well Irrigation under Palar Basin, Tamilnadu". Working Paper: No.179. *Madras Institute of Development Studies*.

———. 2006. "Sustainable Development of Small Water Bodies in Tamil Nadu" Economic *and Political Weekly* 30 June 2006. Vol. XLI No. 26.

———. 2006. (With R. Sakthivadivel). "Land and Water Productivity in Tank Irrigation: A Comparative Analysis". In Sailabala Debi and V.B. Annigeri (eds.) *Sustainable Development and the Indian Economy: Issues and Challenges*. Serials Publications. New Delhi (for CMDR Dharwad). (Co-authored with Dr. R. Sakthivadivel).

———. 2007. "Sustainable Development of Small Water Bodies in Tamil Nadu." In K.A. Rasure (ed.) *Environment and Sustainable Development*. Vol. II. Serials Publications. New Delhi.

———. 2009. (With V. Gandhiraj) "Irrigation in Tamil Nadu – With Special Reference to Tank Irrigation." Monograph. No. 8. *Madras Institute of Development Studies*. August.

———. 2011. Irrigated Agriculture in Tamil Nadu: The Case of Parambikulam Aliyar Project. Simres Publications, Chennai.

———. 2013. "Participatory Irrigation Management under Tanks: The Case of IAMWARM Project." In M. P. Vasimalai & A. Gurunathan (eds.) Governance for Advancing Development (Compendium of Papers Presented at Madurai Symposium 2013), DHAN Vayalagam [Tank] Foundation, Madurai. September 2013: pp. 2.2 to 2.69.

———. 2014. Review of the book: "Transforming Indian Agriculture INDIA 2040 Productivity, Markets and Institutions" by Marco Ferroni (Ed). *Review of Development & Change*. Vol. XIX. No. 2, July-Dec. pp. 117-120

———. 2016. "Performance of Irrigation in Tamil Nadu: A Macro Perspective." Working Paper No. 224. *Madras Institute of Development Studies*. December. Pp. 10-11.

———. 2017. "Tank Irrigation in Tamil Nadu: with Special Reference to Palar Anicut System." In A.N. Sinha Institute of Social Studies Journal, *Journal of Social and Economic Studies, Patna.* . XXVII, No. 1, November. Pp. 46-75.

Siy, R.Y., Jr. 1982. *Community Resource Management: Lessons from the Zanjera*. Quezon City: University of the Philippines Press.

Slater, G. (ed.). 1918. *Some South Indian Villages*. London: Oxford University Press.

Steward, Julian H. 1949. Cultural Causality and Law. Initiation of a Research Trend: Wittfogel's Irrigation Hypothesis. In *Socie-ty and History Essays in Honour of K. A. Wittfogel*. G.L. Ulman. (ed.). New York: Morton.

Tang, Shui Yan. 1992. *Institutions and Collective Action Self- Governance in India*. San Francisco: ICS Press.

Thamodaran, R. *et al.* 1982. An Economic Analysis of Water Management Systems in Southern Tamilnadu: Production Func-tion and Programming Approach. *Indian Journal of Agricultural Economics*. 37(1): 43-56.

Vaidyanathan, A. 1984a. *Organisation and Management of Water Control in China*. V. R. F. Series No. 113. Tokyo: Institute of Developing Economies.

____. 1984b. Organisation of Water Management in Selected Tanks in Tamilnadu. *NABARD Project Study*. Madras: Madras Institute of Development Studies. Unpublished.

____. 1985. Water Control Institutions and Agriculture: A Com-parative Perspective. *Indian Economic Review*. 20(1): 25- 83.

____. 1987. Irrigation and Agricultural Growth. *IJAE*. 42(4): 503-27.

____. 1988. *India's Agricultural Development in a Regional Per-spective*. R.C. Dutt Lectures on Political Economy. Calcutta: Orient Longman Ltd.

____. 1991a. *Integrated Watershed Development: Some Major Issues*. Foundation Day Lecture. Society for Promotion of Wastelands Development. New Delhi: SPWD.

____. 1991b. Critical Issues Facing Indian Irrigation. In Ruth Meinzendick and Mark Svendsen. (eds.). *Future Directions for Indian Irrigation:*

Research and Policy Issues. International Food Policy Research Institute. Washington, D.C.

____. 1992. *Strategy for Development of Tank Irrigation*. Madras: Madras Institute of Development Studies.

____. 1994a. *Second India Studies Revisited: Food, Agriculture and Water*. Madras: Madras Institute of Development Studies.

____. 1994b. Transferring Irrigation Management to Farmers. *Economic and Political Weekly*. 29 (47): 2965-67.

———. 2001 (ed.) *Tanks of South India,* Centre for Science and Environment, New Delhi.

Vaidyanathan, A and S. Janakarajan. 1989. *Management of Irriga-tion and its Effect on Productivity: A Study of Two Surface Irrigation Systems in Tamilnadu*. mimeo. Madras: Madras Insti-tute of Development Studies.

Vander Velde, E.J. 1980. Local Consequences of a Large-Scale Irrigation System in India. In E.W. Coward. (ed.). *Irrigation and Agricultural Development in Asia*. pp. 291-328. New York: Cornell University Press.

Veeraraghavan, A.K. 1940. Dusi. In P.J. Thomas and K.C. Ramak-rishnan. (eds.). *South Indian Villages: A Resurvey*. Madras: University of Madras.

Venkatachalam L. and Kulbhushan Balooni. 2017. Water Transfer from Irrigation Tanks for Urban Use: Can Payment for Ecosystem Services Produce Efficient Outcomes? International Journal of Water Resources Development. July 5.

____. 2018. Payment for Ecosystem Services to Sustain Kudimaramathu in Tamil Nadu. E&PW, Vol LIII. No. 6. 10, February. Pp. 21-23.

Venkatram, B.R. 1985. *Administrative Feasibility of A Tank Irri-gation Authority*. Progress Report No. 72. Pantancheru (AP): ICRISAT.

Von Oppen, M. and K.V. Subba Rao. 1980a. *Tank Irrigation in Semi- Arid Tropical India. Part-I: Historical Development and Spatial Distribution.* Economics Program Progress Report 5. Patancheru (AP): International Crops Research Institute for the Semi-Arid Tropics.

____. 1980b. *Tank Irrigation in Semi-Arid Tropical India. Part- II: Technical Features and Economic Performance*. Economics Pro-gram Progress Report 8. Patancheru (AP): ICRISAT.

Wade, Robert. 1979. The Social Response to Irrigation: An Indian Case Study. *Journal of Development Studies*. 16(1): 9.

____. 1980. Managing the Main Systems. *Economic and Political Weekly*. 15(39): A 102-12.

____. 1982a. The System and Administrative and Political Corrup-tion: Canal Irrigation in South India. *The Journal of Development Studies*. 18(3): 287-325.

____. 1982b. The World Bank and India's Irrigation Reform. *The Journal of Development Studies*. 18(2): 169-180.

____. 1982c. Group Action for Irrigation. *Economic and Political Weekly*. 17(39): A 103-6.

____. 1987a. *Village Republics: Economic Conditions For Collec-tive Action in South India*. Cambridge: Cambridge University Press.

____. 1987b. Managing Water Managers: Deterring Expropriation, or Equity as a Control Mechanism. In W.R. Jordon. (ed.). *Water and Water Policy in World Food Supplies*. pp. 177-83. Texas: A & M University Press.

Wittfogel, Karl. A. 1957. *Oriental Despotism: A Comparative Study of Total Power*. New Haven: Yale University Press.

Government Documents

Adangal Registers of the Villages Served by Kaveripakkam and Dusi-Mamandur Tanks. 1989 and 2016. North Arcot District. Vellore and Thiruvannamalai, Tamil Nadu.

Board of Revenue. PWD. Irrigation. GO 1379. 14-10-1905. Madras.

Board of Revenue. PWD. GO 270I. 24-08-1910. Madras.

Board of Revenue. PWD. GO 250I. 03-06-1914. Madras.

Board of Revenue. PWD. Administrative Report. 1927-28. Madras.

Board of Revenue. PWD. GO 1617I. 19-06-1931. Madras.

Board of Revenue. PWD. GO E-8. 2459. 07-07-1933. Madras.

Board of Revenue. PWD. GO 2404 I. 17-05-1937. Madras.

Government of India. 1881, 1913 and 1981, 2011. Census Registers of the Villages Served by Kaveripakkam and Dusi-Mamandur Tanks. Madras: Government Press.

____. 1883. Note by C. Vincent. *Revised Estimates for the Palar Anicut System*. Calcutta: Government Press.

____. 1903. *Report of the Indian Irrigation Com-mission*. 1901-03. Part II. Calcutta: Government Press.

____. 1981. District Census Handbook, Village and Town Directory. North Arcot District. Madras: Government Central Press.

____. 1988. *CWC*, New Delhi: Ministry of Water Resources.

____. 1989. *Central Board of Irrigation and Power*. Statistics of Minor Irrigation Development. New Delhi: Govern-ment Press.

____. 1991. *National Water Management Project*. Irrigation Wing. Public Works Department. Preliminary Report. Madras: Government Press.

____. 1992. Planning Commission. *Report of the Committee on Pricing of Irrigation Water*. New Delhi: Government Press.

____. 1993. *Water Resources Consolidation Project*. Palar Anicut System (Tamil Nadu). Project Report. Madras: Gov-ernment Press.

____. *Ministry of Agriculture in Brief*. (Various Issues). New Delhi: Government Press.

____. 2017. Ministry of Water Resources. Report of 5th Census of Minor Irrigation Schemes. November. New Delhi.

Government of Tamil Nadu. 1957. *Improvements to Palar Anicut System Below Ranipet*. Public Works Department. Irrigation Re-search Station. I.R.S. Report No. 3/61. Poondi. Madras: Govern-ment Press.

____. 1979. *Report of Tank Irrigation Moder-nisation Project*. mimeo. Two Volumes. Madras: Government Press.

____. 1981. 2016. *District Statistical Handbook*. North Arcot – Vellore, District. Madras: Directorate of Statistics.

____. Various Issues. Tamil Nadu and Economic Appraisal. Evaluation and Applied Research Department, Kuralagam, Chennai.

____. Various Issues. Season and Crop Report. Department of Economics and Statistics. Chennai – 600006.

____. 2001. Tamil Nadu Farmers Management of Irrigation Systems Act–2000 (Act no. 7/2001).

____. 2014. Monitoring and Evaluation Consultancy for "Irrigated Agriculture Modernisation and Water Bodies Restoration and Management" (TN-IAMWARM Project). Final Report. June.

____. 2016. Statistical Handbook of Tamil Nadu. Department of Economics and Statistics. Chennai – 600006.

Cox, Arthur. F. 1894. *Madras District Manuals. North Arcot Dis-trict.* Volumes I and II. Madras: Government Press.

Tamil Water Mamulnamas of Kaveripakkam and Dusi-Mamandur Tanks. 1907. (Originals Written about Fasli 1225 Corresponding to the year 1815 A.D.). District Collectorate. Vellore. Madras: Govern-ment Press.

The Palar Anicut System. 1886. Selections from the Records of the Government of India. Public Works Department. Sl. No. 8. Calcutta: Government Press.

Public Works Department. 1982. Report of Silt Survey in Kaveri-pakkam tank. mimeo. Irrigation Research Station. Poondi. Madras.

Public Works Department. *Palar Anicut System Office Records*. (Undated). Palar Anicut. North Arcot District.

Public Works Department. *Kaveripakkam Section Office Records*. (Undated). Kaveripakkam. North Arcot District.

Public Works Department. *Dusi-Mamandur Section Office Records*. (Undated). Iyenkarkulam. North Arcot District.

Report of the Palar Anicut System. (Undated). Office of the Executive Engineer Vellore Division. North Arcot District.

Report of the Royal Commission on Agriculture in India. 1928. Vol. III. Bombay: Government Central Press.

Village Settlement Registers of the Villages Served by Kaveripak-kam and Dusi-Mamandur Tanks. 1882. Madras: Tamil Nadu Archives.

Village Resettlement Registers of the Villages Served by Kaveri-pakkam and Dusi-Mamandur Tanks. 1913. North Arcot District. Madras: Government Press.

Village Updated Settlement Registers of the Villages Served by Kaveripakkam and Dusi-Mamandur Tanks. 1983. North Arcot Dis-trict. Madras: Government Press.

Tank Survey Photographs

1-2: Discussions with the Executive Engineer of Palar Anicut System, Office at Vellore.

3-4: The Palar River in Floods – Walking and Running Flows Near Vellore – A Blue moon Occurrence.

5-6: Heavy Floods in Palar – Happening once in 10 Years.

7-8: The 7.6 Kilometer Long Inlet channel of KPT – Starts from the Palar Anicut.

9-10: The Palar Anicut built in 1857 is serving as a water diversion anicut to 317 tanks.

11-12: One of the four main channels diversion point at the Palar Anicut.

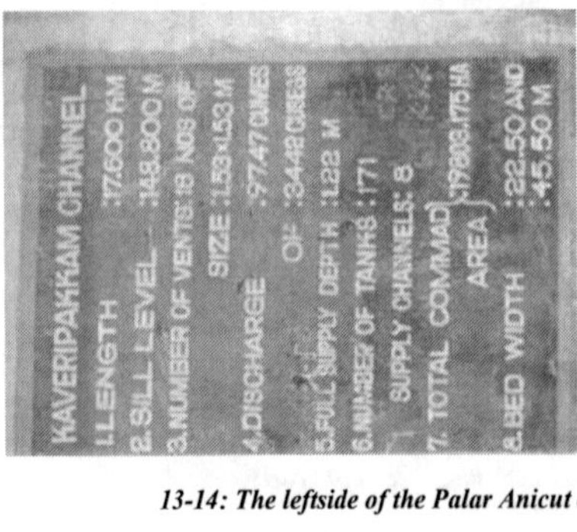

13-14: The leftside of the Palar Anicut off-take point of Kaveripakkam Channel.

15-16: The Closer and inner view of the water diversion sluices at Palar Anicut from the Palar River.

17-18: KPT with Full Storage at the deep Moolai Sluice Police deployment at KPT bungalow for safety.

19-20: A view of KPT when it becomes dry and tank bed cultivation is severe at Gaffoor & Fox Sluices.

21-22: Usual Tank bed / Water spread area cultivation at KPT - Submerged Banana harvested for Sale.

23-24: Well water level at the deep Moolai sluice serving Kattalai village–Head Reach gets more recharge.

25-26: A view of deep Moolai sluice – presently non-functional and water level at Cheri high level sluice.

27-28: Eight KM long and 4-5 mt breadth of Kaveripakkam tank bund protectswater inside the tank bund.

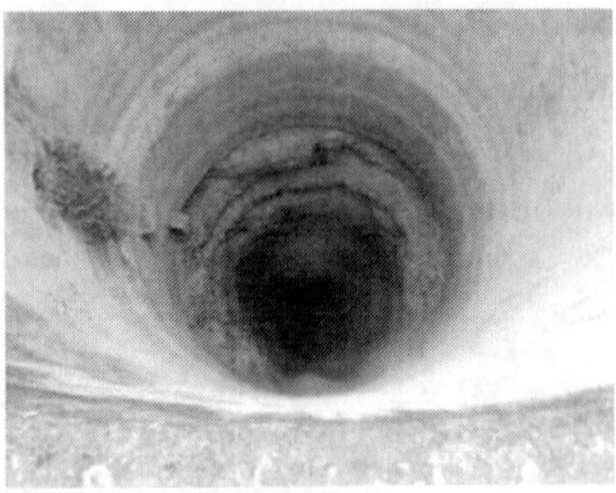

29-30: Elevated Paracheri sluice in KPT- a well close to it–elevated villages always starving–no recharge.

31-32: Extent of Tank bed siltation in KPT close to Surplus Weir – depth goes to over 3-4 metres.

33-34: The two lengthy surplus weirs of KPT – 30 feet elevation from the deep Moolai Sluice.

35-36: The elevated Panniyur sluice close to the surplus weirs and a view of KPT bed at that point.

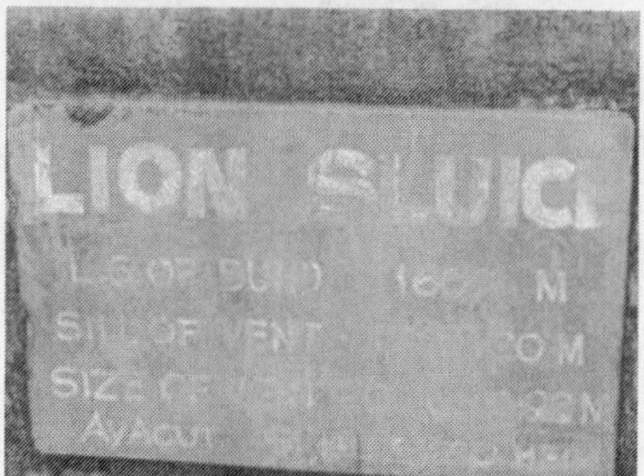

37-38: The Lion Sluice Built by the Order of the Board of Revenue in AD 1835.

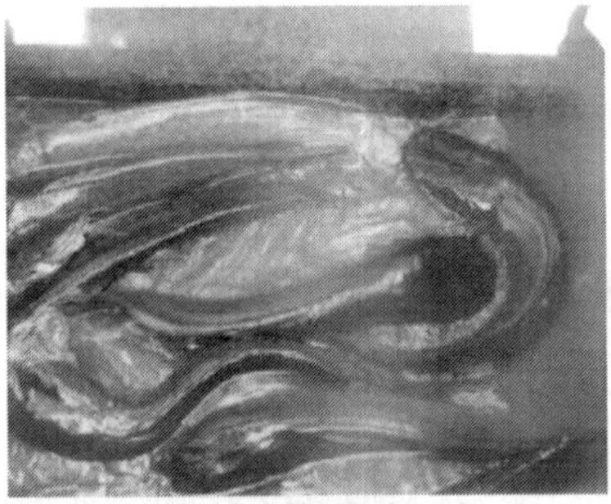

39-40: KPT annual govt. contract fishing income goes more than a Crore in 2015-16 tank supply year.

41-42: The inner and outer view of non-functional Thuraiperumpakkam sluice in KPT in 2015 to 2017.

43-44: Data collection at Cheri VAO office and group discussion with farmers at Cheri village.

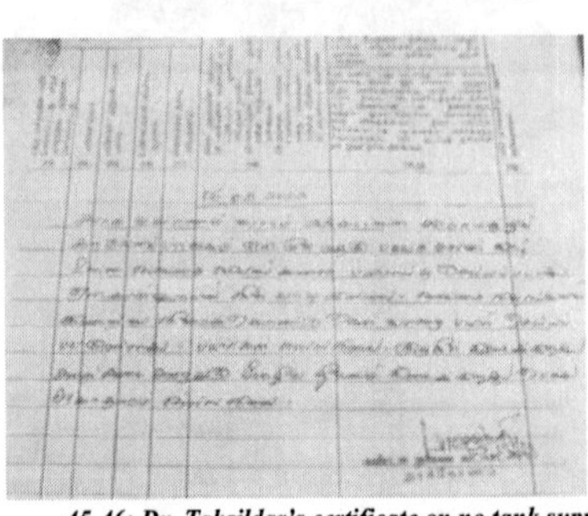

45-46: Dy. Tahsildar's certificate on no tank supply to Kattalai & group discussion with Kattalai farmers.

47-48: Panniyur and Kadaperi villages - group discussion with ayacutdars at VAO's office.

49-50: Data collection at N. Mangalam VAO's office (DMT) and Puthupattu – the lady VAO Officer at KPT.

51-52: Farmer's / HH survey in Alapakkam and Panniyur villages in KPT.

53-54: Paddy and Banana are the major crops in the KPT ayacut villages.

55 to 58: Different phases of paddy crop shown in KPT ayacut.

59-60: Mechanized harvest of Paddy is a common seen in both KPT & DMT – 2 Hrs. per Ha harvest.

61-62: Sugarcane cultivation is diminishing in the KPT ayacut in recent years – low price / water scarcity.

63-64: Sample farmers reported Banana Crop failure in Kattalai village in 2016-17 hurricane - heavy loss.

Index